Addressing the need for new models for the analysis of social network data, Philippa Pattison presents a unified approach to the algebraic analysis of both complete and local networks. The rationale for an algebraic approach to describing structure in social networks is outlined, and algebras representing different types of networks are introduced. Procedures for comparing algebraic representations are described, and a method of analysing the representations into simpler components is introduced. This analytic method, factorisation, yields an efficient analysis of both complete and local social networks.

The first two chapters describe the algebraic representations of the types of networks, and the third chapter covers the ways in which representations of different networks can be compared. A general procedure for analysing the algebraic representations is then introduced, and a number of applications of the approach are presented in the final chapters.

The book should be of interest to all researchers interested in using social network methods.

Structural analysis in the social sciences 7

Algebraic models for social networks

Structural analysis in the social sciences

Mark Granovetter, editor

Other books in the series:

Ronald L. Breiger, ed., *Social Mobility and Social Structure*

John L. Campbell, J. Rogers Hollingsworth and Leon N. Lindberg, eds., *Governance of the American Economy*

David Knoke, *Political Networks: The Structural Perspective*

Kyriakos Kontopoulos, *The Logics of Social Structure*

Mark S. Mizruchi and Michael Schwartz, eds., *Intercorporate Relations: The Structural Analysis of Business*

Barry Wellman and S. D. Berkowitz, eds., *Social Structures: A Network Approach*

The series *Structural Analysis in the Social Sciences* presents approaches that explain social behavior and institutions by reference to *relations* among such concrete social entities as persons and organisations. This contrasts with at least four other popular strategies: (a) reductionist attempts to explain by a focus on individuals alone; (b) explanations stressing the causal primacy of such abstract concepts as ideas, values, mental harmonies and cognitive maps (thus, "structuralism" on the Continent should be distinguished from structural analysis in the present sense); (c) technological and material determinism; (d) explanations using "variables" as the main analytic concepts (as in the "structural equation" models that dominated much of the sociology of the 1970s), where structure is that connecting variables rather than actual social entities.

The social network approach is an important example of the strategy of structural analysis; the series also draws on social science theory and research that is not framed explicitly in network terms, but stresses the importance of relations rather than the atomisation of reductionism or the determinism of ideas, technology or material conditions. Though the structural perspective has become extremely popular and influential in all the social sciences, it does not have a coherent identity, and no series yet pulls together such work under a single rubric. By bringing the achievements of structurally oriented scholars to a wider public, the *Structural Analysis* series hopes to encourage the use of this very fruitful approach.

Mark Granovetter

Algebraic models for social networks

Philippa Pattison
University of Melbourne

Published by the Press Syndicate of the University of Cambridge
The Pitt Building, Trumpington Street, Cambridge CB2 1RP
40 West 20th Street, New York, NY 10011–4211, USA
10 Stamford Road, Oakleigh, Victoria 3166, Australia

First published 1993

Library of Congress Cataloging-in-Publication Data

Pattison, Philippa.

Algebraic models for social networks / Philippa Pattison.

p. cm. – (Structural analysis in the social sciences: 7)

Includes bibliographical references and index.

ISBN 0–521–36568–6

1. Social networks – Mathematical models. I. Title. II. Series.
HM131.P3435 1993
301′.01′51 – dc20

92–34916
CIP

A catalog record for this book is available from the British Library.

ISBN 0–521–36568–6 hardback

Transferred to digital printing 2003

To my parents

Contents

Figures and tables

Figures

Tables

Preface

A class of models for analysing social network data are described in this work. The models are offered in response to two related needs arising from current developments in social science research. Firstly, data on social networks are being gathered much more commonly, a fact that is reflected by the inclusion in 1985 of a set of network questions in the General Social Survey (Burt, 1984). As a result, there is a growing need for a variety of models that will enable the analysis of network data in a number of different forms. Secondly, the role of social networks in the development of social and psychological theory is increasingly prominent and calls for the development of data models attuned to a variety of theoretical claims about the nature of that role.

Arguments for the importance of social networks can be found in both the psychological and sociological domains. Social psychologists have documented their dissatisfaction with the "differential" view of social behaviour embodied in many psychological theories (e.g., Cantor & Kihlstrom, 1981; Fiske & Taylor, 1990; Harre & Secord, 1972; Magnusson & Endler, 1977; Moscovici, 1972) and have argued for an analysis of social behaviour that is more sensitive to the "meaningful" context in which it occurs. One aspect of that context is the network of social relations in which the behaviour in question is embedded, a contextual feature to which empirical studies of some kinds of behaviour have already given explicit recognition (e.g., Henderson, Byrne & Duncan-Jones, 1981).

On the sociological side, the case for the importance of social networks was initiated much earlier, and those studies that demonstrated the salience of social and personal networks have become classics (e.g., Barnes, 1954; Bott, 1957; Coleman, Katz & Menzel, 1957). Indeed, a considerable amount of attention has been devoted to the problems of obtaining information about social networks and representing it in some explicit form (e.g., Fischer, 1982; Harary, Norman & Cartwright, 1965; Henderson et al., 1981; Laumann, 1973; White, Boorman & Breiger, 1976). Moreover, in addition to their role in making social context explicit, social networks have played a significant part in the "aggregation" problem, a role that Granovetter (1973), in particular, has made clear. The aggregation problem is the process of inferring the global, structural implications of local, personal interactions (White, 1970). Granovetter has demonstrated that the problem is not straightforward and has shown in several instances how an understanding of the local social network assists the task of inferring macro level social behaviour (Granovetter, 1973; also, Skog, 1986).

The models for which an analysis is developed in this book have therefore been chosen to be sensitive to these two main themes for the role of social networks in social theory: as an operational form of some aspects of social context and as a vehicle for the aggregation of local interactions into global social effects. The claim is not made that the models selected are unique in filling this role, although it will be argued that their properties are closely aligned with a number of theoretical mechanisms proposed for them.

The starting point for the models is the characterisation of social networks in terms of blockmodels by White et al. (1976) and the subsequent construction of semigroup models for role structure (Boorman & White, 1976; also, Lorrain & White, 1971). In presenting the construct of a blockmodel as a representation for positions and roles from multiple network data, White et al. argued that it was necessary to develop a view of concrete social structure that did not depend on the traditional a priori categories or individual attributes in the sociologists' battery but rather on the networks of relations among individuals. They claimed that blockmodels provide a means for representing and ordering the diversity of concrete social structures, and they showed how the semigroup of a blockmodel provides a representation of its relational structure at a more abstract, algebraic level.

Later, Winship and Mandel (1983) and Mandel (1983) extended the blockmodel framework to include a representation for what they termed "local" roles. In so doing, they decoupled the notion of local role from the global role structure approach of Boorman and White, thus pointing the way to an algebraic characterisation of role in incomplete or egocentred networks.

In this book, I have attempted to develop an integrated method of analysis for these and some related algebraic characterisations of role structure in social networks. I argue that the algebraic description of structure is natural from the perspective of social theory and extremely useful from the perspective of data analysis. In particular, it allows for a general means of analysing network representations into simple components, a property that greatly enhances the descriptive power of the representations. A major theme of the work is that the provision of such a means of analysis is a necessary precursor to adequate practical evaluation of the representations. Moreover, an eventual by-product of this form of analysis should be a catalogue of commonly occurring structural forms and the conditions under which they occur and, hence, a more systematic development of projects initiated by Lorrain and by Boorman and White in their accounts of simple structural models.

The first two chapters describe the algebraic representations adopted for complete and local networks, respectively. The question of which networks possess identical algebraic representations is addressed in chapter 3, together with the more general question of how to compare the algebraic representations of different networks. In chapter 4, a general procedure for analysing the algebraic representations of complete and local networks is described. The task of relating this analysis to aspects of network structure is taken up in chapter 5, where a number of illustrative applications of the overall analytic scheme are also presented. Chapter 6 contains an application of the scheme to complete and local networks measured over time, while chapter 7 presents the algebraic representations that can be constructed for valued network data. Finally, chapter 8 discusses the contribution of the analysis to some important issues for network analysis, including the description of positions and roles, structural models for networks and the comparison of network structures.

The work has benefited from the assistance of many people. Warren Bartlett lent a great deal of encouragement and support in the early stages of the work, and Harrison White and Ronald Breiger have given help in many different ways over a number of years. Many of the ideas developed here have their source in earlier work by Harrison White and François Lorrain and also by John Boyd; the work also owes much to many insightful commentaries by Ron Breiger. I am grateful, too, to Stanley Wasserman for his helpful remarks on two drafts; and I am especially indebted to my family – Ian, Matt and Alexander, my parents and my parents-in-law – for their help and patience.

1

Algebraic representations for complete social networks

Social networks are collections of social or interpersonal relationships linking individuals in a social grouping. The study of social networks has been gaining momentum in the social sciences ever since studies conducted in the 1950s by Barnes (1954), Bott (1957) and others demonstrated the important role of social networks in understanding a number of social phenomena. Social networks have since come to span a diverse theoretical and empirical literature within the social sciences. They have been invoked in a variety of roles in different theoretical contexts and have been conceptualised in a number of ways. The frequency of use of the notion of social network is probably not surprising because an individual's behaviour takes place in the context of an often highly salient network of social relationships. Perhaps more striking is the range of theoretical roles that have been proposed for the social network concept. Social networks have been used to explain various characteristics and behaviours of individuals; they have also been used to account for social processes occurring in both small and large groups of individuals. In addition, they have been viewed as dependent on individual attributes and behaviours, as well as consequences of such aggregate social attributes as the level of urbanisation of a community.

For example, in one type of network research, social scientists have examined the nature of social networks as a function of structural variables such as occupation, stage in life, gender, urbanisation and industrialisation (Blau, 1977; Coates, 1987; Feiring & Coates, 1987; Fischer, 1982; Fischer, Jackson, Stueve, Gerson & McAllister Jones, 1977; Wellman, 1979). In these discussions, the primary focus has been on describing the consequent variation in network characteristics such as the density of ties in a person's local social network, although some authors have expressed the need for more structural concerns (e.g., Friedkin, 1981; Wellman, 1982). Others have considered the interrelationships between social networks and the more traditional sociological categories by examining the distributions of social ties between persons in various categories (e.g.,

Blau, 1977; Fararo, 1981; Fararo & Skvoretz, 1984; Rytina & Morgan, 1982).

In a second type of network research, a more diverse group of researchers have used social networks as a means of explaining individual behaviour. The classic studies of Barnes (1954) and Bott (1957) fall into this class of network studies, as do many more recent investigations (e.g., Kessler, Price and Wortaman, 1985; Laumann, Marsden & Prensky, 1983). A growing body of work, for example, views psychological characteristics such as mental health as dependent in part on features of an individual's interpersonal environment (Brown & Harris, 1978; Cohen & Syme, 1985; Henderson, Byrne & Duncan-Jones, 1981; Kadushin, 1982; Lin, Dean & Ensel, 1986). The network characteristics selected for study in such investigations have included the density of one's local social network (Kadushin, 1982), the availability of attachment in the network (Brown & Harris, 1978) and its perceived adequacy (Henderson et al., 1981). Wellman (1983, 1988) has summarised the essence of this form of network analysis as its emphasis on structural forms allocating access to scarce resources. Social networks provide both opportunities and constraints for social behaviour and are therefore a necessary part of the background information required to explain behaviour (Campbell, Marsden & Hurlbert, 1986; Granovetter, 1985; Marsden, 1983). A variety of behaviours have been considered in this enterprise, including not only indicators of physical, mental, economic and social well-being (e.g., Campbell et al., 1986; Kadushin, 1982; Kessler & McLeod, 1985; Piliksuk & Froland, 1978) but also such diverse behaviours as option trading (Baker, 1984), more general economic behaviours (Granovetter, 1985) and individual decision-making (Anderson & Jay, 1985; Krackhardt & Porter, 1987).

A third type of network research examines the behaviour of a larger group of individuals as a function of the social networks connecting them. The theme in this work is the assessment of the large-scale, global or "macro" effects of individual or "micro" behaviour constrained by the local network structure. Some empirical illustrations of the approach have been conducted in small- to medium-sized groups against a background of a complete mapping of network links between all members of a specified population of persons. Examples include Sampson's (1969) documentation of a "blow-up" in a monastery and Laumann and Pappi's (1976) description of collective decision-making as a function of network links. In each of these cases, the social behaviour of the group was claimed to be understood in terms of information about the structure of the social relationships of its members.

On a larger scale, more direct mappings of the relationship between network characteristics and population parameters have been attempted.

For example, Granovetter (1974) characterised the local personal networks of a sample of individuals and assessed the relationship between network characteristics and aspects of job-finding. Skog (1986) has argued that network processes may underlie long-term fluctuations in national alcohol consumption rates and has observed the need for a greater understanding of the topology of social networks, that is, of the patterns in which network links are distributed in a population. Many social processes occur as the result of micro interactions among persons connected in a network, and the aggregation of these processes across an entire population can clearly depend on the arrangement of links in the network. Skog argued that the Law of Large Numbers may not hold for certain kinds of network structure, so that the assessment of the impact of network topology has far-reaching significance. Granovetter, also, has stressed the need to take social structure into account for a wide variety of social processes, and his arguments have inspired a good deal of empirical study into the role of network structure in information transmission and other processes (e.g., Friedkin, 1980; Granovetter, 1974, 1982; Lin, Dayton & Greenwald, 1978; Lin & Dumin, 1986; Lin, Ensel & Vaughn, 1981; Murray & Poolman, 1982).

A rather different line of work has examined the mutual dependence of individual and network characteristics in small- to medium-sized groups. For example, Breiger and Ennis (1979) have examined the relationship between individual characteristics and properties of the interpersonal environment in which an individual is located (see also Ennis, 1982). In two case studies, they have established a meaningful set of constraints between individual and network features, so adding to our understanding of how particular people come to hold particular network positions. Oliveri and Reiss (1987) viewed characteristics of an individual's personal network as markers for the individual's social orientation and preferences, and they described the networks of mothers and fathers in a sample of families. Leung, Pattison and Wales (1992) have also investigated the relationship between individual and network characteristics by studying the interdependence of the meaning that an individual ascribes to the word *friend* and the network environment in which the ascription is made. Investigations of these latter kinds may be helpful in preventing personal characteristics of an individual and features of the individual's social network becoming a confusing and imperfect proxy for one another (Hall & Wellman, 1985; Wellman, 1982).

The range of theoretical uses of the concept of social network is therefore broad, but it can be argued that most conceptions of the role of social networks fall into one or both of two main classes. The first includes proposals of some kind of link between properties and/or behaviours of an individual and the immediate or extended network environment in which that individual is located. The second class is characterised by the view that social networks define paths for the flow of social "traffic", so that an understanding of social network structure is essential to an understanding of social processes occurring on that network structure. These two views differ in their explanatory emphasis, and there is no necessary inconsistency between them. For instance, it is reasonable to argue for a mutual interdependence of the characteristics of individuals, groups of individuals and the social relationships that connect them.

Nonetheless, it is perhaps surprising that similar features of social networks tend to be evaluated in a variety of network studies. For example, many of the empirical investigations inspired by these theoretical concerns have described social networks in terms of such characteristics as the size of the network, the density of network ties, the centrality of individuals within the network and their integration into a cohesive unit. Many investigators have relied upon a survey approach, constructing a local network of individuals in the immediate network neighbourhood of a randomly selected individual, whereas others have built a more complete picture of the relationships among all persons in a relatively bounded group.

Some evidence suggests, though, that characteristics of social networks may relate to individual and group behaviour in complex ways and that it may not be sufficient to measure features such as the density, size and centrality of a social network, without regard to other structural characteristics (e.g., Friedkin, 1981; Hall & Wellman, 1985). The argument is particularly cogent where social processes are of interest, that is, where the arrangement of network links has been argued to play a substantial role in the development of the process under study (Granovetter, 1973; Skog, 1986). Thus, in the work reported here, I have presented a representation and a means of analysis for social network data that has some structural complexity. The representation allows a unified approach for both complete and local network data and is intended to be sensitive to the two themes just identified for social network research. It is based on the representations developed by Boorman and White (1976), Mandel (1983), White, Boorman and Breiger (1976) and Winship and Mandel (1983), as well as on various developments of them (including those by Breiger & Pattison, 1986; Pattison, 1982; Pattison, 1989; Pattison & Bartlett, 1982). In this chapter, the case in which network data are obtained for each member of a well-defined group is considered. The forms in which such "complete" network data may arise are reviewed, and then the structural representation that is proposed for them is described. Network data in the form of individual-centred local networks are introduced in chapter 2, together with an analogous form of structural representation.

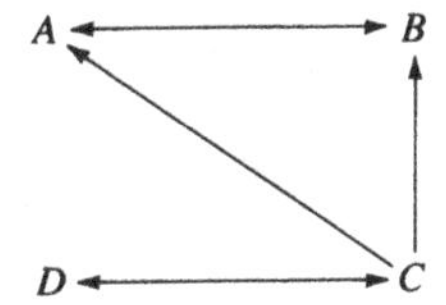

Figure 1.1. A directed graph representation of a friendship network among four members of a work group

Complete network data

The most basic form of social network data can be described as a set of social units, such as individuals, and a collection of pairs of units who are linked by a social relationship of some kind (Freeman, 1989). For example, a Friendship network among a group of individuals belonging to a particular organisation comprises the members of the organisation and the set of pairs of members who are linked by the relation of friendship. An example of such a network for a small, hypothetical work group is shown in Figure 1.1 in the form of a directed graph. Each member of the group is represented as a point, or vertex, of the graph (labelled in Fig. 1.1 by the letters A, B, C and D), and a directed arrow, or edge, links a member to each friend. For example, the link from A to B in Figure 1.1 indicates that A claims B as a friend. The set of group members forms the *vertex* or *node set* X of the graph, and the links defined by pairs of individuals who are friends form the *edge set* of the graph. In Figure 1.1, the vertex set is $X = \{A, B, C, D\}$, and $A \to B$, $B \to A$, $C \to A$, $C \to B$, $C \to D$ and $D \to C$ are the directed edges of the graph. The same network may also be represented in a closely related relational form. The set of organisation members form a set of elements X, and a *relation* F is defined as the set of ordered pairs of members who are linked by a friendship relation. Each ordered pair in the relation F corresponds to a directed edge of the graph of the network. For instance, for the friendship network displayed in Figure 1.1, the relation F may be written as $F = \{(A, B), (B, A), (C, A), (C, B), (C, D), (D, C)\}$.

A third common representation of this kind of network data is a binary matrix. The organisation members again define a set of elements, and these elements may be listed in any order and assigned an integer from 1 to n, where n is the number of members of the group. For instance, A, B, C and D may be assigned the integers 1, 2, 3 and 4, respectively. Then the kth individual in the list may be seen as corresponding to the kth row and the kth column of a square matrix. The cell of the matrix at the intersection of the the ith row and the jth column (i.e., the (i, j)

Table 1.1. *The binary matrix of the friendship network in a small work group*

	1	2	3	4
1	0	1	0	0
2	1	0	0	0
3	1	1	0	1
4	0	0	1	0

cell) may be used to record the presence or absence of a friendship link from the ith individual to the jth individual. The cell is usually defined to have an entry of 1 if the relationship of interest is present (i.e., if the ith individual names the jth individual as a friend) and an entry of 0 if the relationship is absent (i.e., if individual i does not name individual j as a friend). The binary matrix corresponding to the friendship network of Figure 1.1 is displayed in Table 1.1. (The matrix is termed *binary* because each of its entries is either zero or one.)

It may be observed that the diagonal entries of Table 1.1 – that is, the entries in cells (1, 1), (2, 2) and so on – are all zero. Correspondingly, there are no links in Figure 1.1 from any vertex to itself. (A link from a vertex to itself in a directed graph is often termed a *loop*.) In this example, an investigator is likely to be interested only in friendship relations between distinct individuals and may not even wish to consider whether it makes sense to speak of an individual being his or her *own* friend. Indeed, in many studies in which network data are generated, it is assumed that the graphs of the networks have no loops, or, equivalently, that the matrices of the networks have zero diagonals.

In some cases, though, loops and non-zero diagonals may possess useful interpretations. If, for instance, the social units of the network are groups of individuals rather than single persons, then it may be meaningful to regard a group as having a friendship relation to itself as well as to other groups. Certain forms of network analysis may also render the use of loops appropriate (e.g., Arabie & Boorman, 1982; Pattison, 1988). In the treatment of network data developed here, it is not assumed that loops are forbidden, even though they do not occur in some of the examples that are presented.

Symmetric and valued networks. A number of variations on this basic account of a social network have been found useful. For instance, in some cases it is reasonable to assume that every link of a network is reciprocated, that is, that if one individual is linked to a second, then the second is also linked to the first. Networks of close friendships may

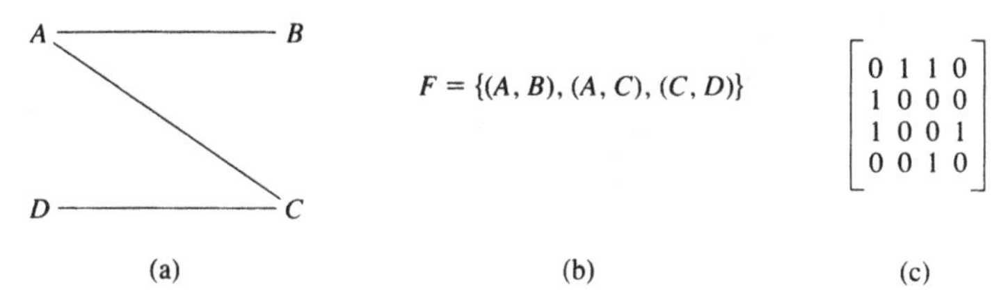

Figure 1.2. Representations for symmetric network relations: (a) (symmetric) graph; (b) set of unordered pairs; (c) symmetric binary matrix

have this character (e.g., Hammer, 1984). The viability of the assumption is essentially an empirical question, but where it is plausible, the representations described earlier may be simplified to some extent. For a reciprocated network relation, the presentation of the network in diagrammatic form need only indicate the presence of an edge and not its direction. The diagram is then termed a *graph*. Also, the relation corresponding to the network need only indicate the pairs of group members who are related, and the ordering of elements within each pair may be ignored. Finally, the binary matrix of the network may be assumed to be *symmetric*, that is, have the property that the entry in cell (i, j) of the matrix is the same as the entry in cell (j, i). For instance, if i names j as a friend, the assumption of symmetry means that j also names i as a friend. In this case, the entries in the row of the matrix corresponding to an individual are identical to those in the column corresponding to the same individual. These three ways of describing a network having reciprocated links are illustrated in Figure 1.2.

Symmetric relations can also arise from networks whose links are *nondirected* rather than directed *and* reciprocated. For instance, it may be useful in some circumstances to define a (nondirected) link to exist between two individuals in a network if one has contact with the other. The relation may not necessarily be reciprocated, but the direction of the link may be irrelevant for some questions. In fact, many network data have been gathered in this form (e.g., Freeman, 1989). They may be presented in exactly the same way as directed, reciprocated relations, that is, in the form of a collection of unordered pairs, a symmetric binary relation, a graph or a symmetric binary matrix. As a result, they are not distinguished here from reciprocated, directed relations, although it should be noted that, for some purposes, distinction may be advisable (Wasserman & Faust, 1993).

In some other cases, it may be possible to make finer distinctions among network links rather than simply determine their presence or absence (also, Wasserman & Iacobucci, 1986). For instance, the links

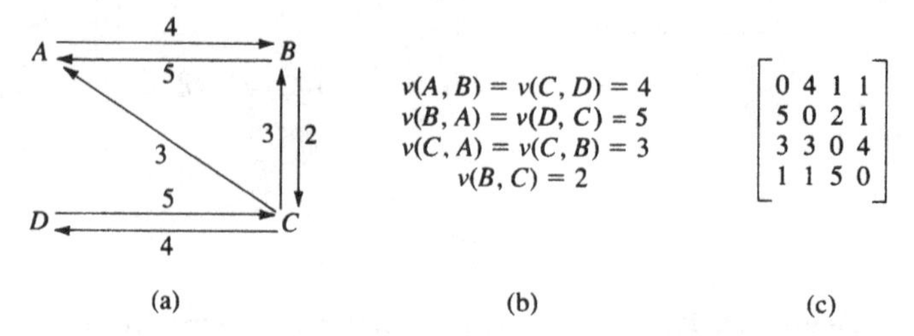

Figure 1.3. Representations for a valued network relation: (a) valued directed graph; (b) valued relation; (c) valued matrix (The strength of friendship links was assessed on a 5-point scale, with 1 = absent, 2 = a little, 3 = somewhat, 4 = quite strong, 5 = strong. Links of strength 1 have been omitted from the directed graph and valued matrix representations.)

may be measured on a numerical scale, with the scale values indicating the strength of the network link, such as the frequency of contact or the strength of friendship. The nature of the numerical scale will depend on the nature of the measurement procedures used to infer network links and on the properties of the measurements themselves (e.g., Batchelder, 1989). We hope, however, that the scale is at least *ordinal*, that is, it faithfully reflects orderings among network links in terms of strength. Where such numerical information is available, the representations require minor modification, as illustrated in Figure 1.3. The graphical representation takes the form of a *valued graph*, in which each directed or nondirected edge has a numerical value attached. The relational form specifies a mapping v from each (ordered or unordered) pair of elements to a possible value of the network link whereas the matrix representation records the value of the link from node i to node j in the cell of the matrix corresponding to row i and column j. For example, the value of the friendship link from A to B in Figure 1.3 is 4, so that the edge directed from A to B in the graph of the network has value 4; the function v assigns the value 4 to the ordered pair (A, B) (i.e., $V(A, B) = 4$), and the entry in cell $(1, 2)$ of the matrix of the network is 4.

In sum, a single network relation for a specified group of persons may be constructed in any of the four ways implied by our description. That is, it may be a symmetric or nonsymmetric binary relation, or a symmetric or nonsymmetric valued relation.

Multiple networks. In many network studies, more than one type of network relation is of interest, and it is necessary to construct more complex representations. For instance, for the small work group whose

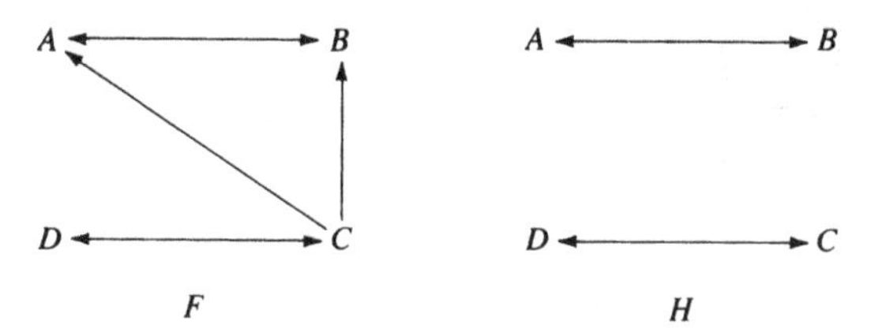

Figure 1.4. A multiple network **W** (F = friendship, H = helping)

Table 1.2. *Binary matrix representation of the multiple network* **W**

Relation	
F	H
0 1 0 0 1 0 0 0 1 1 0 1 0 0 1 0	0 1 0 0 1 0 0 0 0 0 0 1 0 0 1 0

friendship links are displayed in Figure 1.1, information might also be available for a different type of social relationship – for example, who goes to whom for help with work-related problems. This second type of network information is displayed with the first in Figure 1.4 and illustrates a *multirelational* social network. It is a network comprising a single set of network members and more than one type of network relation. In the example of Figure 1.4, two directed graphs are used to present the network. We may also represent the network in terms of two sets of ordered pairs, one set F for friendship links and one set H for helping links. The set F is as before, whereas $H = \{(A, B), (B, A), (C, D), (D, C)\}$. The matrix representation also requires two matrices, one for the friendship relation and one for the helping relation; these are presented in Table 1.2. The multirelational network is labelled **W**, and we may write $\mathbf{W} = \{F, H\}$.

In a representation of this kind, the symbols F and H may actually be used in two distinct ways. As we have just made explicit, each symbol denotes the collection of ordered pairs of elements of X who are linked by a relation of the specified type (either friendship or helping). The symbols will also be used, though, as labels for network links; for instance, we shall say that there is a link of type F from node i to node j if (i, j) is an ordered pair in F. Since the context will always make the intended meaning clear, we shall use the symbols for the relations in both of these ways in what follows.

Table 1.3. *Types of complete network data*

	Single relation		Multiple relations	
	Symmetric	Nonsymmetric	Symmetric	Nonsymmetric
Binary	Symmetric network	Network	Symmetric (multiple) network	(Multiple) network
Valued	Valued symmetric network	Valued network	Valued symmetric (multiple) network	Valued (multiple) network

Each network of a multirelational network may be assessed as a relation that is either symmetric or nonsymmetric, and binary or valued. For simplicity of presentation, we shall describe the overall network in terms of the minimum level of complexity needed to describe each constituent network. Thus, if any of the networks in the multirelational network is valued or nonsymmetric, we present each member of the network in the form appropriate to valued or nonsymmetric relations. For instance, if one relation in a multiple network is binary and another is valued, then we report both relations in valued form. This convention leads to the basic classification of multirelational networks summarised in Table 1.3. The table characterises network data as having either single or multiple relations, and as having constituent networks that are either symmetric or nonsymmetric, and binary or valued. It also identifies the labels to be used for the various forms of network data. For most of the work presented here, the basic form of network data that we shall assume is that of multiple networks, but in chapter 7 we also consider the case of multiple valued networks. The features of these two forms of network data, and the nature of their representations, are summarised in the following two formal definitions.

DEFINITION. Let $X = \{1, 2, \ldots, n\}$ represent a set of social units, and let R_k stand for a relation of some type k (e.g., "is a friend of"), where $k = 1, 2, \ldots, p$. Let $(i, j) \in R_k$ indicate that unit i is R_k-related to unit j (e.g., "i names j as a friend"), where i and j are elements of X. R_k is a *binary relation* on the set X and may be formally described as a set of ordered pairs of elements of X. (General algebraic definitions may be found in Kurosh, 1963; definitions of some basic mathematical terms are also given in Appendix A.) The collection $\mathbf{R} = \{R_1, R_2, \ldots, R_p\}$ of relations on X is termed a *(multiple) network*, and the relations

$R_1, R_2, \ldots, R_p$ are referred to as *primitive relations* or *generators* of **R**. To each binary relation R_k there corresponds a *directed graph* $G(R_k)$, whose vertices are the elements of X and whose edges are defined by

$$i \rightarrow j \text{ is an edge of } G(R_k) \text{ iff } (i, j) \in R_k.$$

(As noted earlier, loops – that is, edges of the form $i \rightarrow i$, for $i \in X$ – are permitted in the generators of **R**.) Relations and their directed graphs are used interchangeably with each other and with their equivalent binary matrix representation. The latter is defined for a binary relation R_k on the set X of n elements as the $n \times n$ square matrix whose entries are given by

$$(R_k)_{ij} \begin{cases} = 1 & \text{iff } (i, j) \in R_k, \text{ or } i \rightarrow j \text{ in } G(R_k) \\ = 0 & \text{otherwise.} \end{cases}$$

A valued network may be formally defined as follows.

DEFINITION. Let X represent a set of n social units, and let $v_k(i, j)$ represent the "strength" of the relationship of type k from unit i to unit j in X. For each k, let V_k represent the relation of type k. V_k can be considered as

1. a valued, directed graph whose nodes are the elements of X and whose edges are defined by the edge of type k directed from node i to node j having value $v_k(i, j)$;
2. a valued relation, assigning the value $v_k(i, j)$ to the ordered pair (i, j); and
3. an $n \times n$ matrix with entries $v_k(i, j)$ (Harary, Norman & Cartwright, 1965).

The collection $\mathbf{V} = \{V_1, V_2, \ldots, V_p\}$ is termed a *(multiple) valued network*.

Several forms of network data that have been considered in the literature are not covered by these general definitions of multiple networks. Two of the most notable forms are those allowing time-dependent network data and those expressing relations among two or more types of social entities.

Time-dependent networks. Network data may be observed at a single point in time or on multiple occasions (e.g., Freeman, 1989; Hallinan, 1978; Wasserman & Iacobucci, 1988). In the case of a single point of observation, the forms of network data just described are appropriate. For multiple observations in time, however, the set of social units and/or the relations of the network may be seen as time-dependent, and each observation point is associated with a multirelational network. Although we can think of the entire collection of relations over all time points as constituting a large multirelational network, we do not, at this stage, accommodate any temporal structure for the collection in the basic representation that we consider. Nonetheless, an attempt to analyse time-dependent data using the structural representations developed here is presented in chapter 6.

Relations among two or more sets of social units. Some of the representational forms that have been proposed in the literature also admit data in the form of relations between two or more types of social units (e.g., Freeman, 1989; Iacobucci & Wasserman, 1990; Wasserman & Iacobucci, 1991; see also the survey in Wasserman & Faust, 1993). A common example of such data are records of the relations among individuals and particular organisations, for instance, individuals and the corporations of which they are directors. The data can be represented in the form of a rectangular binary matrix, whose rows and columns represent the individuals and the corporations, respectively. The (i, j) cell of the matrix has an entry of 1 if the ith individual is a director of the jth corporation and an entry of 0 otherwise. Similar types of data arise if we record the membership of a collection of individuals in any group of formal or informal organisations or the participation of those individuals in some set of activities (e.g., Davis, Gardner & Gardner, 1941; also Homans, 1951). One of the most widely applied class of models admitting these types of data is derived from Atkin's (1977) application of combinatorial topology to social relations (e.g., Doreian, 1980, 1986). A related formalisation has been proposed in the form of bipartite and tripartite graph representations for relations spanning more than one type of social unit, for instance, both persons and groups (Breiger, 1974; Fararo & Doreian, 1984; Wilson, 1982). In part, relations between network members and other entities can be incorporated in binary relational representations by the construction of relations that represent the presence of common relations to other entities, for example, membership in the same group. The representations that result, however, may be more inefficient than those that give explicit recognition to relationships between different types of entities. Moreover, they do not necessarily have the same properties as the more general representations admitting either relations among units of more than one set, or relations that are ternary, quaternary, and so on, or both. Eventually, the methods to be developed may need to be elaborated so as to apply to these more complex representations. Initially, however, attention is confined to binary relational representations because these include the vast majority of the models that have been proposed and applied in the social network literature to date.

Sources of network data

Network data can come from a variety of different sources. In planning a network study, many choices need to be made about how network data are to be collected. In particular, decisions need to be made about who are potential members of the network of interest, that is, where are the "boundaries" of the network; what types of relations are of interest; and how are the network relations to be measured? We briefly consider each of these questions in turn (for more information, see Knoke and Kuklinski, 1982; Marsden, 1990; Wasserman & Faust, 1993).

The boundary of a network

The question of which social units comprise a network clearly depends on the nature and purpose of the network study. In one of the few attempts that have been made to analyse the possible approaches to answering this question, Laumann et al. (1983, 1989) presented a classification of possible frameworks. One of the distinctions they made was between studies adopting a realist or a nominalist perspective. They characterised a *realist* position by an attempt to identify networks whose members possessed some shared subjective awareness of the network as a social entity. Studies of the *nominalist* type were identified by a deliberate choice to define the boundaries of a network according to some research purpose of the investigator, without regard to the subjective status of the network on the part of network members. A cross-cutting distinction was in terms of the focus of the definition of network membership. Membership could be defined in terms of (a) attributes of the potential members of the network, (b) properties of relations among potential network members, (c) activities in which potential members are involved, or (d) some mixture of these. These two distinctions lead to the eight-fold classification of approaches to defining the boundary of a network that Laumann et al. described. The implications of each approach were discussed by Laumann et al., who also presented examples of each approach that have arisen in the literature. A common approach for complete network data appears to be a realist one, defining network membership in terms of attributes of its members, for instance, collections of individuals who belong to the same clearly defined group such as a work group or school class. Occasionally, a nominalist approach is used, as, for example, in the study of relations among community influentials by Laumann and Pappi (1973, 1976).

The boundary problem is particularly important for network studies, as Laumann et al. (1983) argue. It is likely to be especially important for the kinds of representations of social network described here. There have been relatively few systematic investigations of the effects of changes in the boundary of a network on representations of its structure, but it is possible that complex representations exhibit some sensitivity.

Relational content

It was observed earlier that many networks of interest are multirelational, that is, they describe social relations among individuals of more than one type. In designing a network study, how does an investgator determine which social relations are of interest? We might expect the content of relational ties in a network to bear an important relationship to the structure of the network that they define. A network of close friendship ties, for instance, might be expected to manifest a kind of patterning that is different from a network of acquaintance relations. Indeed, it was the predictable nature of these differences that gave force to Granovetter's argument that paths of "weaker" acquaintance relations might be expected to link an individual to a broader social group than paths of "strong" or close ties.

One important distinction among relations often used in network studies is that between social relations perceived by observers or participants of a network and more observable relations of "exchange" among participants (Marsden, 1990). The former type of relations are often termed *cognitive*, being cognitive constructions of the individual(s) constituting the source of the network data. They are exemplified by relations inferred from responses to questions such as "Who are your friends?" and they are often assumed to have some continuity in time, at least over short periods. Actual exchanges, however, are more temporally bound. They are illustrated by network studies that are based on reports or observations of some kind of exchange, such as traces of the frequency and volume of electronic messages among network members.

The question of which type of relational data is more appropriate for a particular network study depends on the purpose of the study (Marsden, 1990). For example, exchange data may be better suited to a study seeking to examine a process of diffusion of information, and cognitive data may be more helpful in understanding the behaviour of a small work group.

Types of relations. Relations in a multiple network are usually labelled according to the type of interpersonal relation that they are intended to signify. The labels are customarily terms of the kind "likes", "is a friend of", "exerts influence upon" and "communicates with". The labels often

Table 1.4. *Relational content in a sample of network studies*

Populations	Relational content	References
Davis–Leinhardt sociogram bank, classrooms, small groups	Positive affect, positive and negative affect	Davis (1970), Davis and Leinhardt (1972), Hallinan (1974), Leinhardt (1972), Leung et al. (1992), Newcomb (1961), Vickers & Chan (1980), Vickers (1981)
Informal groups, self-analytic groups, families	Influence, friendship, esteem, liking, disliking, similarity, closeness, notice taken, notice given	Whyte (1943), Homans (1951), Rossignol & Flament (1975), Breiger & Ennis (1979), Ennis (1982), Kotler & Pattison (1977)
Personal networks	Uniplex, multiplex relations	Boissevain (1974), Mitchell (1969)
Organisations, work groups, novitiates, church groups, prison inmates, inmate addicts	Contacts, communication, friendship, help seeking, work advice, personal advice, most dealings, similar policy, friendship, formal and informal ties, liking, antagonism, helping, job trading, arguing, playing games, instrumental, social ties, affect, esteem, influence, sanction	Arabie (1984), Bernard & Killworth (1977), Thurman (1979), Roethlisberger & Dickson (1939), White (1961), Kapferer (1972), Sampson (1969), Curcione (1975), Herman (1984)
Deaf community, ham radio operators, research specialties	Teletype communications, radio communication, degree of contact and awareness, citations	Killworth & Bernard (1976a) Bernard & Killworth (1977), Breiger (1976), Friedkin (1980), Mullins et al. (1977), Lievrouw et al. (1987), Schott (1987),
Community networks	Business/professional ties, community affairs, social ties	Laumann & Pappi (1976), Laumann et al. (1977)
Corporate structures, social movements	Interlocking directorates, interlocking memberships, corporate relations	Mintz & Schwartz (1981), Mintz (1984), Rosenthal et al. (1985), Mizruchi & Schwartz (1987)
Securities market, world economy	Trading relationships, transnational relations	Baker (1984), Snyder & Kick (1979)

follow the questions of an interview or questionnaire designed to elicit the relational information, such as "Who are your friends?".

Some of the relational types selected by a number of researchers, together with the type of populations to whose interpersonal relations they have been applied, are listed in Table 1.4. The studies presented in

Table 1.4 are only a sample of those that have been conducted, but it is clear from the table that there is considerable variety in the content of the relations that have been considered. It is also the case that the content of relations selected for a study is contingent, at least in part, on the formal approach to network description adopted by the study.

Some early work on network structure considered a single affective or communication relationship defined on a group as the primary relational data (e.g., Bavelas, 1948; Davis and Leinhardt, 1972; Harary, 1959a; Holland and Leinhardt, 1970, 1975, 1978; Katz, 1953; Luce and Perry, 1949). The interpretation of balance theory by Cartwright and Harary (1956) and Flament (1963) pertained to relations of positive and negative affect in small groups and led to the assessment of both positive and negative affective relations in empirical studies. Mitchell and his colleagues (e.g., Kapferer, 1969; Mitchell, 1969) distinguished uniplex and multiplex relations, that is, relations based on a single shared attribute (such as being co-workers) or a number of shared attributes (such as being co-workers as well as kin). More recently, the availability of methods of description for multiple networks has encouraged the use of a wider range of relation types (e.g., Breiger & Ennis, 1979; Herman, 1984; Snyder & Kick, 1979).

Implicit in Table 1.4 is an inferred similarity of certain relational terms. Studies employing questions eliciting sets of friends and those seeking lists of persons "liked", for example, can be expected to generate similar, although not necessarily identical, types of patterns. There is considerable overlap in the meaning of the two corresponding relational questions, whether the overlap is considered largely semantic in origin or an empirical result.

An attempt to identify relatively independent components, or dimensions, in terms of which relational terms might be described, is that of Wish, Deutsch and Kaplan (1976; see also Bales & Cohen, 1979; Wish, 1976). In the spirit of the work of Osgood, Suci and Tannenbaum (1957), Wish et al. undertook to "discover the fundamental dimensions underlying people's perceptions of interpersonal relations" (p. 409) using persons' evaluations of "typical" interpersonal relations as well as of some of their own. On the basis of an INDSCAL analysis (Carroll and Chang, 1970), they identified four major dimensions of relational content, labelled thus: Cooperative and Friendly versus Competitive and Hostile, Equal versus Unequal, Intense versus Superficial and Socioemotional and Informal versus Task-oriented and Formal. The four relational distinctions between liking and antagonism, between equal and discrepant amounts of interpersonal influence, between strong and weak relations and between relations with formal and informal bases, to which the four dimensions may be argued to correspond, are all distinctions that are represented somewhere in Table 1.4. Further, those four distinctions encompass many of the relational terms appearing in Table 1.4; the identified distinctions provide, in many instances, reasonable coverage of relational content. Naturally, some relational distinctions are not important for some groups whereas others not mentioned may be extremely so; the results of Wish et al. (1976) do indicate, however, the nature of typical relational distinctions that individuals make.

Burt (1983) also attempted a different empirical analysis of redundancy in relational terms. He analysed the extent to which similar persons were named by a sample of individuals in response to a variety of survey questions dealing with network relations in the domains of friendship, acquaintance, work, kinship and intimacy. On the basis of his analysis, he argued that five questions could be used to cover a substantial portion of the relational information obtained from the much larger set of survey questions. These five questions dealt with the domains of friendship ("Who are your closest personal friends?" and "Who are the people with whom you socialise and visit more than once in a week?"), acquaintance ("Have you met any people within the last five years who are very important to you but not close friends?"), work ("With whom do you discuss your work?") and kinship ("Have you spent any time during the last year with any of your adult relatives; relatives who are over the age of 21? Who are they?"; Burt, 1983, pp. 67–8).

Network measurement

There are a number of different methods by which network data may actually be obtained. Thus a friendship network of the type portrayed in Figure 1.1 may be constructed using a variety of approaches to the measurement of network links. In a recent survey of network measurement, Marsden (1990) observed that the most common methods relied on self-report measures obtained from surveys and questionnaires but that archival sources, diaries, electronic traces, and observation by a participant or nonparticipant of the network group could also be used. Methods using archival sources are illustrated in the work of Padgett (e.g., Breiger & Pattison, 1986), who traced financial and marriage relations among 116 Florentine families of the fifteenth century from an historical record (Kent, 1978), and in that of White and McCann (1988), who constructed networks of citation among eighteenth century chemists from published scientific papers. Mullins, Hargens, Hecht and Kick (1977) examined relations of awareness, colleagueship, and student–teacher ties in two biological science research fields and related these to citation patterns among the scientists. Some investigators have also used published records of trade and other forms of exchange between nations to examine the structure of network relations among nations (e.g., Breiger, 1981; Snyder & Kick, 1979). Rogers (1987) described networks constructed from traces of electronic communications among individuals, and Bernard and Killworth (1977) analysed the duration and frequency of communications among amateur radio operators over a fixed period. Higgins, McLean and Conrath (1985) describe the use of diary-based communication networks, as well as some of the hazards associated with their use. Finally, two case studies reported by Boissevain (1974) illustrate the use of participant observers as the source of network data.

In many cases using self-report data, a single question is used to define the network links of a particular type. The response to the question by each network member defines the set of persons to whom that person is linked by the relation in question. For example, each of a group of members of an organisation may be asked the questions, "Who are your friends in the organisation?" and "To whom in the organisation do you turn for help when you encounter problems with your work?". Krackhardt (1987), for instance, asked questions such as these in his study of a small organisation. The responses to questions like these may be used to construct a multiple network of the type illustrated in Figure 1.4. This method of obtaining network data is probably the most common. For instance, Laumann and Pappi (1973) identified a group of community influentials in a small West German city. They constructed a list of the identified individuals and asked each of them to indicate on the list (a) the three persons with whom they most frequently met socially, (b) the three persons with whom they had the closest business or professional contact, and (c) the three persons with whom they most frequently discussed community affairs. Nordlie (1958) described two studies in which 17 undergraduates lived together in a fraternity house for a semester. Each week, every student ranked all of the others for "favourableness of feeling", leading to a form of valued network in which the value of a link from one individual to another was the first's favourableness rating of the second. White et al. (1976) created two binary matrices from each valued network by coding the two highest rankings on "favourableness of feeling" as "liking" and the two lowest rankings as "antagonism".

Reliability and validity of network data

No matter what the source of information about network links, it is useful to ask whether the measurement is (a) reliable, that is, subject to only very small amounts of random error, and (b) valid, that is, a measure of the intended relational content, and not something else. The more unreliable a form of measurement, the greater the degree of random error that it exhibits. As Holland and Leinhardt (1973) pointed out some time ago, and as a number of authors have also recently observed, the consequences of measurement procedures and errors in measurement for network data are in need of further empirical and theoretical investigation (e.g., Batchelder, 1989; Bradley & Roberts, 1989; Hammer, 1984; Marsden, 1990; Pattison, 1988).

Batchelder (1989), for instance, has examined from a theoretical perspective the implications of some measurement properties of relational observations for the assessment of network properties. He showed that some important network properties could only be inferred when the measurement of individual network links met certain criteria.

The question of the validity of network data is equally important and has been the subject of some empirical attention. In one set of studies, Kilworth, Bernard and others have reported consistent discrepancies between "cognitive" reports of communication data and observed communications and have concluded that relational data obtained by the usual sociometric methods may have a more limited generalisability than once thought (Bernard & Killworth, 1977; Bernard, Killworth, Kronenfeld & Sailer, 1984; Killworth & Bernard, 1976a, 1979). Others have argued for less pessimistic positions, however, after showing that informant accuracy depends, in part, on the informant's level of interaction (Romney & Faust, 1983) and is biased in the direction of long-term patterns of interaction (Freeman & Romney, 1987; Freeman, Romney & Freeman, 1986). It must be acknowledged, though, that cognitive and exchange network data may be assessing different types of network relations.

In a detailed study of network interview data, Hammer (1984) has also obtained more positive findings. She found good agreement between members of pairs of individuals who were interviewed about the existence of a relationship between them, about the duration of the relationship and about their frequency of contact. She found less agreement, however, on ratings of how well individuals knew each other, and she argued that some known and unknown selection criteria were at work when individuals decided which members of their personal networks to mention in response to a particular interview question (also, Bernard et al., 1990; Sudman, 1988).

In a different type of validity study, Leung et al. (1992) have examined the meaning of the term *friend* and have shown that there is considerable agreement among 15-year-olds about the attributes of the term, despite some small but meaningful variations as a function of their social network position. Such a finding lends essential support to the assumption of White et al. (1976) that "all ties of a given observed type share a common signification (whatever their content might be)" (p. 734).

It seems from all this work that there may be systematic biases at work in the reporting of network data by an individual, and that these need to be taken into account in its interpretation. Marsden (1990) has advocated the use of multiple measures of network relations for improved quality of network data; he has also suggested that simulation might be used to evaluate the effects of different types of measurement error on inferred network properties. Krackhardt (1987) has elaborated the most common observational scheme described above so as to ask *each* individual in a network about relations among all pairs of network members. Such data may also be helpful in formulating questions and methods in the assessment of network data.

So far, we have identified a number of questions to which we need answers in any empirical study of social networks. These include: Who are the members of the network? Which types of relations between network members are of interest? How should they be measured? As we have observed, considerable uncertainty may be attached to any answers that are proposed to these questions, and the ramifications of misspecification at each stage are poorly understood. Indeed, the implications are probably best investigated with a particular representation of a network in mind, and we now consider how to construct a representation of a network that captures some of its structural complexity.

Structure in social networks

One of the challenges facing researchers using social network concepts is that of distilling the significant structural features from social networks represented in one of the forms previously described. The challenge is clearly a difficult one because it requires the resolution of a number of substantial theoretical and empirical questions. For instance, what features of a network are related to various types of individual and group behaviour and what are the mechanisms for these relationships? How should networks be measured? What kinds of interpersonal relations are involved in various social processes?

The question of constructing useful structural models for social networks is probably best addressed on both theoretical and empirical fronts. On the one hand, one can assume a particular structural model and investigate empirically the relationship between social networks represented in this form and other individual or social characteristics of interest. On the other hand, one can evaluate theoretically the consequences of a set of assumptions about structural form, in an attempt to achieve a richer understanding of a particular structural model. The work reported here is of this second kind and is based on the premise that it is useful to know the implications of a set of assumptions about structure before empirical assessments are undertaken.

A number of structural models have been developed for social network data arising in one of the forms described earlier. These models vary in complexity from single indices summarising a particular structural feature of a network to quite complex algebraic and geometrical representations. Each is a means of obtaining simple descriptions from the representation by making use of the structural redundancy that it is presumed to possess. The form of the assumed redundancy is critical in deriving that substantive value and is generally motivated both substantively and practically.

Many of the structural models that have been developed make use of features of directed graphs. So, before the models are described, some of the more widely used constructs for directed graphs are defined.

Directed graphs

A *(directed) graph* comprises a set X of *vertices* or *nodes*, and a set R of *(directed) edges* of the form (x, y), where $x, y \in X$. For example, the graph of Figure 1.1 has node set $X = \{A, B, C, D\}$ and edge set

$$R = \{(A, B), (B, A), (C, A), (C, B), (C, D), (D, C)\}.$$

An *(induced) subgraph* of a (directed) graph on a vertex set X is a subset Y of the nodes from X, together with all of the edges linking elements of Y. The set Y is said to *span* the subgraph. For the directed graph of Figure 1.1, for example, the subgraph induced by the subset $\{A, C, D\}$ of nodes has edges $\{(C, A), (C, D), (D, C)\}$.

A *path* from a vertex x to a vertex y in a directed graph is a sequence of nodes $x = x_0, x_1, \ldots, x_k = y$ such that each pair (x_{j-1}, x_j) of adjacent nodes in the sequence is an edge of the graph. The node x is termed the *source* of the path, and y is termed the *target*. The *length* of the path $x_0, x_1, \ldots, x_k$ is k. For example, there is a path of length 2 from node D to node A in Figure 1.1: the path comprises the nodes D, C and A. There is also a path of length 3 from D to A, comprising the nodes D, C, B and A.

A *path* $x = x_0, x_1, \ldots, x_k = x$ from a vertex x to itself is a *cycle* of length k. The node C in Figure 1.1, for instance, lies on a cycle of length 2. If for two nodes, x and y, in a graph, there is a path from x to y, then y is said to be *reachable* from x. If y is reachable from x, then a *geodesic* from x to y is any path of shortest length from x to y. The length of any shortest path from x to y is termed the *distance* $d(x, y)$ from x to y. If y is not reachable from x, then the distance from x to y is usually defined to be ∞. A graph is termed *strongly connected* if there is a path from each node in the graph to each other node.

For example, in the graph of Figure 1.1, nodes A and B are reachable from all other nodes. The distance from node D to node A is 2 because the path from D to A through C is a geodesic. Nodes C and D are not reachable from either of nodes A and B, so that the graph of Figure 1.1 is not strongly connected.

Paths in a directed graph that ignore the direction of the edges are called semipaths. That is, a *semipath* in a directd graph from node x to node y is a sequence of nodes $x = x_0, x_1, \ldots, x_k = y$ such that (x_{j-1}, x_j) or (x_j, x_{j-1}) is an edge in the graph for each $j = 1, 2, \ldots, k$. A semipath from a vertex to itself is termed a *semicycle*. If there is a semipath from x to y for every pair of nodes x and y in the graph, then the graph is said to be *weakly connected*. There is a semipath in Figure 1.1, for instance, from node A to node D through node C; indeed, all pairs of vertices in the graph are connected by a semipath. As a result, the graph of Figure 1.1 is weakly connected.

Finally, the *indegree* of a vertex x is the number of distinct vertices y for which (y, x) is an edge in the graph. The *outdegree* of x is the number of distinct vertices y for which (x, y) is an edge in the graph. In Figure 1.1, the vertices A, B, C and D have indegree 2, 2, 1 and 1, and outdegree 1, 1, 3 and 1, respectively.

Some analyses for social network data

Table 1.5 contains a summary of the variety of approaches that have been adopted for analysing single and multiple binary representations for networks. Broadly, they may be classified as follows.

1. *Graph indices.* Some analysts have proposed one or more numerical indices to represent particular properties of a network relation. The indices apply most commonly to a single, binary symmetric or directed relation; but some have also been developed for a single valued network (e.g., Peay, 1977a). The indices include (a) the dyad census of the network, that is, the number of mutual or reciprocated links in the network, the number of asymmetric or unreciprocated links and the number of null links (i.e., the number of pairs of unconnected individuals); (b) the degree of hierarchy in the network, that is, the degree to which network members can be fully or partially ordered in terms of connections to other members (Landau, 1951; Nieminen, 1973); (c) the degree of transitivity in the network, that is, the extent to which relations from

Table 1.5. *Some approaches to network analysis*

1 Graph indices	*3 Spatial representations*
hierarchy index	multidimensional scaling and
hierarchisation	clustering of measures of vertex
gross status	similarity for instance:
status differentiation	structural equivalence
influence concentration	automorphic equivalence
graph centrality	regular equivalence
integration	
unipolarity	*4 Collections of subsets*
dimensionality	cliques
strength	*r*-cliques
density	*k*-plexes
degree	*LS* sets
connectivity	clubs
average reachability	clans
mean length of geodesics	nested sets of partitions
connectedness	
dyad census	*5 Blocked matrix and relational models*
balance	balance model
clusterability	clusterability model
transitivity	transitivity model
intransitivity	ranked cluster model
triad census	hierarchical cliques model
	"39+" model
2 Vertex and edge indices	clique structures
status	semilattices
point centrality	blockmodels
strain	stochastic blockmodels
span	
degree	*6 Probabilistic models*
arcstrain	p_1 model
range	stochastic blockmodels
density	biased nets
reachability	Markov graphs

individuals *A* to *B*, and from *B* to *C*, are accompanied by a relation from *A* to *C* (Harary & Kommel, 1979; Peay, 1977a); and (d) the degree to which the network is centralised, with all paths in the network passing through one or more key network members (e.g., Freeman, 1979). Other indices that have been used include (e) the size of the network, that is, the number of network members (e.g., Mitchell, 1969); (f) the density of the network, that is, the ratio of the number of network links that are present to the number of links that are possible (e.g., Barnes, 1969a); (g) the heterogeneity of the network, defined in terms of the variation among members in the number of other members to whom they are connected (Snijders, 1981); and (h) the dimensionality of the network, referring to the dimensionality of some space in which the network may be embedded (Freeman, 1983; Guttman, 1979). Holland and Leinhardt (1970, 1978) suggested the *triad census* for the network as a useful all-purpose collection of indices representing network structure. The triad census of a network records the number of occurrences in a network of each of 16 possible forms of relation among three network members. Some more specific indices can be constructed from the triad census; for instance, the degree of transitivity of the network is reflected by the frequency of occurrence of transitive triads. For multiple networks comprising relations of positive and negative affect, indices assessing the degree to which the network conforms to the predictions of balance theory have been proposed (e.g., Cartwright & Harary, 1956; Peay, 1977a).

2. *Vertex and edge indices.* Indices representing properties of each vertex or edge in a graph have also been proposed. For example, a number of indices have been constructed to assess the "status" or "prestige" of each member of a network, that is, the extent to which a person receives interpersonal ties from other network members (taking into account the positions of those members) (e.g., Harary, 1959b; Katz, 1953; Langeheine & Andresen, 1982; Nieminen, 1973). The centrality of a vertex has been conceptualised in a number of ways (Freeman, 1979), for instance, in terms of its "closeness" to other vertices, the extent to which it lies on shortest paths between other vertices, and the number of other vertices to which it is directly connected (Freeman, 1979; also, Bolland, 1988; Donninger, 1986; Gould, 1987). Stephenson and Zelen (1989) have adopted an information theory approach to measuring centrality. Faust and Wasserman (1993) present a survey and synthesis of indices of centrality and prestige. The *point strength* of a vertex is defined as the increase in the number of components in the graph when the vertex is removed (Capobianco & Molluzzo, 1980), and the *arcstrain* of an edge is a measure of its participation in triads not conforming to the theory of structural balance (Abell, 1969).

3. *Representations based on similarity of network members.* Some investigators have attempted to represent the structure of a social network by mapping the network members into a multidimensional space. The mapping is usually designed so that the proximity of network members in the multidimensional space reflects their closeness, or similarity, in the network. The closeness or similarity of individuals in the network can be conceptualised in a number of ways. In a network comprising a single symmetric relation, the graph distance between nodes is often the starting point of the analysis (e.g., Doreian, 1988a). In networks containing multiple relations, definitions of vertex similarity are usually based on some measure of similarity of network position

(Pattison, 1988). A number of models of network position have been proposed (Pattison, 1988), but three important ones are based on the notions of structural equivalence (Lorrain & White, 1971), automorphic equivalence (Borgatti, Boyd & Everett, 1989; Pattison, 1980; Winship, 1988) and regular equivalence (White & Reitz, 1983, 1989). These three notions provide increasingly abstract definitions of what it means for individuals to hold the "same" position in a network.

Two individuals *A* and *B* are *structurally equivalent* if they have exactly the same network links to and from other network members. That is, if individual *C* is related to *A* by some relation *R*, then *C* must also be related to *B* by the relation *R*. Similarly, if *A* is related to a person *D* by relation *T*, then *B* must also be linked to *D* by *T*. The condition is illustrated in Figure 1.5a, where *A* and *B* are structurally equivalent, as are *C* and *D*. For instance, *A* is related by *R* to *A* and *B* and by *T* to *C* and *D*, and both *C* and *B* are linked by *R* to *A*. Exactly the same relations hold for *B*; that is, *B* is related to *A* and *B* by *R* and to *C* and *D* by *T*, and both *A* and *B* are linked to *B* by *R*.

Automorphic equivalence relaxes the condition by allowing automorphically equivalent individuals *A* and *B* to be linked by a relation *R* to the same kinds of individuals, rather than to the same individuals. In particular, *A* and *B* are *automorphically equivalent* if there exists a re-labelling of the vertices of the network, with *B* re-labelled by *A*, so that there is a link of type *R* from *A* to *C* in the original network if and only if there is also a link of type *R* from the vertex re-labelled *A* to the vertex re-labelled *C*. This condition also holds for links of type *R* from *D* to *A*: such a link exists if and only if there is also a link of type *R* from the vertex re-labelled *D* to the vertex re-labelled *A*. Figure 1.5b illustrates the definition. The vertices *A*, *B*, *C*, and *D* may be re-labelled *B*, *A*, *D* and *C* (see Fig. 1.5c), and it may readily be verified that the conditions on the re-labelling of links hold. That is, the network displayed in Figure 1.5c has precisely the same links as the network displayed in Figure 1.5b.

The notion of regular equivalence relaxes the condition of automorphic equivalence further. Two individuals *A* and *B* are *regularly equivalent* if (a) whenever *A* has a link of type *R* to an individual *C*, then *B* has a link of type *R* to some individual *D* who is regularly equivalent to *C*, and similarly for the links of type *R* from *B* to *D*; and if (b) whenever an individual *C* is linked to *A* by *R*, then some individual *D* who is regularly equivalent to *C* is linked by *R* to *B*. In Figure 1.5d, *A* and *B* are regularly equivalent, and so are *C* and *D*.

For each of these ways of defining the similarity of network vertices, a distance measure can be constructed that assesses the extent to which any pair of vertices are equivalent with respect to the specified

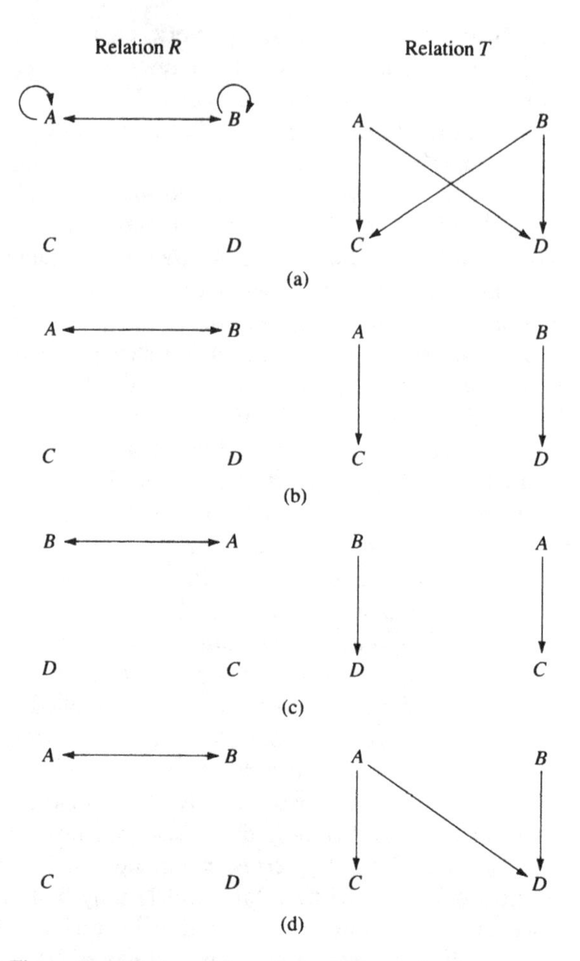

Figure 1.5. Structural, automorphic and regular equivalence: (a) Partition $(AB)\,(CD)$ is a structural equivalence; (b) partition $(AB)\,(CD)$ is an automorphic equivalence; (c) re-labelling nodes in (b) establishes an automorphism ($A \rightarrow B$, $B \rightarrow A$, $C \rightarrow D$, $D \rightarrow C$); (d) partition $(AB)\,(CD)$ is a regular equivalence

definition (e.g., Pattison, 1988; White & Reitz, 1989). Thus, a matrix of distances between vertices with respect to a particular type of equivalence may be computed; alternatively, a matrix of path distances between vertices of a single network may be calculated. The distance matrix may then be subjected to a general-purpose algorithm for obtaining a metric or nonmetric mapping of the distance matrix into some multidimensional space. The space itself may be metric, for instance, Euclidean, or it may be an ultrametric space constructed from a hierarchical cluster analysis (Arabie, 1977; Breiger, Boorman & Arabie, 1975; Breiger & Pattison, 1986; Burt, 1976; Doreian, 1988a,b; Faust, 1988).

4. *Collections of subsets.* Procedures for identifying cohesive subsets in a network have also attracted a great deal of attention. Intuitively, a *cohesive subset* is regarded as a subset of network members for whom network links within the subset are somewhat denser than network links connecting members of the subset to individuals outside it. Attempts at more explicit definitions of a cohesive subset include definitions permitting nonoverlapping and overlapping subsets of X, such as cliques, clubs, clans, r-cliques, k-plexes, and LS sets (Alba, 1973; Arabie, 1977; Batchelder & Lefebvre, 1982; Luce, 1950; Mokken, 1979; Peay, 1980; Seidman, 1983; Seidman & Foster, 1978). Other ways of defining subgroups have also been investigated. For instance, Batchelder and Lefebvre (1982) examined the "stratification" of the graph of a single symmetric network, namely, the division of the nodes of the graph into two distinct nonempty subsets such that a tie exists between every pair of nodes from distinct sets. Partitions of network members reflecting positional similarities in terms of network relations have also been of interest. For instance, each of the notions of vertex similarity just described – namely, structural equivalence, automorphic equivalence, and regular equivalence – lead to partitions of the set of network members. For each equivalence notion, we can partition the network members so that every class of the partition contains a set of individuals who are equivalent to each other. For example in Figures 1.5a,b,d the partition (A, B) (C, D) divides the vertex set into classes that are structurally equivalent, automorphically equivalent, and regularly equivalent, respectively.

5. *Blocked matrix and relational models.* Some representations of network structure have been expressed in a relational form. In addition to various algebraic models for particular types of network data – such as the balance model for positive and negative relations (Cartwright & Harary, 1956), clique structures for a single relation (Boyle, 1969), the semilattice model for informal organisations (Friedell, 1967) and triad-based relational models (e.g., Johnsen, 1985) – this group of representations contains blockmodels for multiple networks (e.g., Arabie, Boorman & Levitt, 1978; White et al., 1976). A blockmodel for a multiple network may be thought of as a hypothesis specifying (a) "blocks" of individuals who possess identical network relations as well as (b) the nature of the relations between these blocks. The notion of identical network relations is made explicit in the definition of structural equivalence of persons. As outlined earlier, two persons in a multiple

network are said to be structurally equivalent if they possess exactly the same set of relationships to and from every other network member. More precisely, whenever one of a structurally equivalent pair of persons stands in a relationship of a given type to a third person, then so does the other bear a relationship of that type to that third person. This requirement governs relations directed both from and towards members of a structurally equivalent pair.

DEFINITION. Let $\mathbf{R} = \{R_1, R_2, \ldots, R_p\}$ be a multiple network on a set X. Elements i and j of X are *structurally equivalent* if, for any $m \in X$,

1 $(i, m) \in R_k$ iff $(j, m) \in R_k$, for any $k = 1, 2, \ldots, p$; and
2 $(m, i) \in R_k$ iff $(m, j) \in R_k$, for any $k = 1, 2, \ldots, p$.

Individuals who are structurally equivalent are argued to hold the same position in the network (White et al., 1976), and it may be shown that the members of a multiple network may be partitioned into groups, or blocks, of persons such that all individuals assigned to the same block are structurally equivalent. The block containing a set of structurally equivalent people then becomes a representative of the (identical) social position that each block member holds (White et al., 1976).

The other main feature of the blockmodel is the collection of relations among the blocks of individuals. The interblock relations are represented as binary relations among the set of blocks and serve to define the social "roles" associated with the positions represented by the blocks. The relations among the blocks may be inferred from the relations among members of the blocks: one block bears a relationship of a given type to another if the members of the first hold that relationship to members of the second. That this rule for inferring relations among blocks is not ambiguous follows from the properties of structural equivalence.

Blockmodels may be derived from sets of observed relational data or they may be specified independently of data, on theoretical grounds. In general, a blockmodel is defined as specifying

1 an assignment of persons (elements) to positions, or blocks, and
2 the relationships between the blocks.

That is, the blockmodel indicates which persons hold the same social positions and whether a relationship of a given type between two positions is present or absent. Like the original network data, the blockmodel may be represented as a set of multiple binary relations, with the relations defined among positions or blocks rather than among individual people. An empirical collection of multiple, binary relations yields on exact fit to a given blockmodel if

1 persons assigned to the same block are, indeed, structurally equivalent, and

2 relations among blocks inferred by the procedure just outlined are in agreement with those specified by the blockmodel.

Such a fit of data to a blockmodel is termed by Breiger et al. (1975) a *fat fit* and is, of course, difficult to obtain empirically. The criteria for determining the fit of a set of data to a blockmodel have accordingly been relaxed to allow approximate fits. Two simple approximations are a *lean fit* (Breiger et al., 1975) and a fit to an α-blockmodel (Arabie et al., 1978). Both of these forms relax the requirement that if one block has a relation of some type to another, then every individual in the first block has that relation to every individual in the second. For a lean fit it is only required that at least one individual in the first block is related to at least one individual in the second block, whereas for an α-blockmodel the proportion of pairs of individuals from the two blocks for which the relation holds must be at least α.

DEFINITION. Let $\mathbf{T} = \{T_1, T_2, \ldots, T_p\}$ be a blockmodel on a set B of blocks (so that $\mathbf{T}$ is a multiple network on B), and let $\mathbf{R} = \{R_1, R_2, \ldots, R_p\}$ be a multiple network on a set $X = \{1, 2, \ldots, n\}$. Let f be a mapping from the set X of members of the multiple network $\mathbf{R}$ onto the set B of blocks in $\mathbf{B}$. For each block $b \in B$, let n_b denote the number of elements i in X for which $f(i) = b$. Then T is

1 a *fat-fit* to $\mathbf{R}$ if $(b, c) \in T_k$ iff $(i, j) \in R_k$, for all $i, j \in X$ such that $f(i) = b$ and $f(j) = c$; $k = 1, 2, \ldots, p$;
2 a *lean-fit* to $\mathbf{R}$ if $(b, c) \in T_k$ iff $(i, j) \in R_k$, for some $i, j \in X$ for which $f(i) = b$ and $f(j) = c$; $k = 1, 2, \ldots, p$;
3 an *α-blockmodel* for $\mathbf{R}$ if $(b, c) \in T_k$ iff $(i, j) \in R_k$, for at least $\alpha n_b n_c$ pairs of elements (i, j) for which $f(i) = b$ and $f(j) = c$; $k = 1, 2, \ldots, p$ (Arabie et al., 1978); and
4 an *$(\alpha_1, \alpha_2, \ldots, \alpha_p)$-blockmodel* for $\mathbf{R}$ if $(b, c) \in T_k$ iff $(i, j) \in R_k$, for at least $\alpha_k n_b n_c$ pairs of elements (i, j) for which $f(i) = b$ and $f(j) = c$; $k = 1, 2, \ldots, p$.

A blockmodel, and multirelational data sets for which it is a fat fit, a lean fit and an α-blockmodel ($\alpha = 0.5$), are illustrated in Table 1.6. In the second panel of Table 1.6, the density of each network relation for each pair of blocks in the blockmodel is presented. The density of a relation for a pair of blocks is the proportion of existing to possible links among block members on the relation and may be defined formally as follows:

DEFINITION. Let $\mathbf{R} = \{R_1, R_2, \ldots, R_p\}$ be a multiple network on the set X, and let f be a mapping from X onto a set B of blocks. If b, c are blocks in B, and if n_b and n_c are the number of elements mapped by f onto b and c, respectively, then the *density* of the pair (b, c) for the relation R_k

Table 1.6. *A blockmodel and multiple networks for which it is a fat fit, a lean fit and an α-blockmodel* ($\alpha = 0.5$)

Blockmodel B		Network **U** for which **B** is a fat fit		Network **V** for which **B** is a lean fit		Network **W** for which **B** is an α-blockmodel (α = 0.5)	
R_1	R_2	R_1	R_2	R_1	R_2	R_1	R_2
1 0	0 1	111 00	000 11	011 00	000 10	011 00	000 10
0 1	0 0	111 00	000 11	000 00	000 11	100 10	001 01
		111 00	000 11	110 00	000 00	110 00	100 11
		000 11	000 00	000 11	000 00	100 01	000 00
		000 11	000 00	000 00	000 00	001 10	100 00

Corresponding density matrices

U		V		W	
R_1	R_2	R_1	R_2	R_1	R_2
1.00 0.00	0.00 1.00	0.44 0.00	0.00 0.50	0.56 0.17	0.22 0.67
0.00 1.00	0.00 0.00	0.00 0.50	0.00 0.00	0.33 0.50	0.17 0.00

is $m^k_{bc}/n_b n_c$, where m^k_{bc} is the number of pairs (i, j) for which $f(i) = b$, $f(j) = c$ and $(i, j) \in R_k$; $i, j \in X$. If there are no loops permitted in R_k, then the density of the pair (b, b) for the relation R_k is defined as $m^k_{bb}/n_b\ (n_b - 1)$ (Breiger et al., 1975).

Given this definition, it may be seen that a fat-fit blockmodel possesses pairs of blocks whose densities for relations in **R** are either 0 or 1, and that $(b, c) \in T_k$ if and only if the pair (b, c) of blocks has density 1 for the relation R_k. The block $(b, c) \in T_k$ for a lean-fit blockmodel if the pair (b, c) for the relation R_k has density greater than 0. For an α-blockmodel, the density corresponding to interblock links must be at least α; and for an $(\alpha_1, \alpha_2, \ldots, \alpha_p)$-blockmodel, the pair (b, c) must have density for the relation R_k of at least α_k.

As noted, a blockmodel may be specified independently of data to which it pertains, or it may be derived from it. An example of the first approach is provided by Breiger's (1979) account of the blockmodels predicted by various theories of the nature of community power structures. In this case, the algorithm BLOCKER (Heil and White, 1976) may be used to assess whether (and if so, in what ways) a given set of data provides a lean fit to a specified blockmodel. Permutation tests (Hubert, 1987) may also be used to assess the fit of a blockmodel to network data.

In the case of deriving a blockmodel from the data, analysis has typically been accomplished by the use of a hierarchical clustering algorithm, such as CONCOR (Breiger et al., 1975) or STRUCTURE (Burt, 1976). More recently, Arabie, Hubert and Schleutermann (1990) have investigated the identification of blockmodels using the Bond-Energy approach. In the case of hierarchical clustering schemes, a matrix of distances or similarities among people is generated, with persons being at zero distance or possessing maximum similarity if, and only if they are structurally equivalent. The resulting matrix is clustered to yield a division of the group of persons into subgroups of almost structurally equivalent people. (The usual problems attendant on cluster analysis, such as the selection of the most appropriate proximity measure and method of clustering and the most useful number of clusters in the obtained solution, are also issues for blockmodel analysis; these matters are discussed in some detail by Arabie & Boorman, 1982; Burt, 1986a; Faust, 1988; Faust & Romney, 1985; Faust & Wasserman, 1992; Pattison, 1988.) From the subgroups, or clusters, thus obtained, some permutation of the group members is inferred and imposed upon the original relational data matrices. The permuted data matrices may then be partitioned according to the division of the group into clusters, or blocks, so that the density of relational ties present within any submatrix may be computed. Use of a cutoff-density of α then yields an α-blockmodel for the data: interblock relations of a given type having a density greater than α are coded as being present, those with a density of α or less, as absent. (For example, the blockmodel **B** in Table 1.6 is an α-blockmodel for the network shown on the right-hand side of Table 1.6, with $\alpha = 0.5$.) More detailed accounts of blockmodelling and the theoretical and practical issues that it raises have been given by Arabie and Boorman (1982), Arabie et al. (1978), Breiger (1976, 1979), Breiger et al. (1975), Faust & Wasserman (1992), Wasserman and Faust (1993) and White et al. (1976).

6. *Probabilistic models.* Probabilistic network models express each edge in a single or multiple network as a stochastic function of vertex and network properties (e.g., Fararo, 1981, 1983; Fararo & Skvoretz, 1984; Holland & Leinhardt, 1981; Holland, Laskey & Leinhardt, 1983; Wasserman & Galaskiewicz, 1984). For instance, Holland and Leinhardt's (1981) p_1 model for a single network relation expresses the probability of a network tie between two members in terms of "density" and "reciprocity" parameters for the network and "productivity" and "attractiveness" parameters for each member.

The preceding six categories of representation are not mutually exclusive: for instance, blockmodels can sometimes be considered as belonging to categories 3 and 4 as well as to 5. Moreover, some representations have now been constructed so as to take advantage of the structural features of more than one category; for example, stochastic blockmodels propose that each edge of a blockmodel is a stochastic function not only of vertex and edge properties but also of relations between blocks (Anderson, Wasserman & Faust, 1992; Fienberg, Meyer & Wasserman, 1985; Wasserman & Anderson, 1987; Wang & Wong, 1987).

The representations listed in Table 1.5 also vary considerably in the generality of the form of redundancy assumed to underlie them. Blockmodels, for example, subsume balance models as a special case, and overlapping clique models admit nonoverlapping cliques as particular instances. A number of graph, vertex and edge indices may be derived, or at least approximated, from knowledge of a blockmodel representation for a given set of network data, and approximate spatial models may often also be constructed. Indeed, a substantial empirical literature attests to the high level of generality represented by the blockmodel approach (e.g., Breiger, 1976, 1979; Mullins et al., 1977; Snyder & Kick, 1979; Vickers & Chan, 1980; White et al., 1976) even though such possibilities as overlapping blocks are proscribed by them. Moreover, promising developments in the area of stochastic blockmodels mean that the advantages associated with probabilistic representations may eventually come to be associated with this class of models (Holland et al., 1983; Wang & Wong, 1987; Wasserman & Anderson, 1987).

Properties of a structural representation

For a particular research problem, an investigator using social networks needs to select one or more representations of network structure from the large array summarised in Table 1.5. We have noted that the representations differ in their generality, so that in the absence of any clear theoretical direction about an appropriate form, it is wise to choose a more general representation. Moreover, we shall argue that to obtain a general representation sympathetic to the two main themes for network research that we have outlined, we should select a representation that admits multiple relations and that is sensitive to paths in networks.

Multiple relations. It is now widely recognised that different types of interpersonal connections operate in different ways to explain various network phenomena (e.g., information flow; Granovetter, 1973; Kapferer, 1972; Lee, 1969; Mitchell, 1969). Few of the representations of Table 1.5 allow for the simultaneous consideration of multiple relations. Indeed, the distinctive feature and rationale for many of the blocked matrix and relational representations listed under the fifth category of Table 1.5 is their capacity for joint representation of multiple relations. A goal of blockmodel construction, for instance, is the description of groups of persons who have identical relations with other network members across a number of different *types* of relations.

There are also grounds for making qualitative distinctions between network ties on the basis of their strength. Granovetter (1973) argued that a strong tie between a pair of individuals is involved in a different set of social processes than a "weak" one and correspondingly is interrelated with other ties in a different way. He reviewed a body of evidence suggesting the power of weak ties to penetrate social boundaries that are impermeable to stronger ties, and he attributed that power to the transmission potential resulting from the open-ended, far-reaching nature of the networks that they form. Similar conclusions followed from Lee's (1969) study of networks of communication involved in finding an abortionist, wherein weak ties provided the paths for the process of information-seeking and strong ties were effectively excluded from such paths. Thus, in some circumstances it may prove useful to treat strong and weak social ties as distinct, rather than construct a single, valued relation whose values signify the strength of the tie. Clearly, this strategy is available only if a multirelational representation of the data is being entertained.

Paths in networks. To some degree, a blockmodel can be seen as providing a formal representation for the local or interpersonal environment of a person in a network. The block to which a person is assigned and the relations of the block with other blocks in the model idealise the position and associated role of the person in the network. Thus, at least one of the major theoretical requirements for a network representation is satisfied by a blockmodel. The other main requirement is the need to specify the paths for the flow of social traffic implied by the network and hence to permit the calculation of the large-scale effects of local network processes.

It is important to note first that the structural constraints implicit in a network of interpersonal connections have important implications for social processes. Network connections are avenues for the potential flow of such social phenomena as information, influence, attitudes, hindrance and uncertainty. Consider, for example, a network in which person B is an acquaintance of person A and person C is, in turn, an acquaintance of person B. Then C is an acquaintance of an acquaintance of A: that is, there is an acquaintance path of length 2 from A to C. If A passes information to B, then it is claimed that C is a potential recipient of that information from B. In this sense, the path of length

2 from *A* to C provides one possible channel for the spread of information from *A* to C.

A number of empirical demonstrations of social processes flowing along paths in networks have been made, including Bott's (1957) account of conjugal role performance, Barnes's (1954) analysis of social behaviour in a Norwegian fishing village and the study of "hysterical contagion" in a textile plant by Kerckhoff, Back and Miller (1965; see also Boissevain, 1974; Boissevain and Mitchell, 1973; Kapferer, 1969; Mitchell, 1969). Similar demonstrations have been made for the spread of infectious diseases (e.g., Klovdahl, 1985). Some observers have even documented social processes in action on previously measured networks. Laumann and Pappi (1973, 1976), for example, showed how divisions of opinion on local issues among members of a community elite were consistent with properties of the interpersonal network, as did Doreian (1988a) in a political network of a Midwestern county. Kapferer (1969) recorded how a person's social network could be used to mobilise support in a crisis.

Others such as Granovetter (1973, 1974) have established the significance of network paths in understanding global social processes. In his study of job finding, Granovetter (1974) reviewed a range of studies undertaken in the United States, covering a wide selection of occupational groups, which demonstrated that approximately 60% of persons found their jobs through personal contacts, principally friends and relatives. Less than 20% obtained employment by formal means, that is, by responding to advertisements and working through employment agencies.

Such data illustrate the degree to which "local" events, occurring within the local structural framework specified by the interpersonal network, can have significant structural implications on a global scale. Granovetter (1974) noted the disjunction between the micro-level treatments of the topic of job-finding, that is, those studies which offered plausible psychological and economic accounts of the motives of job-finding and the macro-level analyses, which concerned the statistics of persons flowing between various occupational categories. He argued that the structure of an individual's personal network largely determined the information available to a person and hence the possible courses of action open to that person. Thus, he claimed, consideration of networks of personal contacts is a potentially crucial link in the integration of these two analytic levels.

The cognition of compound ties. Diffusion and contagion studies (e.g., Coleman, Katz and Menzel, 1957; Kerckhoff et al., 1965) illustrate the real part played by paths in networks in the operation of social processes.

A more subtle kind of reality is afforded them in some instances in the form of an awareness of the paths by those whom they link. Kinship relations provide some common examples: the path formed by tracing the relation "is a brother of" followed by "is a mother of" is equivalent to the relation "is a (maternal) uncle of". The path describes a compound relation, "mother's brother", which is perceived, in general, and given a single term of reference in recognition of its salience. Similarly, one is cognisant, in general, of those to whom one bears a relationship such as "daughter's sister's son". In principle, moreover, any composite relation formed from the kin generator relations ("is a wife of" and its transpose "is a husband of", and "is a father of" and its transpose "is a child of") may be given meaning by those using such terms. In a similar way, the composition of the relation "is a boss of" with "is a friend of" is often perceived by both boss and friend, as is "is a friend of" with itself. One's friend's friends often comprise a listable set, as do one's friend's enemies, boss's boss's bosses, and so on. Perhaps a more convincing argument for the conscious recognition of those to whom one bears a compound tie is provided by studies such as that of Mayer (1977) in which it was shown how a person in a given network position sets out to manipulate those to whom that person bears a compound relation, by calling upon intermediaries to influence those one step closer to the target persons. Similarly, Granovetter's (1974) results demonstrated the relatively high frequency with which employment is learnt of and procured through the contacts of one's acquaintances. Whyte (1943) documented a related phenomenon for Cornerville:

> According to Cornerville people, society is made up of big people and little people – with intermediaries serving to bridge the gap between them. The masses of Cornerville people are little people. They cannot approach the big people directly but must have an intermediary to intercede for them. They gain this intercession by establishing connections with the intermediary, by performing services for him and thus making him obligated to them. The intermediary performs the same functions for the big man. The interactions of big shots, intermediaries and little guys builds up a hierarchy of personal relations based upon a system of reciprocal obligations. (pp. 271–2)

An obvious restriction on the extent to which individuals are aware of their compound relations is imposed by the length of corresponding network paths. Compound relations may not be traced indefinitely by the parties involved. The maximum length of traceable compounds might be expected to vary as a function of structural constraints operating.

Thus, for example, the ranks institutionalised in the army and the grade structure of the public service make the tracking of quite long compounds of the form "superordinate's superordinate's superordinate's . . ." a relatively easy task in their defining populations. In comparison, the tracing of "friend's friend's friend's . . ." in seemingly less-structured populations seems much more difficult, as the results of Bernard and Killworth (1973, 1978) suggest.

The representations with which we deal here make no particular assumptions about whether compound relations are transparent to the individuals that they link. The extent to which such relations are transparent, however, is a relevant consideration in the construction of social process models and requires further empirical study in the longer-term project of building network process models with plausible structural assumptions.

In sum, there is some evidence for the claim that paths in networks can give rise to extremely powerful social phenomena. The description of the form they may take is therefore essential to models of the social processes with which they are associated. It might be expected that the structural constraints presumed to obtain within a given social system (as determined by the methods listed in Table 1.5) would have a substantial impact on its path geometry and so on associated social process models. Despite the critical nature of such constraints for process models, however, most to date have assumed them to be extremely strong. As Boorman (1975) observed, "formidable difficulties arise when one attempts to deal in generality with networks of arbitrary topology" (p. 220). Some limited progress has been made in the development of contagion and diffusion models. For example, models characterised by the well-mixedness assumption, that is, by the assumption that the probability of contact between any pair of individuals from a specified pair of states is a constant, have been extended by the introduction of statistical biases towards symmetry, transitivity and circularity (Fararo & Skvoretz, 1984; Foster & Horvath, 1971; Foster, Rapoport & Orwant, 1963; Rapoport, 1957, 1983; Rapoport & Horvath, 1961). A more general basis for the exploration of such biases has been described by Frank and Strauss (1986; also Strauss & Freeman, 1989).

Renewed interest in stochastic models promises new advances in models for the flow of processes along network paths. To lay the groundwork for such advances, it is helpful to consider what formal operations are available for the representation of network flows.

An algebra for complete social networks

The key construction for the representation of network flows is widely recognised to be some form of composition operation (e.g., Ford &

Fulkerson, 1962; Lorrain, 1972; Peay, 1977b). Different network representations have been associated with different definitions of a composition operation; for valued networks, Peay (1977b) has outlined the various substantive interpretations that may be associated with a number of them (also Doreain, 1974). A widely used operation for binary networks is that of relational composition (Boorman and White, 1976; Breiger & Pattison, 1978; Lorrain, 1975; Lorrain & White, 1971).

The approach based on relational composition was introduced in different formal terms and with somewhat different theoretical emphases in Lorrain and White (1971) and Lorrain (1972, 1975) and in Boorman and White (1976). The algebraic construction is, however, essentially the same in each instance and operates upon a collection of multiple networks. The networks may refer either to relations among individuals or to relations among other social units, such as blocks in a blockmodel. Two principles underlie the construction: (a) the salience of compound relations or paths in networks and (b) a natural means of comparing them. In what follows, we refer to the social units of the network as individuals, but the definitions apply equally well to relations among other types of social units, such as blocks or groups of individuals.

Compound relations and network paths

Consider three individuals A, B and C in the multiple network **W** of Figure 1.4. It can be seen that individual C names B as a friend, and B seeks help from individual A. Thus, A is a helper to a friend of C, that is, A is one of C's friend's helpers. We say that C stands in the *compound* relation FH to A, or that C is *connected* to A by the compound relation, or compound path, FH. In fact, the compound relation or compound role FH links C to A via a path of length 2, the first step of which is a relationship of type F and the second, a relationship of type H. The relationship exists irrespective of any other relations enjoyed by C and A: the defining feature of the compound is the existence of some person to whom C bears the relation F and who, in turn, is related by H to A. More than one such person may link C and A in this way: the strategy of recording only the existence of a path from C to A is a structural one; it is not designed to reflect the likelihood of a path from A to C of this type being activated, only that activation is possible. The set of all pairs of persons in a network connected by the compound relation FH define a new network relation among members of the network. The relation has the same form as the original collection of binary relations and may be represented as a directed graph, a binary relation or a binary matrix.

Now, to construct the compound relation FH, we may reason as follows. Individual i is connected to individual j by the compound

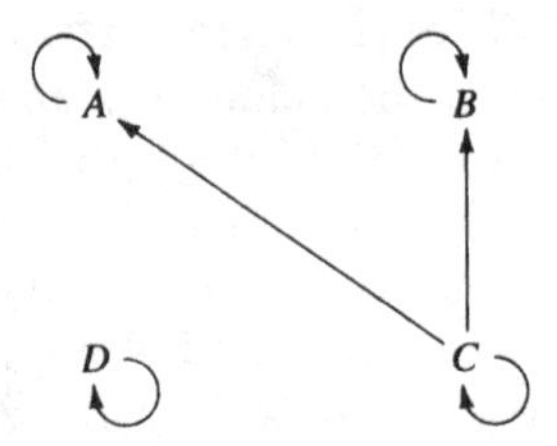

Figure 1.6. The compound relation *FH*

relation *FH* if a friend of i seeks help from j, that is, if j is a helper to a friend of i, or j is one of i's friend's helpers. For instance, if we consider the elements of the network in turn, we see that *A* is related to *B* by *F*, who is related to *A* by *H*. Thus, *A* is related to itself by the relation *FH*. Element *B* is similarly related to itself by *FH*, whereas *C* is related to *A*, *B* and *D* by *F* and hence to itself, *A* and *B* by *FH*. Element *D* is related to *C* by *F*, and *C* is related only to *D* by *H*, so that the relation *FH* links *D* only to itself. Thus, we obtain the compound relation

$$FH = \{(A, A), (B, B), (C, A), (C, B), (C, C), (D, D)\},$$

which is shown in Figure 1.6 in directed graph form.

The matrix of the compound relation *FH* may also be constructed directly from the matrices of the relations *F* and *H*. The construction relies on the operations of *inclusive or*, represented by the symbol $\vee$, and *and*, represented by the symbol $\wedge$. These operations act on the matrix entries of 0 and 1, as follows:

$$0 \vee 0 = 0, \quad 0 \vee 1 = 1, \quad 1 \vee 0 = 1, \quad 1 \vee 1 = 1$$

$$0 \wedge 0 = 0, \quad 0 \wedge 1 = 0, \quad 1 \wedge 0 = 0, \quad 1 \wedge 1 = 1.$$

Both the *inclusive or* and *and* operations are associative. That is, for the *and* operation, $(x \wedge y) \wedge z = x \wedge (y \wedge z)$, where each of x, y and z is either 0 or 1, and we may write both expressions as $x \wedge y \wedge z$ without ambiguity. Similarly, for the *inclusive or* operation, $(x \vee y) \vee z = x \vee (y \vee z)$, and both expressions may be written as $x \vee y \vee z$. The *Boolean product* of the matrices of relations *F* and *H* can then be expressed as

$$FH_{ij} = (F_{i1} \wedge H_{1j}) \vee (F_{i2} \wedge H_{2j}) \vee \cdots \vee (F_{in} \wedge H_{nj}).$$

That is, cell (i, j) of *FH* has a unit entry if, for any element k, the (i, k) cell of *F* and the (k, j) cell of *H* both have unit entries; otherwise, cell (i, j) of *FH* contains an entry of zero. Thus, the Boolean product operation leads to a product matrix *FH* that is also a binary matrix of the same size as *F* and *H*.

The expression of the Boolean product in the preceding form makes clear its relationship to ordinary matrix multiplication, which may be expressed in the form:

$$(FH)_{ij} = F_{i1}H_{1j} + F_{i2}H_{2j} + \cdots + F_{in}H_{nj}.$$

That is, in the ordinary matrix product of F and H, the (i, j) cell of the product matrix is the sum of products of entries in cells (i, k) of F and (k, j) of H, and because both F and H are binary, this is the number of elements k for which both cells have a unit entry. In the Boolean product, the operations of multiplication and summation in this expression are replaced by those of *and* and *inclusive or*, so that the outcome is 1 if there is at least one element k for which the (i, k) cell of F and the (k, j) cell of H are both 1 and 0 otherwise. In other words, an entry in the Boolean product of F and H is 1 if and only if it is at least 1 in the ordinary matrix product of F and H; otherwise, it is zero.

We can construct an infinite set of compound relations from F and H, each corresponding to a possible string of the relations F and H. For example, from the relations F and H, we can construct the compound relations

$$FF,\ FH,\ HF,\ HH,\ FFF,\ FFH,\ FHF,\ FHH, \qquad \text{and so on.}$$

Some of these compound relations are displayed in Figure 1.7 in directed graph form and in Table 1.7 in binary matrix form. The first four compound relations are constructed from two primitive relations in the network and define paths in the network of length 2. For instance, FF records the existence of paths comprising two friendship links among members of the network. Similarly, FH records the existence of paths of length 2 whose first link is a friendship relation and whose second link is a help-seeking relation. Paths of length 3 may be constructed by finding the composition of a compound relation corresponding to paths of length 2 and a primitive relation. For instance, the relation FHF may be constructed by finding the composition of the relation FH and the relation F. In matrix terms, for instance, we obtain

$$\begin{bmatrix} 1\,0\,0\,0 \\ 0\,1\,0\,0 \\ 1\,1\,1\,0 \\ 0\,0\,0\,1 \end{bmatrix} \begin{bmatrix} 0\,1\,0\,0 \\ 1\,0\,0\,0 \\ 1\,1\,0\,1 \\ 0\,0\,1\,0 \end{bmatrix} = \begin{bmatrix} 0\,1\,0\,0 \\ 1\,0\,0\,0 \\ 1\,1\,0\,1 \\ 0\,0\,1\,0 \end{bmatrix}.$$

The same relation may be constructed by finding the composition of the relation F with the relation (HF) because it can readily be established that the order of calculation of compounds in a string is not important; we do not need to distinguish, say, $(FH)F$ from $F(HF)$ and so can write both as FHF. This property of the composition operation is termed

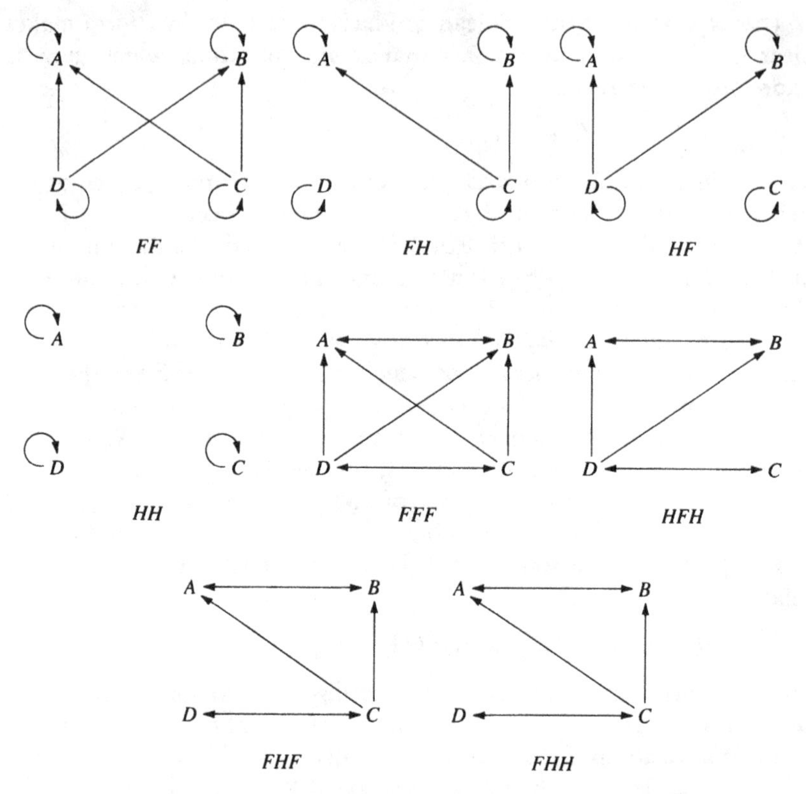

Figure 1.7. Some compound relations for the network **W**

Table 1.7. *Some compound relations for the network* **W** *in binary matrix form*

FF	*FH*	*HF*	*HH*	*FFF*	*FFH*	*FHF*	*FHH*	*HFF*	*HFH*
1000	1000	1000	1000	0100	0100	0100	0100	0100	0100
0100	0100	0100	0100	1000	1000	1000	1000	1000	1000
1110	1110	0010	0010	1101	1101	1101	1101	1101	0001
1101	0001	1101	0001	1110	1110	0010	0010	1110	1110

associativity. Note, though, that the order of each relation in the string is important; so *FHF* is not necessarily the same relation as *HFF*.

The operation of forming compound relations is termed *composition*, or *concatenation*; according to Lorrain and White (1971, it represents "the basic logic of interlock" (p. 54) in the system of relationships. Its

Table 1.8. *The blockmodel* $\mathbf{N} = \{L, A\}$

L	A
1 1 0 0	1 0 1 1
1 1 0 0	1 0 1 0
0 0 1 0	1 0 0 1
0 0 1 1	1 0 0 0

construction permits one to take into account "the possibly very long and devious chains of effects propagating within concrete social systems through links of various kinds" (Lorrain and White, 1971, p. 50).

In other words, compound relations are claimed to define the paths along which social processes flow: those whom they link may or may not be aware of them. An awareness of compound relationships by the individuals concerned, however, is not essential to their usefulness; Lorrain (1972) argued that "any concatenation of social relationships is itself a social relationship, whether perceived or not" (p. 9).

Formally, the composition of relations may be defined as follows:

DEFINITION. The *composition* of two binary relations R and T on a set X is given by

> $(i, j) \in RT$ iff there exists some $k \in X$ such that $(i, k) \in R$ and $(k, j) \in T$; $i, j, \in X$.

RT is termed a *compound relation.* (Composition is also referred to as *multiplication* and the compound relation RT as the *product* of R and T.) If $\mathbf{R} = \{R_1, R_2, \ldots, R_p\}$ is a network defined on a set X, and if $T_h \in \mathbf{R}$ for each $h = 1, 2, \ldots, m$, then $T_1T_2 \cdots T_m$ is a compound relation of *length m*. In particular, $(i, j) \in T_1T_2 \cdots T_m$ if and only if there exists a sequence $i = k_0, k_1, \ldots, k_m = j$ of elements of X, such that $(k_{h-1}, k_h) \in T_h$; $h = 1, 2, \ldots, m$. We say that there is a *labelled path* of length m from i to j, with the labels $T_1, T_2, \ldots, T_m$, respectively.

As noted earlier, the definition applies to multiple networks defined on any kind of social unit. In particular, it applies to networks defined among individuals, as well as to blockmodels constructed from them. As an example of the latter case, we construct compound relations for the blockmodel reported in Table 1.8. The blockmodel is labelled **N** and possesses two relations defined on four blocks. The blockmodel was constructed by White et al. (1976) for relations of Liking (L) and Antagonism (A) among the members of the fraternity reported by Nordlie (1958; also Newcomb, 1961). All primitive and compound relations of length 2 and 3 generated by L and A are reported in Table 1.9.

Table 1.9. *Primitive relations and compound relations of lengths 2 and 3 for the blockmodel* **N**

L	*A*	*LL*	*LA*	*AL*	*AA*	*LLL*
1100 1100 0010 0011	1011 1010 1001 1000	1100 1100 0010 0011	1011 1011 1001 1000	1111 1110 1111 1100	1011 1011 1011 1011	1100 1100 0010 0011
LLA	*LAL*	*LAA*	*ALL*	*ALA*	*AAL*	*AAA*
1011 1011 1001 1000	1111 1111 1111 1111	1011 1011 1011 1011	1111 1110 1111 1100	1011 1011 1011 1011	1011 1011 1011 1011	1011 1011 1011 1011

Comparing paths in networks and the Axiom of Quality

The second principle underlying the representation that we construct is the notion that it is useful to make comparisons among the collection of all primitive and compound relations defined on a network. If one relation U links pairs of individuals who are all also linked by a relation V, then we can think of the relation U as being *contained* in the relation V. For instance, if individuals in a work group only approach their friends for help with work-related problems, then we would represent this state of affairs using the ordering

$$H \leq F,$$

where F and H denote the Friendship and Help relations, respectively. For instance, this ordering holds in the network **W** displayed in Figure 1.4 because each individual seeks help from an individual who is also named as a friend.

In formal terms, this notion leads to a partial ordering among binary relations on a set:

DEFINITION. Let W and V be binary relations on a set X. Define

$$W \leq V$$

if $(i, j) \in W$ implies $(i, j) \in V$, for all $i, j \in X$.

It is easy to establish that the relation $\leq$ has the properties of a mathematical *quasi-order*, that is, it is *reflexive* ($W \leq W$, for any binary relation W on X) and *transitive* ($W \leq V$ and $V \leq U$ imply $W \leq U$). We may also define the converse relation $\geq$ by

$W \geq$ iff $V \leq W$.

Moreover, from the relations $\geq$ and $\leq$, an equality relation can be constructed:

$$W = V \text{ iff } W \leq V \text{ and } V \leq W.$$

From the relations $\leq$ and $=$, in turn, the relation $<$ may be defined, using

$$W < V \text{ iff } W \leq V \text{ but not } W = V.$$

The relation $>$ is defined similarly.

Now, if $W = V$, then W and V link precisely the same pairs of individuals. Boorman and White (1976) termed the equation of two such relations the Axiom of Quality:

DEFINITION (Axiom of Quality). Let W and V be binary relations on X. Define

$$W = V \text{ iff } W \leq V \text{ and } V \leq W.$$

For example, the relation FFF and the relation F from the network **W** are equated by the Axiom of Quality. Indeed, we can show that

$$F = FFF = FHF = HHF = FFFFF$$

all hold, as do many other equations to F. Similarly, $FF = FFFF = FFFH$, and $HHH = H$; also $HHHH = HH$, $HFHF = HF = HFHH$ and so on. Now the collection of compound relations generated by a given collection of binary relations among members of a network is infinite in number. The Axiom of Quality states that any two relations (compound or otherwise) that define exactly the same set of connections among persons are to be equated. Thus, because there are only a finite number of distinct binary relations that can be defined on a finite set X, there are also only a finite number of distinct relations (compound or otherwise) in the set generated by a given collection of multiple, binary relations. In fact, at most there are 2^k, where $k = n^2$ and n is the number of elements in X.

We can find the set of distinct binary relations by an iterative process. We begin with the distinct primitive relations in the network and then construct all compound relations corresponding to paths of length 2 in the network. Beginning with F and H, for instance, we construct FF, FH, HF and HH. We then compare these binary relations (FF, FH, HF and HH) with the primitive relations and with each other and select any that are distinct. Table 1.7 indicates that, in this case, all of the relations are distinct. We then form a compound relation of length 3 from *each* of these compound relations of length 2 and *each* of the primitive relations and again examine the list for new distinct relations. That is, we construct *FFF*, *FFH*, *FHF*, *FHH*, *HFF*, *HFH*, *HHF* and *HHH*. In this list, only *FFF* and *HFH* are distinct for the network **W**. The other relations are equal to a relation already generated: that is, $FFH = FFF$, $FHF = F$, $FHH = F$, $HFF = FFF$, $HHF = F$ and $HHH = H$. In the next round, therefore, we construct compound relations of length 4 from distinct compound relations of length 3 and the primitive relations. Hence, we compute *FFFF*, *FFFH*, *HFHF* and *HFHH* and find that no new distinct relations are obtained. We have thus found the eight distinct relations that may be generated from the relations *F* and *H* in the network **W**.

The partially ordered semigroup of a network

The set of distinct relations that we find by this process defines an algebraic structure termed a *partially ordered semigroup*. The composition operation acts as a binary operation satisfying an algebraic identity known as the associative law (defined later). The equations among (compound) relations, that is, the outcome of identifying all those relations defining the same set of interpersonal connections in the group, are translated into equations of the semigroup, encoding the specific structural interrelations among the given collection of relations. The semigroup equations and partial orderings record the relations between the network relations and represent relational structure in a network in a way that does not refer to specific sets of network links.

In fact, the distinct primitive and compound relations of the network, together with the composition operation, possess the properties of both a semigroup and a partially ordered semigroup.

DEFINITION. A *semigroup* is a set of elements *S* and an associative binary operation on *S*. The *binary operation* is a mapping of ordered pairs of elements of *S* onto a single element of *S* and is usually represented as a product, with the pair of elements (U, V) from *S* being mapped onto the single element *UV* of *S*. The binary operation is associative, that is, it has the property

$$(UV)W = U(VW);$$

for any $U, V, W \in S$ (e.g., Clifford & Preston, 1961).

In the case of the network relations, *S* is the set of distinct primitive and compound relations and the binary operation on *S* is the composition operation. For the network **W** of Figure 1.4 the set *S* is $\{F, H, FF, FH, HF, HH, FFF, HFH\}$. The composition operation is defined for any pair of members of *S*, and the result of computing the composition of any two elements in *S* is again an element in *S* (because *S* contains all distinct primitive and compound relations). The outcome of all possible compositions of pairs of elements in S may be recorded in a table, termed the *multiplication table* of the semigroup. The left half of Table 1.10 contains the multiplication table for the semigroup of the network **W**. The multiplication table has rows and columns corresponding to the primitive and compound relations in S, and the table entry corresponding to a particular row and column records the distinct relation obtained when the composition of the row relation and the column relation is computed. For instance, the entry in the first row and column contains the composition of F and F, namely FF; the entry in the first row and second column contains the result FH of composing the first row relation F with the second column relation H. Similarly, the entry in the fourth row and second column, F, is the result of composing FH with H and corresponds to the equation $FHH = F$, which was previously found to hold.

As well as possessing the structure of a semigroup under the composition operation, the set S of distinct relations obtained from a network may be shown to give rise to a partially ordered semigroup.

DEFINITION. A semigroup S together with a partial ordering $\leq$ among its elements is a *partially ordered semigroup* if $U \leq V$ implies $WU \leq WV$, and $UW \leq VW$, for any $W \in S$ (e.g., Fuchs, 1963).

It may be established that if $U \leq V$ holds for two binary relations on a set X, then $WU \leq WV$ and $UW \leq VW$ also both hold for any binary relation W on X. Hence, the distinct relations generated by a multiple network **R** also constitute a partially ordered semigroup $S(\mathbf{R})$. The partially ordered semigroup may be presented in the form of two tables, a *multiplication* table, as before, and a *partial order* table.

The partial order table also has rows and columns corresponding to the distinct relations generated by the network. The entry in the ith row and jth column is 1 if the ith distinct relation is greater than or equal to the jth distinct relation and zero otherwise. The partial order table for the partially ordered semigroup $S(\mathbf{W})$ of the network **W** is presented in the right half of Table 1.10. For instance, the ordering relation

$$FF \geq FH$$

for the network **W** is indicated by the unit entry in the third row and fourth column of the partial order table for the relations of the network **W**.

A partial ordering may also be conveniently represented by a partial order diagram (e.g., Birkhoff, 1967; also known as a Hasse diagram, e.g., Kim & Roush, 1983), constructed as follows.

DEFINITION. Let S be a partially ordered set (i.e., a set S and a partial ordering $\leq$) and let $s, t \in S$. The element s *covers* the element t if $t < s$,

Table 1.10. *The multiplication table and partial order for the partially ordered semigroup* $S(\mathbf{W})$

Multiplication table

	F	*H*	*FF*	*FH*	*HF*	*HH*	*FFF*	*HFH*
F	*FF*	*FH*	*FFF*	*FFF*	*F*	*F*	*FF*	*FH*
H	*HF*	*HH*	*FFF*	*HFH*	*F*	*H*	*FF*	*FH*
FF	*FFF*	*FFF*	*FF*	*FF*	*FF*	*FF*	*FFF*	*FFF*
FH	*F*	*F*	*FF*	*FH*	*FF*	*FH*	*FFF*	*FFF*
HF	*FFF*	*HFH*	*FF*	*FF*	*HF*	*HF*	*FFF*	*HFH*
HH	*F*	*H*	*FF*	*FH*	*HF*	*HH*	*FFF*	*HFH*
FFF	*FF*	*FF*	*FFF*	*FFF*	*FFF*	*FFF*	*FF*	*FF*
HFH	*HF*	*HF*	*FFF*	*HFH*	*FFF*	*HFH*	*FF*	*FF*

Partial order

	F	*H*	*FF*	*FH*	*HF*	*HH*	*FFF*	*HFH*
F	1	1	0	0	0	0	0	0
H	0	1	0	0	0	0	0	0
FF	0	0	1	1	1	1	0	0
FH	0	0	0	1	0	1	0	0
HF	0	0	0	0	1	1	0	0
HH	0	0	0	0	0	1	0	0
FFF	1	1	0	0	0	0	1	1
HFH	0	1	0	0	0	0	0	1

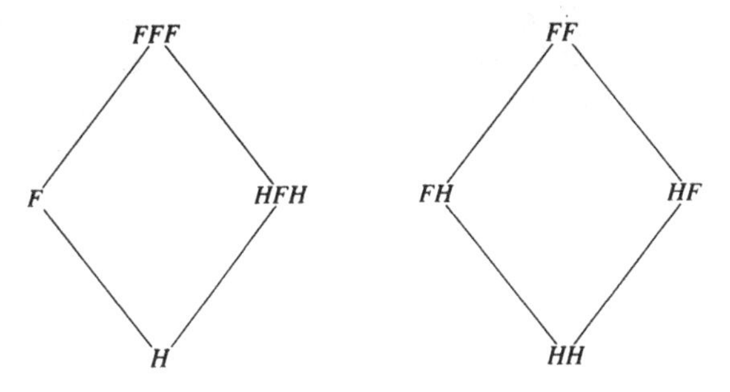

Figure 1.8. Hasse diagram for the partial order of $S(\mathbf{W})$

and $t \leq w \leq s$ implies $w = s$ or $w = t$. The *Hasse diagram* or *partial order diagram* of S is a graph whose vertices are the elements of S and in which an edge is drawn between s and t, with s above t, if s covers t.

In other words, the Hasse diagram of a partial order suppresses any orderings which are implied by other orderings. Consider, for example, the Hasse diagram for the partial ordering of the semigroup $S(\mathbf{W})$ of the network $\mathbf{W}$ which is presented in Figure 1.8. It can be seen from Table 1.10 that $FFF > F$ and also that $F > H$. Hence the ordering $FFF > H$ is implied by the transitivity of the partial order and no link is drawn directly from FFF to H in the Hasse diagram. There is, however, a path from FFF to H in the diagram, with FFF drawn above H, so that the ordering $FFF > H$ may be inferred. There are no relations x distinct from FFF and F, however, for which $FFF > x$ and $x > F$. That is, FFF covers F and the Hasse diagram contains a direct link from FFF to F. The direction of the links is indicated by their position on the page: any relation *above* a second relation and connected to it is greater than the second relation. It may be observed by examining Figure 1.8 that *all* orderings appearing in Table 1.10 can be inferred from the Hasse diagram; for instance, FFF is greater than all those elements to which it is connected and lie below it, namely, F, HFH and H (and, of course, $FFF \geq FFF$ also holds).

As a second example, consider the blockmodel $\mathbf{N} = \{L, A\}$ of Table 1.8. It generates the distinct relations L, A, LA, AL, AA and LAL, which are presented in Table 1.9 in matrix form. Equations generated by equating compound relations with the same directed graph include

$$LL = L,\ LAA = AA,\ ALA = AA,\ AAL = LAL \text{ and } AAA = AA,$$

whereas orderings include

$$L \leq LAL,\ AL \leq LAL,\ A \leq LA \leq AA \leq LAL \text{ and } A \leq AL.$$

The equations and orderings record interrelations among the relations. The equation $LL = L$, for example, claims that each block likes those blocks liked by its friends, a claim that may be verified by examining the L matrix in Table 1.8. Other equations may be similarly interpreted in relational terms; the entire set of equations and orderings generated by a given network provides a comprehensive statement of relational interrelating in the network.

Some applications of the semigroup representation have made systematic use of the accompanying partial order (e.g., Fennell & Warnecke, 1988; Light & Mullins, 1979), some have used it occasionally (e.g., Boorman & White, 1976) and others have effectively ignored it (e.g., Bonacich & McConaghy, 1979; Breiger and Pattison, 1978; Lorrain, 1975). Clearly, the information conveyed by the partially ordered semigroup contains the information encoded in the abstract semigroup. Indeed, the latter is obtained from the former by dropping the partial order table. At this stage, no consensus has emerged on the role of the partial order in the representation, but there are at least two reasons for including it. Firstly, the comparison relation carries important substantive information: it is useful to know, for instance, that the collection of friends of one's business associates is contained in the set of one's business associates. Secondly, the representation for local networks presented in chapter 2 may be compared more directly with the partially ordered semigroup representation for entire networks. Using the partially ordered semigroup for entire networks therefore leads to a more unified treatment of structure in local and entire networks.

In sum, the definition of the network **R** as the "raw data" for a representation of social structure has been argued to be an appropriate relational basis. In some cases, that relational basis may also be usefully summarised in the form of a blockmodel **B**. It has then been proposed that the composition operation, applied to either the original network relations among network members or to relations among blocks in a blockmodel, yields a suitable formalisation for the tracing of paths in networks. The comparison of these paths or compound relations yields an algebraic structure, the partially ordered semigroup $S(\mathbf{R})$ for observed network relations, or $S(\mathbf{B})$ for blockmodel relations, that represents the interrelatedness of relations in **R** or **B**.

The decision about whether the partially ordered semigroup should be constructed from the original network relations or from a blockmodel derived from the network relations depends on a number of factors. Firstly, it depends on the level of structural detail required by the analysis: the detail is generally much greater for original data than for a derivative blockmodel. Secondly, and perhaps most importantly, the measurement characteristics of the network data need to be considered. A number of researchers have suggested that the blockmodel for a multiple network may be more robust than the network itself (e.g., Pattison, 1981; White et al., 1976), so that any derived algebraic structure for the blockmodel is likely to possess a greater degree of robustness as well. For instance, overlooking a member of a network, or mis-specifying a network relation, may have no effect on the blockmodel of the network and hence on the blockmodel semigroup; yet the partially ordered semigroup of the network data may be affected (also Boyd, 1991). Thirdly, there are practical considerations. The semigroup of a blockmodel is generally very much smaller than the semigroup of an observed network, and as a result it is usually easier to describe its equations and orderings. Although techniques developed in chapter 4 go some way to resolving this descriptive problem, most reported semigroup analyses have been for blockmodels rather than raw network data (e.g., Boorman & White, 1976; Breiger & Pattison, 1978).

Some support for the usefulness of the semigroup representation comes from studies of models of certain kinds of kinship systems, of which the semigroup representation is a generalisation. White (1963) showed that relations determining the clan of wife and child in certain marriage class systems characteristic of some Australian aboriginal tribes lead to the construction of permutation group models. He was able to classify possible kinship structures according to their marriage and descent rules and provide the foundation for a comparison of the various structures. Boyd (1969) extended the representation by proposing that various algebraic transformations relating different group representations be used to assist the latter task, especially that part of it directed to the analysis of the evolution of kinship structures over time. Boyd reviewed evidence supporting the usefulness of permutation group models and claimed that by using the mathematical analytical power afforded by the theory of groups, in particular, and algebraic systems, in general, one could investigate more fully both the relationships among various systems of the same type (e.g., the marriage systems of various tribes) and the interdependencies of such systems with those of a different type (e.g., componential analyses of their kinship terms).

An algorithm for semigroup construction

In practice, the finite semigroup S of a network $\mathbf{R} = \{R_1, R_2, \ldots, R_p\}$ may be constructed by the following algorithm. Let W_1 be the set of primitive relations in $\mathbf{R}$,

$$W_1 = \{R_1, R_2, \ldots, R_p\},$$

and assume that the R_i are distinct. Multiply each element of W_1 on the right (or dually, on the left) by R_1 to form the collection $\{R_1R_1, R_2R_1, \ldots, R_pR_1\}$. Each time a new compound relation is computed, compare its

Table 1.11. *Generating the semigroup of the blockmodel* **N**

World length	Distinct compound relations	Equations
1	$W_1 = \{L, A\}$	
2	$W_2 = \{LA, AL, AA\}$	$LL = L$
3	$W_3 = \{LAL\}$	$ALL = AL$
		$AAL = AA$
		$LAA = AA$
		$ALA = AA$
		$AAA = AA$
4	$W_4 = \emptyset$	$LALL = LAL$
		$LALA = AA$

directed graph or binary matrix with those of preceding relations, that is, with relations in W_1 or compound elements generated thus far. If the new compound is distinct, enter it into the set W_2 of new distinct relations formed from two primitive relations; if it is not distinct, record the equation between the new compound relation and some preceding relation. Similarly, multiply each element of W_1 on the right by $R_2, R_3, \ldots, R_p$, storing distinct compound relations in W_2 and noting equations as they are generated. W_2 then contains all distinct compound relations formed from two members of **R** (i.e., all distinct compound relations of length 2). If W_2 is nonempty, multiply each element of W_2 by the generator R_1, then R_2, and so on, up to R_p, placing distinct compound relations of length 3 in W_3 and retaining equations among relations thus generated. Continue the process, multiplying elements of W_k (distinct compound relations of length k) by the generators $R_1, R_2, \ldots, R_p$ until, for some word length m, W_m is an empty set; that is, until no new distinct compound relations can be generated from W_{m-1}. That such a word length exists and is finite is guaranteed by the finiteness of S (because there are no more than 2^k distinct relations in S, where $k = n^2$ and n is the number of elements in X). Then the distinct compound relations in

$$W_1 \cup W_2 \cup \cdots \cup W_{m-1}$$

are the elements of the semigroup S, and equations generated in the process of implementing the algorithm constitute a finite set of defining equations for the semigroup. Once distinct semigroup elements have been identified, they may be compared to construct the corresponding partial ordering.

For example, the blockmodel $\mathbf{N} = \{L, A\}$ of Table 1.8 yields, when subjected to the algorithm, the results set out in Table 1.11. The multiplication table of the semigroup is presented in Table 1.12. The as-

Table 1.12. *Multiplicaton table for the semigroup* $S(\mathbf{N})$

	Symbolic form						Numerical form						
	L	*A*	*LA*	*AL*	*AA*	*LAL*		1	2	3	4	5	6
L	*L*	*LA*	*LA*	*LAL*	*AA*	*LAL*	1	1	3	3	6	5	6
A	*AL*	*AA*	*AA*	*LAL*	*AA*	*LAL*	2	4	5	5	6	5	6
LA	*LAL*	*AA*	*AA*	*LAL*	*AA*	*LAL*	3	6	5	5	6	5	6
AL	*AL*	*AA*	*AA*	*LAL*	*AA*	*LAL*	4	4	5	5	6	5	6
AA	*LAL*	*AA*	*AA*	*LAL*	*AA*	*LAL*	5	6	5	5	6	5	6
LAL	*LAL*	*AA*	*AA*	*LAL*	*AA*	*LAL*	6	6	5	5	6	5	6

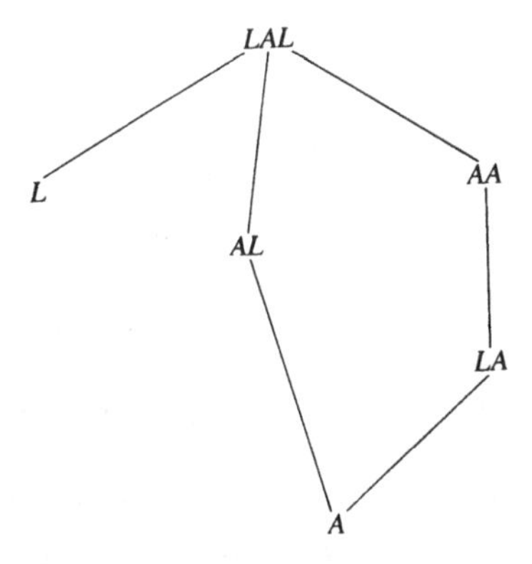

Figure 1.9. Hasse diagram for the partial order of $S(\mathbf{N})$

sociated partial ordering is displayed in Hasse diagram form in Figure 1.9. The multiplication table is presented in Table 1.12, in symbolic form in a square table and in an equivalent numerical form. The two forms are equivalent and, as the algorithm described earlier suggests, a reduced form of each table, termed a right multiplication table, is sufficient to specify the semigroup. That is, the first p columns of the entire multiplication table (or, equivalently, the first p rows) are required to characterise the semigroup. A C program, PSNET, that implements the algorithm is available from the author on request. The convention adopted by the program and in the presentation of semigroup tables later, is that equations among the generators of the semigroup $\{R_1, R_2, \ldots, R_p\}$ are recorded outside the semigroup table. Thus, if $R_i = R_j$, with i not equal to j, then R_i is included as a semigroup generator whereas R_j is deleted and the equation $R_i = R_j$ noted separately.

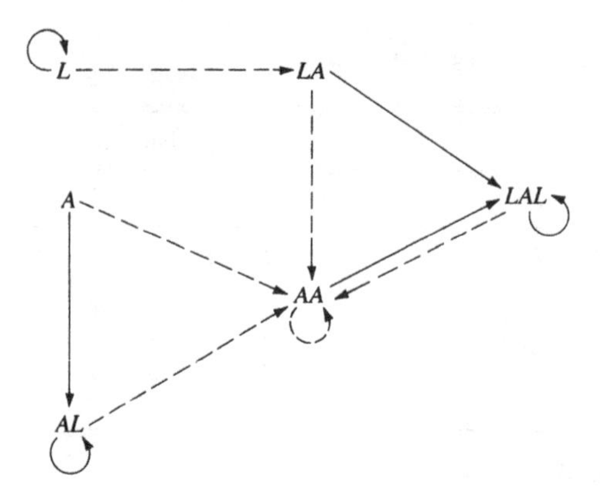

Figure 1.10. The Cayley graph of the semigroup $S(\mathbf{N})$ (solid and dashed edges are L and A, respectively)

Semigroup presentation. The program PSNET stores the multiplication table of the semigroup of a network in the form of an edge table and a word table. Following an algorithm described by Cannon (1971), it constructs the *Cayley graph* of the semigroup (Grossman and Magnus, 1964). Distinct compound or primitive relations, generated in the manner described by the semigroup construction algorithm, are vertices of the Cayley graph. Labelled edges of the graph record the result of post-multiplying distinct compound relations by the generators. Thus, the edge directed from the distinct compound relation T and labelled by the generator R_i is directed toward V, the distinct compound resulting from the composition of T and R_i. That is, the edge corresponds to the equation

$$TR_i = V.$$

The Cayley graph of the semigroup of Table 1.12 is presented in Figure 1.10. The *edge table* of the semigroup records the target of each labelled edge emerging from each vertex (distinct compound relation) and corresponds to the first p columns of the multiplication table of the semigroup. The *word table* lists, for each distinct relation, the vertex from which an edge was first directed to it and the label of that edge. It thus readily permits the reconstruction of a minimal path from the generators to that vertex and speeds multiplication when only the abbreviated multiplication table (that is, the edge table) is stored. (As a result, it is particularly useful when large semigroups are under investigation.) Edge and word tables for the semigroup presented in Table 1.12 are shown in Table 1.13, together with the partial order of the semigroup. The latter form will be used frequently throughout the book.

Table 1.13. *Edge and word tables and partial order for the semigroup S*(**N**)

Semigroup element	Word table			Edge table generators		Partial order
	Word	Node	Generator	1	2	
1	L	—	—	1	3	1 0 0 0 0 0
2	A	—	—	4	5	0 1 0 0 0 0
3	LA	1	2	6	5	0 1 1 0 0 0
4	AL	2	1	4	5	0 1 0 1 0 0
5	AA	2	2	6	5	0 1 1 0 1 0
6	LAL	3	2	6	5	1 1 1 1 1 1

Network semigroups and the free semigroup of generator relations. Finally, we present a slightly more formal approach to defining the partially ordered semigroup of a multiple network. The purpose of the presentation is to allow us to draw the close links between the algebraic construction presented in this chapter for complete networks and that for local networks developed in chapter 2.

DEFINITION. Let $\mathbf{R} = \{R_1, R_2, \ldots, R_p\}$ be a set of primitive relations and let $FS(\mathbf{R})$ denote the collection of all finite strings of the form $T_1T_2 \cdots T_k$, where $T_i \in \mathbf{R}$. Define a binary operation on the strings in $FS(\mathbf{R})$ by

$$(T_1T_2 \cdots T_k)\ (U_1U_2 \cdots U_l) = T_1T_2 \cdots T_kU_1U_2 \cdots U_l \qquad (1.1)$$

where $T_i \in \mathbf{R}$, $U_j \in \mathbf{R}$, $i = 1, 2, \ldots, k$; $j = 1, 2, \ldots, l$. The operation is termed *juxtaposition* and is clearly associative. The collection of relations $FS(\mathbf{R})$, together with the binary operation of (1.1) define the *free semigroup on* **R**. The compound relation $T_1T_2 \cdots T_k$, where $T_i \in \mathbf{R}$ $(i = 1, 2, \ldots, k)$ is called a *word* of $FS(\mathbf{R})$ of *length k*. $R_1, R_2, \ldots, R_p$ are termed the *generators* of $FS(\mathbf{R})$.

As we have already seen, each compound relation in the free semigroup $FS(\mathbf{R})$ corresponds to a binary relation on X. The ordering on binary relations which was introduced earlier may also be seen as an ordering relation on $FS(\mathbf{R})$, with

$$W \leq V$$

if and only if $(i, j) \in W$ implies $(i, j) \in V$, for all $i, j \in X$. The relation $\leq$ is actually a *quasi-order* on $FS(\mathbf{R})$, that is, a reflexive and transitive binary relation on $FS(\mathbf{R})$. It leads naturally to an equivalence relation E on $FS(\mathbf{R})$ with the definition

$$(W, V) \in E \text{ iff } W = V,$$

and it also induces an ordering on the classes of E by

$$e_1 \leq e_2 \text{ if and only if } W \leq V \text{ for all } W \in e_1,\ V \in e_2.$$

It may readily be seen that this induced partial ordering on the classes of E is precisely the same as the partial ordering defined on distinct semigroup elements which was introduced earlier.

Moreover, the partial ordering has the property of being preserved by multiplication on the right and left and hence yields a partially ordered semigroup on the classes of E. Thus, there is a one-to-one correspondence between the elements of the partially ordered semigroup $S(\mathbf{R})$ of a network $\mathbf{R}$ on a set X and the classes of an equivalence relation E defined on the free semigroup $FS(\mathbf{R})$ of generator relations. Consequently, operations and orderings among the elements of S can also be viewed as operations and orderings on the classes of E.

Summary

It has been argued in this chapter that social networks play an important substantive role in the social sciences. In particular, they provide a means of describing some salient features of the immediate and extended "social environment" of an individual, and they also allow the tracing of paths along which social processes are likely to flow. In constructing a formal representation for social networks, it was claimed that a suitable representation should possess two important properties. Firstly, it should be multirelational and so encompass different types of network relations in the description of an individual's social environment. Secondly, it should be concerned with the description of different kinds of network paths.

The partially ordered semigroup of a network was introduced as an algebraic construction fulfilling both of these requirements. The partially ordered semigroup of a network records the relationships between all possible labelled paths in the network and, as such, provides a representation of its "relational structure". It may be constructed either from network data describing links between individuals or from network data describing links between aggregate social units, such as blocks in a blockmodel. The choice of the relational data from which to construct the algebra depends on the purpose of the network study.

In the next chapter, similar arguments are offered in support of an algebraic representation of relational structure in local, or personal, networks. That is, an algebra is constructed that is both multirelational and path-based and which is the local network analogue of the partially ordered semigroup of a complete network. The problem of describing and analysing these algebraic structures for networks is then considered in chapters 3, 4 and 5.

2

Algebraic representations for local social networks

Although a large number of significant network studies have involved complete networks of the kind discussed so far, there is also considerable interest in the analysis of personal, or ego-centred, networks arising from survey research (e.g., Feiring & Coates, 1987; Fischer, 1982; Fischer et al., 1977; Henderson et al., 1981; Laumann, 1973). An indication of the extent of this interest was the decision to include a set of network questions in the General Social Survey (*GSS*) from 1985 (Burt, 1984). As a result of that decision, it is now easier for social scientists to address a variety of questions about the links between network and other personal and social characteristics using *GSS* data (e.g., Bienenstock, Bonacich & Oliver, 1990; Burt, 1986b, 1987a, 1987b; Burt & Guilarte, 1986; Huang & Tausig, 1990; Marsden, 1986, 1988).

It is usual in survey studies to sample a number of individuals from some population of interest and to enquire about characteristics of their "local" networks. For instance, the *GSS* network questions seek information about those people with whom an individual has discussed an important personal matter in the preceding six months, as well as information about the interrelations of those people. Symmetric relational data for two types of relations in the local network are obtained: (a) who is acquainted with whom (or, equivalently, who is a stranger to whom) and (b) who is "especially close" to whom (Burt, 1984). For partial network data in this form, the semigroup representation introduced in chapter 1 is not necessarily appropriate. An implicit assumption of semigroup construction is the existence of a meaningful network boundary. That is, we assume that we have a complete record of the relationships of a given individual, a complete record of the relationships of all those to whom he or she is related, and so on. (We assume here, as before, that the types of relationships under consideration belong to a prespecified primitive set.) In the case of personal or ego-centred networks, the record is truncated at some point so that only a partial network may be constructed. The form of this partial network varies from one study to another but, most commonly, has taken one of the forms illustrated in Figure 2.1.

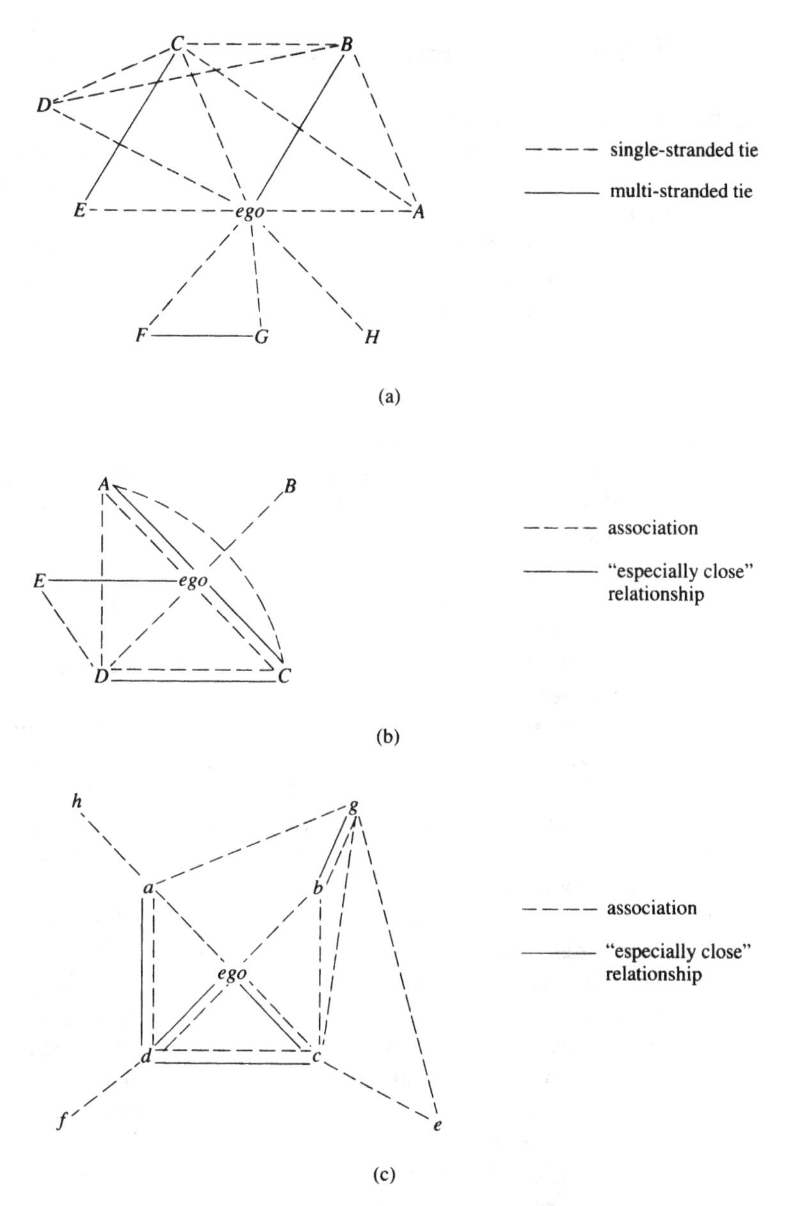

Figure 2.1. Some partial networks: (a) a complete first-order zone; (b) a restricted first-order zone ($N = 5$); (c) a two-stage snowball sample L (restricted second-order zone, $N = 5$)

Types of local networks

Figure 2.1a is typical of the literature on personal or ego-centred networks instigated by Barnes (1954) and Bott (1957) and developed by Mitchell, Barnes, Kapferer, Boissevain and others (e.g., Barnes, 1969a; Boissevain, 1974; Kapferer, 1969; Mitchell, 1969). The figure represents the *first-order zone* of an identified individual (*ego*), that is, the set of persons to whom *ego* has relations of some specified kind, together with all of the relations obtaining among them. The set of individuals to whom *ego* is linked by any relation is assumed to be unrestricted in size in this type of network and may be generated by using a "free choice" format for questions eliciting network information. Of major interest in such partial networks have been the size and composition of the first-order zone of the network, its density (the degree to which the associates of *ego* are associated with one another) and the distribution of ties of various kinds, such as uniplex (or single-stranded) and multiplex (or multi-stranded) relations.

The partial network illustrated in Figure 2.1b is the same in form to that of Figure 2.1a except that the size of the first-order zone has been limited to a fixed number N of individuals (excluding *ego*, the respondent). Such networks may result from a "fixed choice" format for network questions, such as "Name up to five people to whom you feel close". Typically, N is small, such as 5 (Burt, 1984; Fischer, 1982) or 6 (Wellman, 1979) or 9 (Kadushin, 1982), and may represent a small fraction of the size of the actual first-order zone. The latter has been estimated by Pool and Kochen (1978) to be in the range of 500 to 2,000 for the acquaintance networks of many individuals; whereas for networks of persons with whom personal matters are discussed, Fischer (1982) reported first-order zones ranging in size from 2 to 67 with a mean of 19. Some similar measures, such as density, also have usually been of interest for networks of the form in Figure 2.1b, and a number of useful results have followed from a classification of individuals in the first-order zone in terms of other attributes (such as whether individuals belong to the same neighbourhood or not; Wellman, 1979).

Figure 2.1c illustrates the snowball, or star, sampling procedure (Frank, 1979; Goodman, 1961). Starting with a single individual, such as the one labelled *ego* in Figure 2.1c, we initiate a k-stage sampling process. At each stage, for every individual added in the previous stage of sampling, up to N of their associates are added to the sample. For example in Figure 2.1c, the individuals labelled a, b, c and d are added at the first stage of sampling, together with all reported relations of association and close relationships. At the second stage, individuals e, f, g and h are added, together with any associations and close relations among the new and old members of the network identified so far. Goodman developed procedures for estimating some parameters for the entire network from the sample data, in the case in which each individual names exactly N associates; others have developed the methods further (e.g., Frank, 1979). Note that snowball sampling is distinct from other methods of network sampling discussed by Granovetter and others (e.g., Erickson & Nosanchuk, 1983; Frank, 1971; Granovetter, 1976). In some of these other sampling schemes, pairs of individuals are sampled at random from the entire population; and in other schemes, a random sample of individuals, augmented by all of their network links, may constitute the basic sampling unit. (In the latter case, information about network links existing among sampled members of the network is complete. In principle, therefore, the same methods for representing relational structure in the sample as in the entire network may be applied in this case, but it will be demonstrated in chapter 5 that the structure in the sample bears no necessary relationship to the structure in the whole network population. The question of the usefulness of the representation for the case in which network boundaries are arbitrary therefore remains an open one.)

The k-stage process of the snowball sampling scheme leads to the construction of a local network of an individual that has been termed a kth-order zone. The *second-order zone* of an individual comprises the individual's first-order zone augmented by the first-order zone of each of its members. That is, it comprises persons to whom the individual is linked by a path of length no greater than 2 and all the network links between them. More generally, the *kth-order zone* of an individual is the $(k-1)$th-order zone augmented by the first-order zone of every person in the $(k-1)$th-order zone. It therefore contains the collection of persons to whom the individual is linked by a path of length k or less, and all of their interrelations. For instance, Figure 2.1c represents the second-order zone of the individual labelled *ego*. The first-order zone of *ego* is the subgraph of Figure 2.1c spanned by the vertices *ego*, *a*, *b*, *c* and *d*.

Clearly, the various notions of partial network represented in Figure 2.1 are not all distinct. For instance, form Figure 2.1c with N very large and k equal to 1 is similar to form (a), form (c) with fixed N and k equal to 1 is identical to form (b), and form (b) with N large approximates form (a). Moreover, the ideal hypothetical case in which both N and k are unbounded is that assumed, in principle, for an entire network. Note that the decision to restrict N is usually made on practical grounds (e.g., Burt, 1984), and where the effects of restriction have been studied, the practice has not been recommended (e.g., Holland & Leinhardt, 1973).

Relations among the various types of partial and complete networks are summarised in Table 2.1. The form of partial network illustrated in Figure

Table 2.1. *Types of local network*

Order of local network, k	Maximum outdegree of each node, N	
	Unrestricted	Restricted
Fixed at 1	Unrestricted first-order zone (Fig. 2.1a)	Restricted first-order zone (Fig. 2.1b)
Fixed at $k > 1$	Unrestricted kth-order zone (k-stage snowball sample)	Restricted kth-order zone (k-stage snowball sample (Fig. 2.1c)
Unbounded	Complete network	Entire network with restricted outdegree

2.1b can be seen as a restricted case of all of the other network forms, and we assume that this figure represents the minimum amount of information available about a partial network. We shall refer to all types of partial and complete networks that have an identified focal individual or other social unit as the *local network* of the individual or unit. The identified individual or unit will often be labelled *ego*. The methods to be described have been developed to apply to all of these forms of local network.

As for complete networks, personal or ego-centred networks can also be distinguished in terms of the source from which information about ties in the network has been obtained. In some sampling schemes, *ego* is the source of all relational information present in the network (e.g., in the General Social Survey, Burt, 1984); in others, individuals named by *ego* are approached and asked to identify their perceived relations with persons in their own personal network, and so on. The latter methodology, for example, may be applied in the snowball sampling scheme of Figure 2.1c. In other cases, a participant observer is the source of information for all relational ties (e.g., Roethlisberger & Dickson, 1939). In every case, the relational data obtained may be represented in one of the forms illustrated in Figure 2.1, and the various methods of obtaining information about ties are not distinguished in this representation. Clearly, though, knowledge of this aspect of network methodology may bear heavily on the interpretation of the representation obtained from the data and needs to be kept in mind.

As we observed in chapter 1, further clarification of the effects of obtaining network data from various sources is likely to follow from studies using the "cognitive social structures" introduced by Krackhardt (1987). He advocates the gathering of information about all ties of specified kinds among all individuals in a specified group from *each* member of the group. With such data, one may begin to examine the relationships among representations constructed in different ways; these investigations will be made more practical by the availability of analytic methods developed in chapters 4 and 5.

It is worth stressing that the multiple network representations discussed in chapter 1 assume networks in which both N and k are unbounded. Implicitly, it was supposed that relational data had been gathered for every member of a bounded population of persons of interest. The question of whether a population X of persons is ever sufficiently circumscribed, in practice, to make this assumption reasonable, is an empirical one. Does the failure to include some individual(s) in a population under study lead to substantial changes in the obtained representation of the network? The question can be made more precise using definitions introduced later; and like the question of the effects of methodological variations in data collection, it can also be tackled more easily using the analytic methods to be derived for them.

Representing local networks

Each local network relation can be presented in the same three forms as a complete network relation, that is, as a directed graph, a binary relation and a binary matrix. The vertices or elements of the representation are the members of *ego*'s kth-order zone (for some specified value of k), and an edge or relation is directed from vertex a to vertex b if a stands in the given type of relation to b. We shall adopt the convention of listing *ego* as the first network element in the matrix version of the representation, so that the first row and column of the matrix represent relations expressed by and towards the focal individual of the network.

Any given network relation may be symmetric or nonsymmetric and binary or valued; moreover, local networks may be constructed using single or multiple network relations. Thus, each of the forms of representation for complete networks summarised in Table 1.3 has a local network analogue, and we shall generally assume local network data in the form of multiple, binary, nonsymmetric relations. The case of valued relations is considered in chapter 7.

The local network **L** presented in Figure 2.1c in the form of a graph possesses multiple, binary, symmetric relations. It is presented in Table 2.2 in matrix form, with the elements *ego*, *a*, *b*, *c*, *d*, *e*, *f*, *g* and *h* assigned the numbers 1 to 9, respectively. The relational representation is in terms of two symmetric binary relations, A (representing "association") and C (representing "especially close" relations), whose *unordered* pairs are given by

Table 2.2. *The local network* **L** *in binary matrix form*

Element	Relation C	Relation *A*
	1 2 3 4 5 6 7 8 9	1 2 3 4 5 6 7 8 9
1	0 0 0 1 1 0 0 0 0	0 1 1 1 1 0 0 0 0
2	0 0 0 0 1 0 0 0 0	1 0 0 0 1 0 0 1 1
3	0 0 0 0 0 0 0 1 0	1 0 0 1 0 0 0 1 0
4	1 0 0 0 1 0 0 0 0	1 0 1 0 1 1 0 1 0
5	1 1 0 1 0 0 0 0 0	1 1 0 1 0 0 1 0 0
6	0 0 0 0 0 0 0 0 0	0 0 0 1 0 0 0 1 0
7	0 0 0 0 0 0 0 0 0	0 0 0 0 1 0 0 0 0
8	0 0 1 0 0 0 0 0 0	0 1 1 1 0 1 0 0 0
9	0 0 0 0 0 0 0 0 0	0 1 0 0 0 0 0 0 0

$$A = \{(ego, a), (ego, b), (ego, c), (ego, d), (a, d), (a, g), (a, h), (b, c), (b, g), (c, d), (c, e), (c, g), (d, f), (e, g)\}$$

and

$$C = \{(ego, c), (ego, d), (a, d), (b, g), (c, d)\}.$$

In the presentation that follows, the description of a representation for local networks is introduced first for network data that assume some completeness, that is, where neither N nor k has been restricted. The representation is then generalised to the case in which the partial nature of the relational information is included in the representation, that is, for which restrictions on N or k or both are given explicit recognition.

An algebra for local social networks

In an innovative departure from blockmodel analysis, Mandel (1983; also Winship and Mandel, 1983; Breiger & Pattison, 1986) developed a method of characterising the local relational structure associated with a social position that is distinct from that implicit in the blockmodel approach (White et al., 1976). The procedure is mindful, however, of the same theoretical constraints as those underlying the construction of the partially ordered semigroup of the network. That is, it preserves the distinctions between relations of different kinds, it is concerned with variation in patterns of relations associated with different positions and it emphasizes the role of compound relationships in the flow of social processes (Boorman & White, 1976; Lorrain & White, 1971).

The approach has been described in varying forms by Breiger and

Pattison (1986), Mandel (1978, 1983), Pattison (1989) and Wu (1983), and all of these forms rely on the constructions originally proposed by Mandel (1978). In each case, the essential construction is the same and may be presented as a local network analogue of the partially ordered semigroup for a complete multiple network.

Paths in local networks

The major difference between a complete network and a local network is that the latter has an identified individual, or *ego*, as the central focus of the network. The links present in the local network are included because of their relationship to this identified individual. The relationship may be a *direct* one, in the sense that the link belongs to the individual's first-order zone, or it may be *indirect* and hence a part of some *k*th-order zone for the individual. The representation for structure in local networks takes account of this focus on the identified individual in the local network by constructing only those paths in the local network that have *ego* as their source. That is, we trace a number of compound relations, or paths, in the local network, but we restrict attention to those paths beginning at the identified individual. Each path of a given type links *ego* to a subset of elements in the network, and we may associate that subset with the given path type. Then, as for complete networks, we may make comparisons among the paths that we have constructed, and the comparisons lead to the construction of an algebraic structure. In particular, we may think of one path type as *contained* within another if the subset of individuals associated with the first is contained in the subset of individuals associated with the second. Two path types are *equivalent* if they are associated with exactly the same subsets of individuals. Thus, in this way, we derive a set of statements about the equalities and orderings among paths in the local network that emanate from *ego*, and these statements define an algebraic structure for the network. The algebraic structure is a *localised* representation of network structure in that is deals only with relations emanating from the focal individual of the network.

Consider, for example, the snowball network L of Figure 2.1c. The relation A links *ego* to a, b, c and d, so we may associate the relation A with the subset $\{a, b, c, d\}$. The set of paths labelled C emanating from *ego* have c and d as their endpoints; hence the relation C is associated with the vertex subset $\{c, d\}$. The set associated with C is clearly contained in the set associated with A, so that we may say that $C < A$. That is, in *ego*'s local network, close relations are a subset of acquaintance relations. Paths of length 2 labelled AA link *ego* to every member of the network S; thus, AA is associated with the subset $\{ego, a, b, c, d, e, f, g, h\}$ and the orderings

Table 2.3. *Paths of length 3 or less in the network* **L** *having* ego *as source*

Path label	Subset	Binary vector
ego * *C*	{*c*, *d*)	000110000
ego * *A*	{*a*, *b*, *c*, *d*)	011110000
ego * *CC*	{*ego*, *a*, *c*, *d*}	110110000
ego * *CA*	{*ego*, *a*, *b*, *c*, *d*, *e*, *f*, *g*}	111111110
ego * *AC*	{*ego*, *a*, *c*, *d*, *g*}	110110010
ego * *AA*	{*ego*, *a*, *b*, *c*, *d*, *e*, *f*, *g*, *h*}	111111111
ego * *CCC*	{*ego*, *a*, *c*, *d*}	110110000
ego * *CCA*	{*ego*, *a*, *b*, *c*, *d*, *e*, *f*, *g*, *h*}	111111111
ego * *CAC*	{*ego*, *a*, *b*, *c*, *d*, *g*}	111110010
ego * *CAA*	{*ego*, *a*, *b*, *c*, *d*, *e*, *f*, *g*, *h*}	111111111
ego * *ACC*	{*ego*, *a*, *b*, *c*, *d*}	111110000
ego * *ACA*	{*ego*, *a*, *b*, *c*, *d*, *e*, *f*, *g*, *h*}	111111111
ego * *AAC*	{*ego*, *a*, *b*, *c*, *d*, *g*}	111110010
ego * *AAA*	{*ego*, *a*, *b*, *c*, *d*, *e*, *f*, *g*, *h*}	111111111

$A < AA$ and $C < AA$ both hold. Similarly, paths of type *AC* are associated with the subset {*ego*, *a*, *c*, *d*, *g*}, and so on. Table 2.3 shows those to whom *ego* is linked in **L** by all paths of length 3 or less. Each path type is represented in the form of a subset, as well as in the form of a binary vector. The latter has entries corresponding to the elements of the network and has a unit entry corresponding to an element whenever there is a path of the specified type from *ego* to the element concerned. All other entries are zero. The binary vector recording the presence or absence of relations of type *R* from an element *x* to other network elements is denoted by $x * R$. Thus, for instance, the binary vector corresponding to the relation *AC* for *ego* (network element 1) is $1 * AC = [1\ 1\ 0\ 1\ 1\ 0\ 0\ 1\ 0]$.

Comparing paths in local networks

We now examine systematically the subsets associated with paths of length 3 or less in **L** for possible equations and orderings. (We see later how to deal with longer paths.) We can see that a number of equations and orderings hold among these subsets, including the equations $CCC = CC$, $CCA = AA = CAA = ACA = AAA$ and $CAC = AAC$, as well as the orderings $C < A$, $C < CC$, $C < CA$ and so on. As Table 2.3 makes clear, only eight distinct subsets are associated with paths of length 3 or less in **L**, namely, those associated with *C*, *A*, *CC*, *CA*, *AC*, *AA*, *CAC* and *ACC*. Indeed, the number of distinct subsets associated with paths of any length in a local network is necessarily finite because there are only a finite number of distinct subsets of a finite set. In particular, for a local network of n elements, there are at most 2^n distinct subsets. Thus, we may group all possible paths for a local network into a finite number of classes.

Moreover, we may infer certain properties for these classes of paths. Consider, for instance, the equation represented by $CCC = CC$. It corresponds to the statement that paths comprising three close relations link *ego* to exactly the same set of individuals as paths comprising two close relations (namely, *ego*, *a*, *c* and *d*). Now consider more complex paths that we may construct from *CC* and *CCC* by adjoining on the right an additional string of primitive relations, that is, a string of *A* and *C* relations. Suppose that we select the string *ACACC*. The process of adjoining this string to the right of *CC* or *CCC* to obtain the strings *CCACACC* or *CCCACACC* has a direct interpretation in terms of the tracing of paths in the network. The path *CC* links *ego* to a collection of individuals in the local network. The path *CCACACC* also links *ego* to a subset of individuals, and each path passes through one of the individuals to whom *ego* is linked by the path *CC*. Indeed, we can think of the path *CCACACC* as comprising two sections, one path labelled *CC* from *ego* to a member of the subset associated with *CC*, and one path labelled *ACACC* from the latter member to a member of the subset associated with *CCACACC*. Now, since *CC* and *CCC* link *ego* to the same subset of individuals, the paths *CCACACC* and *CCCACACC* must also link *ego* to the same individuals. Thus, from the equation

$$CC = CCC$$

the equation

$$CCACACC = CCCACACC$$

must follow. Indeed, the equation

$$CCX = CCCX$$

must hold for any string *X* of the primitive relations *A* and *C*.

More generally, we can show that this property holds for any equation among paths. That is, if $R = T$ is an equation, then so is $RX = TX$, for any string of relations *X*. Or, to put it another way, whenever a pair of paths belong to the same class, then so does any pair constructed from them by adjoining a string on the right. As a result, we can search for distinct subsets associated with paths in a local network by adjoining primitive relations to distinct representatives of classes. For instance, the distinct subsets associated with paths of length 3 in **L** have representative paths of *CAC* and *ACC*. By constructing the paths *CACC*, *CACA*, *ACCC* and *ACCA*, we will discover any new and distinct

Table 2.4. *Right multiplication table for the local role algebra of* ego *in the network* **L**

		Generator	
Element	Class	C	A
1	C	CC	CA
2	A	AC	AA
3	CC	CC	AA
4	CA	CAC	AA
5	AC	ACC	AA
6	AA	CAC	AA
7	CAC	CAC	AA
8	ACC	AC	AA

subsets of vertices associated with paths of length 4 in **L**. In fact, we find that these four paths may be equated with CAC, AA, AC and AA, respectively, so that there are only eight distinct classes of paths of any length for the local network **L**.

We can record the class containing the result of adjoining primitive relations to representatives of distinct classes in the form of a right multiplication table. For example, we label the eight distinct classes for the snowball network **L** by the strings C, A, CC, CA, AC, AA, CAC and ACC and then list them as the rows of a table. The columns of the table may be assumed to stand for the primitve relations C and A, respectively (see Table 2.4). The class that results when the ith class has the jth primitive relation adjoined on the right is represented in row i and column j of the table. For instance, the outcome of adjoining C on the right of CC is recorded in row 3 and column 1. To find the class that is obtained by adjoining $ACACC$ on the right of CC, we first find (a) the class resulting from adjoining A to CC, then (b) the class resulting from adjoining C to the outcome of (a), and so on. From Table 2.4, for instance, we see that $(CC)A = AA$, that $(AA)C = CAC$, that $(CAC)A = AA$, that $(AA)C = CAC$ and that $(CAC)C = CAC$, and therefore that $(CC)(ACACC) = CAC$.

We refer to the process of adjoining a string of primitive relations to the right of a path label as *right multiplication*. The property just described means that equations are preserved by right multiplication: if $R = T$ is an equation, then $RX = TX$ must also be an equation for any string of primitive relations X. We may also establish that orderings are preserved by right multiplication. If $R \leq T$ is an ordering in a local network, then *ego* is linked by R to a subset of those to whom *ego* is linked by T. Now if a path RX links *ego* to an individual z, then there must

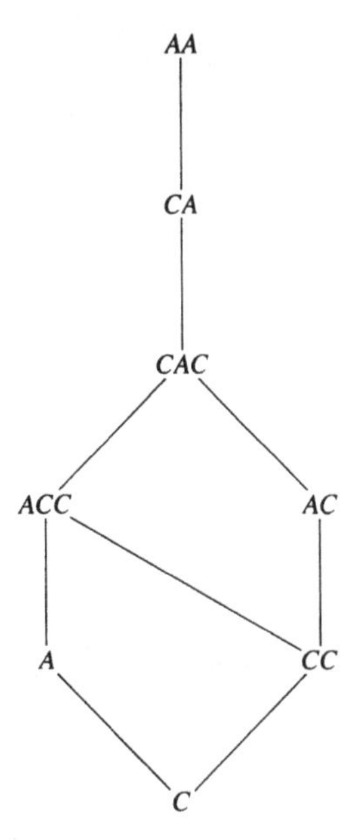

Figure 2.2. Partial ordering for the local role algebra of the network **L**

be a path labelled R from *ego* to some individual x, and a path labelled X from x to z. Therefore, there must be a path labelled T from *ego* to x, and hence a path labelled TX from *ego* to z. In other words, from $R \leq T$, it follows that $RX \leq TX$, for any string of relations X.

The local role algebra of a local network

As a result of this property, we may present the orderings among paths of the local network emanating from *ego* as a partial ordering on the classes of equivalent paths. This partial ordering is displayed for the classes of the local network **L** in Figure 2.2. The right multiplication table in Table 2.4 and the partial order in Figure 2.2 represent all of the orderings and equations among paths from *ego* in the network **L**. Together they define an algebraic structure termed a *local role algebra*, and this is the proposed representative of structure in the local network.

We can formalise these constructions in the following way.

DEFINITION. Let $X = \{1, 2, \ldots, n\}$ represent the set of members of a local network of some individual and suppose that the identified individual corresponds to element 1. Let $\mathbf{R} = \{R_1, R_2, \ldots, R_p\}$ be a set of primitive relations defined on X. The *free semigroup* $FS(\mathbf{R})$ on $\mathbf{R}$ is the set of all strings $U_1U_2 \cdots U_k$ of finite length of elements of $\mathbf{R}$ (i.e., $U_i \in \mathbf{R}$, $i = 1, 2, \ldots, k$), together with the operation of *juxtaposition*, given by

$$(U_1U_2 \cdots U_k)\,(V_1V_2 \cdots V_h) = U_1U_2 \cdots U_kV_1V_2 \cdots V_h,$$

where $U_1U_2 \cdots U_k$ and $V_1V_2 \cdots V_h$ are strings of elements of $\mathbf{R}$ (i.e., $U_i, V_j \in \mathbf{R}$, $i = 1, 2, \ldots, k$, $j = 1, 2, \ldots, h$). Each string of relations A from $\mathbf{R}$ corresponds to a labelled path on the local network beginning at the identified element; the collection of individuals to whom the identified element is linked by such a path may be denoted by $1 * A$. We may define a relation $\leq$ on the paths by

> $A \leq B$ iff for each $x \in X$ there is a path labelled B from 1 to x whenever there is a path labelled A from 1 to x, that is, iff $1 * A$ is a subset of $1 * B$,

where $A, B \in FS(\mathbf{R})$. We may also define an equivalence relation on the paths of the local network by

$$A = B \text{ iff } A \leq B \text{ and } B \leq A.$$

Then it may be demonstrated that the number of distinct classes of the equivalence relation is finite, and that if C_A denotes the class containing A we may define a partial ordering on the classes according to $C_A < C_B$ if and only if $A < B$ for some $A \in C_A$ and $B \in C_B$. Furthermore, we may define a right multiplication operation

$$C_AR = C_{AR}$$

($R \in \mathbf{R}$) on the classes that satisfies

$$C_A = C_{A'} \quad \text{implies} \quad C_AR = C_{A'}R.$$

The free semigroup $FS(\mathbf{R})$, together with the relation $\leq$, comprise the *local role algebra* Q_1 for element 1 in the local network. The local role algebra may be presented in the form of a finite right multiplication table and a partial ordering on the distinct equivalence classes of the local role algebra.

An algorithm for constructing a local role algebra

In practice, the local role algebra for an element labelled 1 whose network is described by a set of relations $\mathbf{R} = \{R_1, R_2, \ldots, R_p\}$ on the set $X = \{1, 2, \ldots, n\}$ may be constructed as follows.

1 Construct the collection $\{1 * R_1, 1 * R_2, \ldots, 1 * R_p\}$ and let W_1 be a set of distinct binary vectors in the collection. Record the

Table 2.5. *Constructing the local role algebra of* ego *in the network* **L**

i	W_i	Equations generated
1	$\{1 * C, 1 * A\}$	
2	$\{1 * CC, 1 * CA, 1 * AC, 1 * AA)$	
3	$\{1 * CAC, 1 * ACC\}$	$(1 * CC)C = 1 * CC$
		$(1 * CC)A = 1 * AA$
		$(1 * CA)A = 1 * AA$
		$(1 * AC)A = 1 * AA$
		$(1 * AA)C = 1 * CAC$
		$(1 * AA)A = 1 * AA$
4	$\emptyset$	$(1 * CAC)C = 1 * CAC$
		$(1 * CAC)A = 1 * AA$
		$(1 * ACC)C = 1 * AC$
		$(1 * ACC)A = 1 * AA$

equality corresponding to any vector $1 * R_j$ omitted from the list. (For instance, if $1 * R_j = 1 * R_i$, then place $1 * R_i$, say, in W_1 and record the equation $1 * R_j = 1 * R_i$ outside the table.) Set $k = 1$.

2 Let $W_{k+1} = \phi$. Multiply each binary vector in W_k on the right by each matrix in **R**. If the operation of multiplying $1 * A$ by R_i yields a binary vector that is different from all binary vectors in $W = W_1 \cup W_2 \cup \cdots \cup W_k \cup W_{k+1}$ that have been constructed so far, then place $1 * AR_i$ in W_{k+1} and record the outcome as $(1 * A)R_i = 1 * AR_i$. Otherwise, $1 * AR_i$ is equal to an existing vector $1 * B$ in W, and the result may be recorded as the equation $(1 * A)R_i = 1 * B$. Continue until all multiplications have been performed for all vectors in W_k.
3 Set $k = k + 1$. If $W_k = \phi$, go to step 4; otherwise, return to step 2.
4 Construct the right multiplication table for the local role algebra from the equations recorded in step 2. The rows of the table correspond to elements in $W = W_1 \cup W_2 \cup \cdots \cup W_k$; these may be numbered by integers from 1 to, say, w. The columns of the table correspond to the p relations in **R**. The entry in row h and column i of the table records the label of the vector in W, which is the result of multiplying the hth vector in W by the ith relation in **R**.
5 Construct the partial ordering among the vectors in W.

A C program RANET that implements the algorithm for constructing the local role algebra of an element is available from the author on request. The outcome of applying the algorithm to the local network **L** of Figure 2.1c is displayed in Table 2.5. The algorithm begins with

Table 2.6. *The blockmodel network* **N**

L	A
1 1 0 0	1 0 1 1
1 1 0 0	1 0 1 0
0 0 1 0	1 0 0 1
0 0 1 1	0 0 1 1

Table 2.7. *The local role algebra for block 1 in the network* **N**

Right mult. table				
		Generator		
Element	Class	L	A	Partial order
1	L	1	2	1 0 0
2	A	3	2	0 1 0
3	AL	3	2	1 1 1

$W_1 = \{1 * C, 1 * A\}$ and when $k = 1$, it generates the list of distinct vectors $W_2 = \{1 * CC, 1 * CA, 1 * AC, 1 * AA\}$ and no equations. When $k = 2$, we obtain $W_3 = \{1 * CAC, 1 * ACC\}$, and the equations that are generated include $(1 * CC)\ C = 1 * CC$, $(1 * CC)\ A = 1 * AA$, $(1 * CA)\ A = 1 * AA$ and so on.

The local role algebra for a block in a blockmodel may be obtained in exactly the same way. Consider, for example, block 1 in the blockmodel network $\mathbf{N} = \{L, A\}$ shown in Table 2.6; the local role algebra for block 1 generated by the algorithm is shown in Table 2.7.

The local role algebra of a subset in a local network

Wu (1983) pointed out that it is also possible to define a similar construction that begins with a collection of identified vertices in a local network, rather than with a single identified vertex. The construction permits us to describe the local role algebra of social units intermediate in size between a single social unit (a local role algebra) and an entire network (a partially ordered semigroup). Instead of confining attention to paths having a single identified vertex as their source, one deals with paths beginning at each of a number of vertices in a selected subset of the network. Then one may construct labelled paths corresponding to

Table 2.8. *The local role algebra for the subset {1, 2} in the network* **N**

	Right mult. table			
		Generator		
Element	Class	L	A	Partial order
1	L	1	3	1 0 0 0 0
2	A	4	3	0 1 0 0 0
3	LA	5	3	0 1 1 0 0
4	AL	4	3	1 1 0 1 0
5	LAL	5	3	1 1 1 1 1

strings A of relations in **R** and compare the paths in the following way. If, for every vertex x in the identified subset of vertices Y, the presence of a path labelled A from x to a vertex z implies the presence of a path labelled B from x to z, then we say that $A \leq B$. The resulting set of comparisons among paths yields the local role associated with the subset Y. Under this definition, $A \leq B$ holds in the local role algebra Q_Y for the subset Y if and only if it holds for the local role algebra Q_x of each single element x in Y. If Y is equal to the set X of *all* elements in the local network, then $A \leq B$ in Q_X if and only if $A \leq B$ in the semigroup $S(\mathbf{R})$ of the network **R**.

Consider, for instance, the network **N** of Table 2.6. The local role algebra for block 1 of the network is displayed in Table 2.7, and the local role algebra corresponding to the subset $Y = \{1, 2\}$ of blocks is shown in Table 2.8. The latter local role algebra was constructed using a modification of the algorithm presented earlier for the local role algebra of a single element. The modification is achieved by replacing each binary vector $1 * A$ by a corresponding binary submatrix $Y * A$ throughout. The submatrix $Y * A$ has rows corresponding to elements of Y and columns corresponding to elements of X. The (i, j) entry of the matrix is 1 if there is a path labelled A from the ith element of Y to the jth element of X, and 0 otherwise. For instance, the submatrices $Y * A$ for distinct members of Q_Y are shown in Table 2.9.

A local role algebra may also be defined when the subset Y is equal to the whole set X on which the local network is defined. For instance, the local role algebra corresponding to the subset $X = \{1, 2, 3, 4\}$ is shown in Table 2.10. Comparison with Table 1.13 reveals that the right multiplication table and partial order for the local role algebra Q_X are identical to those for the semigroup of the network. We establish later that, in general, the local role algebra defined on the whole of a network is equivalent to the partially ordered semigroup of the network.

Table 2.9. *Distinct submatrices in the local role algebra for the subset* {1, 2} *of the network* **N**

Submatrix				
L	*A*	*LA*	*AL*	*LAL*
1100	1011	1011	1111	1111
1100	1010	1011	1110	1111

Table 2.10. *Local role algebra for the subset* {1, 2, 3, 4} *of the network* **N**

	Right mult. table			
		Generator		
Element	Class	*L*	*A*	Partial order
1	*L*	1	3	1 0 0 0 0 0
2	*A*	4	5	0 1 0 0 0 0
3	*LA*	6	5	0 1 1 0 0 0
4	*AL*	4	5	0 1 0 1 0 0
5	*AA*	6	5	0 1 1 0 1 0
6	*LAL*	6	5	1 1 1 1 1 1

DEFINITION. Let Y be a subset of X comprising n_Y elements, and let $Y * A$ denote the $n_Y \times n$ submatrix of the matrix of A whose rows correspond to elements of Y. Then the collection $\{Y * A;\ A \in FS(R)\}$ defines a *local role algebra for the subset* Y *of* X with the binary operation of juxtaposition as before, and the ordering in Q_Y being given by

$$Y * B \leq Y * A \text{ iff } (Y * B)_{ij} \leq (Y * A)_{ij} \text{ for all } i \in Y,\ j \in X.$$

This definition is equivalent to the earlier one for the local role algebra of an element when $Y = \{1\}$ is a single network element.

Incoming paths. The construction of a local role algebra, just outlined, relies exclusively on comparison among paths emanating from *ego*. An analogous construction may be developed for paths having *ego* as their target rather than their source. The development is achieved by replacing each network relation R_i in the preceding derivation by its *converse*, that is, by the relation R_i' defined by $(x, y) \in R_i'$ if and only if $(y, x) \in R_i$, for $x, y \in X$. Further, if the collection of converse relations $\mathbf{R}' =$

$\{R_1', R_2', \ldots, R_p'\}$ is added to the set $\mathbf{R} = \{R_1, R_2, \ldots, R_p\}$ to give an augmented set of network relations $\mathbf{R}'' = \{R_1, R_2, \ldots, R_p, R_1', R_2', \ldots, R_p'\}$, then one generates a local role algebra based on comparisons among semi-paths having *ego* as their target or their source. (Recall that a *semipath* in a network is a sequence of nodes $x_0, x_1, \ldots, x_k$ such that (x_{j-1}, x_j) or (x_j, x_{j-1}) is an edge in some network relation $W_j \in \mathbf{R}$, for each $j = 1, 2, \ldots, k$. The node x_0 is the *source* of the semipath, and x_k is its *target*. The *label* of the semipath is $W_1 W_2 \cdots W_k$.) Mandel (1978), for instance, reported some applications of local role algebras with the latter augmented set of network relations. Of course, if all the network relations are symmetric, as in the example just described, then the augmentation of the network by converse relations makes no difference to the algebraic structure that is generated.

Role algebras

The local role algebra of an element in a local network or of a subset of the local network is an instance of a general algebraic structure that may be termed a *role algebra* (Pattison, 1989). As we see later, it is important to characterise the properties of these structures when we come to make comparisons among them. In fact, a characterisation may be constructed from the following observations about the properties of local role algebras drawn from the earlier discussion.

Firstly, the elements of local role algebras are labelled paths corresponding to strings of generator relations in **R**. Secondly, the composition of a pair of paths corresponds to the juxtaposition of their associated strings, so that the free semigroup $FS(\mathbf{R})$ comprising all strings of finite length and the juxtaposition operation is generated. Thirdly, an ordering relation may be defined on the labelled paths by

$$U \leq V \text{ iff } (1, x) \in U \text{ implies } (1, x) \in V,$$

for any $x \in X$; $U, V \in FS(\mathbf{R})$; and the ordering relation is both reflexive and transitive and hence a quasi-order. Fourth, the orderings among paths are preserved by multiplication on the right, that is,

$$U \leq V$$

implies

$$UW \leq VW,$$

for any $W \in FS(\mathbf{R})$.

These properties are captured in the general definition:

DEFINITION. A *role algebra* $[W, f, Q]$ is a set W, a binary operation f on W and a binary relation Q on W satisfying

1. $[W, f]$ is a free semigroup, with the operation f presented in the form $f(s, t) = st$, for any $s, t \in W$;
2. Q is a quasi-order (i.e., Q is reflexive and transitive); and
3. for any $s, t, u \in W$, $(s, t) \in Q$ implies $(su, tu) \in Q$.

The definition may be applied to local role algebras by setting W to be the elements in the free semigroup $FS(\mathbf{R})$, that is, all finite length labelled paths constructed from the generator set $\mathbf{R}$; f to be the composition operation for labelled paths; and Q to be the quasi-order

$$(U, V) \in Q \text{ iff } V \le U.$$

In fact, in what follows, all role algebras considered pertain to the same set W and the same binary operation f, so it is convenient to denote the role algebra $[W, f, Q]$ simply by Q. Two notable examples of role algebras have already been introduced. The first is the local role algebra for a particular element of a network, and the second is the local role algebra for a subset of network elements.

For the local role algebra for element 1 of a network, let

$$(A, B) \in Q_1 \text{ iff } 1 * B \le 1 * A.$$

It may be verified that Q_1 is both reflexive and transitive, and hence is a quasi-order, and that inclusions are preserved under right multiplication (i.e., $1 * B \le 1 * A$ implies $1 * BC \le 1 * AC$ for any $C \in FS(\mathbf{R})$; Mandel, 1978), so that all the axioms for a role algebra are satisfied. As observed earlier, the local role algebra may be presented in a finite form as a right multiplication table and partial order on the set of equivalence classes associated with Q_1.

The local role algebra associated with a subset Y of individuals in a local network is also a role algebra. As before, the binary operation f is the composition of relations, and the binary relation Q is inclusion of submatrices:

$$(A, B) \in Q_Y \text{ iff } Y * B \le Y * A.$$

When Y is equal to the set of all elements X of the local network, the associated local role algebra Q_X is clearly also a role algebra, in which the binary relation Q is defined by

$$(A, B) \in Q_X \text{ iff } B \le A$$

(i.e., if and only if $(i, j) \in B$ implies $(i, j) \in A$, for all $i, j \in X$). In fact, when Y is equal to the whole of the set X on which the local network is defined, the associated local role algebra satisfies the additional property that, for any $A, B \in Q$, $(A, B) \in Q$ implies $(CA, CB) \in Q$ for all $C \in FS(\mathbf{R})$. That is, as well as being preserved by multiplication on the right, equations are preserved by multiplication on the left. This property is formalised in the following definition.

DEFINITION. A *two-sided role algebra* is a role algebra $[W, f, Q]$ for which

$$(s, t) \in Q \text{ implies } (us, ut) \in Q,$$

for any $u \in W$, $s, t \in W$.

Thus, a two-sided role algebra is one for which inclusions are preserved by multiplication on both the right and the left: $s \leq t$ implies $su \leq tu$ and $us \leq ut$, for any u. It may be seen that the role algebra for an entire set X is a two-sided role algebra; moreover, it is straightforward to show that the partially ordered semigroup of a network and its two-sided local role algebra are equivalent in the following sense.

THEOREM 2.1. *Let* $\mathbf{R} = \{R_1, R_2, \ldots, R_p\}$ *be a network on* X. *For any* A, B $\in$ FS(*R*), *the ordering* B $\leq$ A *holds in the partially ordered semigroup* S(*R*) *if and only if* (A, B) $\in Q_X$, *the two-sided local role algebra of the set* X.

Relations among role algebras: The nesting relation

Given that we can generate a local role algebra from an element or from a set of elements in a network, it is natural to ask how the resulting constructions compare. One means of comparing role algebras is by comparing their quasi-orders: Is an ordering among the relations present in one quasi-order always present in another? If this is the case for *every* ordering present in the first quasi-order, then we may describe the second quasi-order as *nested* in the first (Mandel, 1978). That is, role algebras with the same set W and the same binary operation f may be partially ordered by the nesting relation.

DEFINITION. A role algebra $[W, f, Q]$ is *nested* in a role algebra $[W, f, T]$ if, for any $s, t \in W$, $(s, t) \in T$ implies $(s, t) \in Q$; that is, if Q contains T. (Mandel, 1983.) We write

$$[W, f, Q] \leq [W, f, T]$$

or, more simply,

$$Q \leq T.$$

For example, the local role algebra for the element 1 of **N** is nested in the local role algebra for the subset {1, 2} of **N**. The equations and orderings characterizing Q_1 are

$$1 * LL = 1 * L, \quad 1 * LA = 1 * A, \quad 1 * AA = 1 * A,$$
$$1 * ALL = 1 * AL, \quad 1 * ALA = 1 * A, \quad 1 * L < 1 * AL,$$
$$1 * A < 1 * AL$$

and those characterising $Q_{\{1,2\}}$ are

$$1 * LL = 1 * L, \quad 1 * AA = 1 * LA, \quad 1 * LAA = 1 * LA,$$
$$1 * ALL = 1 * AL, \quad 1 * ALA = 1 * LA, \quad 1 * LALL = 1 * LAL,$$
$$1 * LALA = 1 * LA, \quad 1 * L < 1 * AL, \quad 1 * L < 1 * LAL,$$
$$1 * A < 1 * LA, \quad 1 * A < 1 * AL, \quad 1 * A < 1 * LAL,$$
$$1 * LA < 1 * LAL, \quad 1 * AL < 1 * LAL.$$

Because the orderings present in $Q_{\{1,2\}}$ are also true in Q_1, then Q_1 is nested in $Q_{\{1,2\}}$. Indeed, more generally, the local role algebra Q_Y of a subset Y of X is nested in the local role algebra $Q_{Y'}$ for the subset Y' of X whenever Y is a subset of Y'. In particular, Q_Y is nested in Q_X for *every* subset Y of X.

As a result, the local role algebra for a subset Y of X may be expressed in terms of the (two-sided) local role algebra of X in which it is nested, and this form of expression is often useful. We use the local role algebra of Y to define an equivalence relation on the finite classes of the local role algebra of X, and then an induced ordering on those classes. That is, we define

DEFINITION. Let **R** be a network on a set X, and let Y be a subset of X. Denote by Q_Y the local role algebra of Y and by S the partially ordered semigroup of **R**, that is, the two-sided role algebra on X. Define the equivalence relation r_Y on the elements of S by

$$(s, t) \in r_Y \text{ iff } (s, t) \in Q_Y \text{ and } (t, s) \in Q_Y.$$

It may be shown that r_Y has the property that

$$(s, t) \in r_Y \text{ implies } (su, tu) \in r_Y, \text{ for all } s, t, u \in S.$$

As a result, r_Y is termed a *right congruence* on the partially ordered semigroup S of the network **R** (Mandel, 1978; Wu, 1983). The classes of r_Y may also be partially ordered by

$$s^* \le t^* \text{ iff } (t, s) \in Q_Y \text{ for some } s \in s^*, t \in t^*;$$

where s^*, t^* are classes of r_Y.

It follows that the role algebra of a subset Y of X may be presented as a quasi-order on the elements of the partially ordered semigroup S of **R**.

Table 2.11. *Quasi-orders on S*(**N**) *corresponding to the role algebras* Q_1 *and* $Q_{\{1,2\}}$

		Modified right mult. table	
		L	*A*
Q_1	1 0 0 0 0 0	1	2
	0 1 1 0 1 0	4	2
	0 1 1 0 1 0	4	2
	1 1 1 1 1 1	4	2
	0 1 1 0 1 0	4	2
	1 1 1 1 1 1	4	2
$Q_{\{1,2\}}$	1 0 0 0 0 0	1	3
	0 1 0 0 0 0	4	3
	0 1 1 0 1 0	6	3
	1 1 0 1 0 0	4	3
	0 1 1 0 1 0	6	3
	1 1 1 1 1 1	6	3

For instance, the local role algebra for block 1 in the network **N** gives rise to the right congruence (1) (2, 3, 5) (4, 6) on the classes of the local role algebra for $X = (1, 2, 3, 4)$ (Table 2.10); the quasi-order on *S*(**N**) corresponding to the local role algebra Q_1 is presented in the upper panel of Table 2.11 (left-hand side). The right congruence on the role algebra for *X* corresponding to the local role algebra for the subset {1, 2} of blocks is (1) (2) (3, 5) (4) (6), and the corresponding quasi-order on *S*(**N**) is shown in the lower panel of Table 2.11. As expected, the local role algebra for block 1 is nested in the local role algebra for the subset of {1, 2}, and this is nested, in turn, in the local role algebra for *X*.

Because each local role algebra for a subset *Y* of a local network *X* is nested in the two-sided local role algebra generated by *X* itself, it is often convenient to compute the local role agebra Q_Y from the partially ordered semigroup *S*(**R**) of the entire local network. The computation may be performed in two steps, which is illustrated for the local role algebra of the subset (1, 2} of the network **N** of Table 2.6. The distinct relations in *S*(**N**) are displayed in Table 2.12; the right multiplication table and partial order for *S*(**N**) are shown in Table 2.10. The first step is to examine the submatrices corresponding to the subset *Y* of interest in each relation *R* of *S*(**N**) for possible additional equations and orderings not already represented in Table 2.12. In the case of the subset {1, 2}, this amounts to comparing the first two rows of relations in *S*(**N**), and we find the additional orderings

Table 2.12. *Distinct relations in the semigroup S*(**N**) *of the network* **N**

Relation					
L	*A*	*LA*	*AL*	*AA*	*LAL*
1100	1011	1011	1111	1011	1111
1100	1010	1011	1110	1011	1111
0010	1001	1001	1111	1011	1111
0011	1000	1000	1100	1011	1111

$$Y * AA \leq Y * LA \quad \text{and} \quad Y * L \leq Y * AL.$$

From the first ordering, the equation

$$Y * AA = Y * LA$$

follows because the ordering

$$Y * LA \leq Y * AA$$

holds in $S(\mathbf{N})$.

The second step is to impose the new orderings on the right multiplication table and partial order for $S(\mathbf{N})$. The new orderings applied to the partial order of $S(\mathbf{N})$ lead to the quasi-order presented in the lower panel of Table 2.11. From this quasi-order, we see that the third and fifth elements of $S(\mathbf{N})$ belong to the same class of r_Y, so that the right congruence corresponding to the subset Y is (1) (2) (3, 5) (4) (6). The right multiplication table for $S(\mathbf{N})$ may also be modified to reflect the equation of the third and fifth elements, as in the right-hand side of the lower panel of Table 2.11. The right multiplication table and partial order tables for Q_Y are then obtained by rewriting these tables, so that the fifth element is combined with the third, and the classes are relabelled with consecutive integers: the result is Table 2.8. The same intermediate tables for the local role algebra Q_1 of block 1 are presented in the upper panel of Table 2.11.

Presentation of role algebras

Just as local role algebras can be presented as a right multiplication table and partial order on a finite set of equivalence classes of $FS(\mathbf{R})$, so can a role algebra, in general, be presented in this way. We define an equivalence relation e_Q on $FS(\mathbf{R})$ according to

$$(s, t) \in e_Q \text{ iff } (s, t) \in Q \text{ and } (t, s) \in Q.$$

The partial order Q^* can then be defined on the equivalence classes of e_Q by

$$(s^*, t^*) \in Q^* \text{ iff } (s, t) \in Q$$

for some $s \in s^*$, $t \in t^*$, where s^* and t^* are classes of e_Q. The partial order Q^* on the finite set of right congruence classes of the role algebra corresponds to the quasi-order Q on the infinite set of elements of W. The right congruence relation e_Q on $FS(\mathbf{R})$ may be represented in the form of a right multiplication table with the rows of the table indexing the right congruence classes of e_Q and the columns corresponding to members of $\mathbf{R}$. The (i, j) entry of the table records the class containing the product of any element in the ith class with the jth member of $\mathbf{R}$. The role algebra Q can then be represented in finite form by the finite quasi-order Q^* and the finite right multiplication table for e_Q.

In all of the examples to be discussed, the quasi-order Q^* will be presented rather than Q, and for convenience it will be identified with Q. The right multiplication table for the right congruence relation e_Q will be referred to simply as the right multiplication table for the role algebra.

Local role algebras and role-sets

Mandel (1978, 1983) described the local role algebra of an individual constructed from the individual's local network as an abstract characterisation of the individual's local role. A different approach to the representation of an individual's local role has been presented by Winship and Mandel (1983). The relationship between the two approaches may be described in the context of the *relation plane* of an individual.

DEFINITION. Let $\mathbf{R} = \{R_1, R_2, \ldots, R_p\}$ be relations defined on a set $X = \{1, 2, \ldots, n\}$ representing the members of the local network of the element 1. The *relation plane* RP_1 of element 1 is a binary matrix of dimension $\infty \times n$, whose rows correspond to the elements of $FS(\mathbf{R})$, listed in some fixed order, and whose columns correspond to the elements of X. The ith row of the relation plane RP_1 is the *relation vector* $1 * R$ for the ith relation R in $FS(\mathbf{R})$. The *role-relation* R_{1j} is the jth column of the relation plane and is a binary vector whose ith entry is 1 if element 1 is related to element j by the ith relation in $FS(\mathbf{R})$, and 0 otherwise.

The relation plane RP_1 may also be presented as a finite binary matrix of dimension $w \times n$, where w is the number of distinct classes in the local role algebra of element 1. In this case, the ith row of the matrix corresponds to the ith distinct class of Q_1. The two presentations are equivalent because one form can be constructed from the other using

Table 2.13. *Relation plane for* ego *in the network* L

Relation	Element								
	ego	*a*	*b*	*c*	*d*	*e*	*f*	*g*	*h*
C	0	0	0	1	1	0	0	0	0
A	0	1	1	1	1	0	0	0	0
CC	1	1	0	1	1	0	0	0	0
CA	1	1	1	1	1	1	1	1	0
AC	1	1	0	1	1	0	0	1	0
AA	1	1	1	1	1	1	1	1	1
CAC	1	1	1	1	1	0	0	1	0
ACC	1	1	1	1	1	0	0	0	0

Table 2.14. *Relation plane for block 1 in the network* **N**

Relation	Block			
	1	2	3	4
L	1	1	0	0
A	1	0	1	1
AL	1	1	1	1

the information contained in the right multiplication table for the local role algebra of the element. The relation planes of *ego* in the network L and block 1 in the blockmodel network **N** are displayed, in finite form, in Tables 2.13 and 2.14, respectively.

The difference between the characterisations of local role described by Mandel (1983) and Winship and Mandel (1983) is that they are based on complementary subdivisions of the relation plane. Mandel (1983) used orderings among the collection of relation vectors, or rows of RP_1, to define a local role algebra. Winship and Mandel's (1983) approach to local role definition, on the other hand, was through the collection of n role-relations for element 1, that is, through the columns of RP_1.

DEFINITION. Let RP_1 be the relation plane of element 1 and let $\{R_{1j}: j \in X\}$ be the collection of role-relations for element 1. The *role-set* for element 1 is the set of distinct role-relations in $[R_{1j}: j \in X\}$ (Winship & Mandel, 1983).

Winship and Mandel (1983) characterised the role of an element by its role-set. Mandel (1978) referred to the role-set definition of local role as a concrete version of local role because it involves the actual role-

Table 2.15. *The role-set for block 1 in the network* **N**

Role-relation	
Label	Vector
R_{11}	(1 1 1)
R_{12}	(1 0 1)
R_{13}, R_{14}	(0 1 1)

relations associated with the element. The local role algebra for an element is, on the other hand, an abstract version. The local role algebra is derived from a collection of relation vectors but does not refer to them explicitly, and a given role algebra is consistent with a number of realizations in the relation plane.

For example, the local role algebra for block 1 of X is characterised by the right multiplication table and partial order of Table 2.6. The representation is in terms of relationships between abstract relations (L, representing Liking, and A, representing Antagonism) and makes no reference to particular blocks in the set X. Instead, it contains information of the form LA contains L; that is, if block 1 likes some other block, then it also likes someone who is antagonistic toward that block. The role-set representation, on the other hand, shown in Table 2.15, lists the relationships that block 1 has with each other block in the set X. The role-relation $R_{12} = (1\ 0\ 1)$, for instance, indicates that in relation to block 2, block 1 has relations of types L and AL. Block 1 has the same role-relations with respect to blocks 3 and 4 – namely, $R_{13} = R_{14} = (0\ 1\ 1)$ – in which relations of type A and AL are present, but not L.

Partial networks and partial role algebras

In the case of incomplete partial networks, we do not necessarily possess useful information about long paths in the network. For instance, in a network constructed from the *k*th-order zone of some *ego*, we may have accurate information about all paths emanating from *ego* of length k or less but much less certain information about paths of length greater than k. Paths having lengths $k + 1$ or greater can reach outside of *ego*'s *k*th-order zone, whereas paths of length $k + 2$ or more can go outside of the *k*th-order zone and then return to terminate within it. Because we have no record of these longer paths from *ego*'s local network, it may be useful to construct a representation of relational structure for *ego* that depends

Table 2.16. *Truncated relation plane of order 2 for block 1 in the network* **N**

Relation	Block 1	2	3	4
L	1	1	0	0
A	1	0	1	1
LL	1	1	0	0
LA	1	0	1	1
AL	1	1	1	1
AA	1	0	1	1

only on paths of length k or less. Indeed, it may sometimes also be useful to restrict the length of paths under consideration for other reasons. For instance, the strategy may be used to avoid the computational problems that can arise when large local networks possess a large number of distinct relation vectors. Mandel (1978, 1983), for example, considers paths of length 3 or less in many of his illustrative applications of role algebras.

A representation of *ego*'s relational structure that depends only on paths of length k or less may be constructed from a "truncated" relation plane for an individual. The truncated plane contains only those relation vectors corresponding to paths of length no greater than k. It can be defined as follows.

DEFINITION. Let RP_1 be the relation plane for element 1 in a local network $\mathbf{R} = \{R_1, R_2, \ldots, R_p\}$ on a set $X = \{1, 2, \ldots, n\}$ of dimension $\infty \times n$. The *truncated relation plane of order k* is the submatrix of RP_1 whose rows correspond to paths in the local network of length k or less. The truncated relation plane has dimension $K \times n$, where $K = p + p^2 + \cdots + p^k$.

Because the truncated relation plane is necessarily finite when k is finite, it may be presented both in an expanded form, showing the relation vector for each path of length k or less, and in a compressed form, displaying only distinct relation vectors. The expanded form is used in Tables 2.16 and 2.17. Table 2.16 shows the truncated relation plane of order 2 for block 1 in the network **N**; its relation vectors indicate the paths of length 2 or less from block 1 to other blocks in **N**. Similarly, Table 2.17 presents the truncated relation plane of order 2 corresponding to paths of length 2 or less for *ego* in the network **L**.

Now the major difference between the full and truncated relation planes constructed from a local network is that not all compound paths are defined in the truncated case. Thus, the composition operation of a

Table 2.17. *Truncated relation plane of order 2 for* ego *in the network* **L**

Relation	Element: *ego*	*a*	*b*	*c*	*d*	*e*	*f*	*g*	*h*
C	0	0	0	1	1	0	0	0	0
A	0	1	1	1	1	0	0	0	0
CC	1	1	0	1	1	0	0	0	0
CA	1	1	1	1	1	1	1	1	0
AC	1	1	0	1	1	0	0	1	0
AA	1	1	1	1	1	1	1	1	1

local role algebra needs to be replaced by a *partial* composition operation in an algebra based on a truncated relation plane. The ordering relation, however, is defined for all relation vectors in the truncated relation plane. Further, the orderings are necessarily preserved by right multiplication whenever the right products are defined. That is, if we find that

$$U \leq V$$

holds for a pair of relation vectors in the truncated relation plane of order k for element 1 from X, and if the relation vectors for UW and VW both belong to the truncated plane as well, then it is necessarily the case that

$$UW \leq VW$$

holds. These considerations suggest the following definition of an algebra for a truncated relation plane:

DEFINITION. Let $\mathbf{R} = \{R_1, R_2, \ldots, R_p\}$ be a local network for element 1 on a set X. Define S_k to be the collection of all strings of elements of $\mathbf{R}$ of length k or less (that is, each $U \in S_k$ may be written as $U_1U_2 \cdots U_m$, where $U_j \in \mathbf{R}$, for each $j = 1, 2, \ldots, m$ and $m \leq k$). The *partial local role algebra for element 1 of order k*, denoted Q_1^k, may be defined as

1 the set S_k; together with
2 the ordering relation

$$U \leq V$$

if and only if

$$(1, x) \in U \text{ implies } (1, x) \in V,$$

for any $x \in X$; U, $V \in S_k$; as well as

Table 2.18. *Partial local role algebra Q_1^2 for block 1 in the network* **N**

	Right mult. table			
		Generator[a]		
Element	Class	1	2	Partial order
1	L	1	2	1 0 0
2	A	3	2	0 1 0
3	AL	*	*	1 1 1

[a] An asterisk in the right multiplication table indicates a product that is not defined.

3 the partial binary operation on S_k for which

$$(U_1 U_2 \cdots U_m)(V_1 V_2 \cdots V_l) = U_1 U_2 \cdots U_m V_1 V_2 \cdots V_l, \text{ if } m + l \leq k,$$

and is undefined if $m + l > k$.

We write $(V, U) \in Q$ if and only if $U \leq V$, and it may easily be shown that Q is reflexive and transitive and hence a quasi-order. Further, if $(V, U) \in Q$ and both VW and UW are defined, then it follows that $(VW, UW) \in Q$; for any $W \in S_k$.

For example, the partial local role algebra Q_1^2 generated from the truncated relation plane of order 2 for block 1 in network **N** is shown in Table 2.18. A partial role algebra can be presented in the form of a right multiplication table and a partial order on the distinct relation vectors corresponding to members of S_k. For instance, for block 1, it can be seen that the truncated relation plane of order 2 (Table 2.16) contains only three distinct relation vectors, namely (1 1 0 0), (1 0 1 1) and (1 1 1 1) corresponding to L, A and AL, respectively. Hence the right multiplication and partial order tables for the partial algebra pertain to products and orderings among the elements L, A and AL. Undefined products in the right multiplication table of the partial algebra are indicated with an asterisk. Comparing Tables 2.18 and 2.7 shows the effect of truncation at path length 2 on the derived algebraic representation of local role. The two representations share the equations and orderings

$$1 * LL = 1 * L, \quad 1 * LA = 1 * A, \quad 1 * AA = 1 * A,$$
$$1 * L < 1 * AL, \quad 1 * A < 1 * AL,$$

whereas the representation based on the "full" relation plane for element 1 has the additional equations

Table 2.19. *Partial local role algebra* Q_1^2 *for* ego *in the network* **L**

Element	Class	Generator[a] C	Generator[a] A	Partial order
1	C	3	4	1 0 0 0 0 0
2	A	5	6	1 1 0 0 0 0
3	CC	*	*	1 0 1 0 0 0
4	CA	*	*	1 1 1 1 1 0
5	AC	*	*	1 0 1 0 1 0
6	AA	*	*	1 1 1 1 1 1

(Right mult. table: Class, Generator C, Generator A.)

[a] An asterisk in the right multiplication table indicates a product that is not defined.

$$1 * ALL = 1 * AL, \qquad 1 * ALA = 1 * A,$$

namely, those pertaining to paths in the network of length greater than 2. In the case of this local role algebra, the effect of truncation of the relation plane at order 2 does not appear to be particularly great.

The partial local role algebra Q_1^2 of order 2 for *ego* in the network **L** is presented in Table 2.19. In this case, comparison of the full representation for *ego* in the network **L** with its partial counterpart constructed from the truncated relation plane of order 2 yields some common orderings, such as $ego * C < ego * A$, $ego * C < ego * CC$ and so on, but no common equations. All of the equations in the "full" local role algebra for *ego* in network **L** involve paths of length 3 or more, and none of these are represented in the partial local role algebra of order 2. There is much greater overlap, however, between the orderings and equations of the "full" local role algebra for *ego* in **L** and the partial local role algebra Q_1^3 of order 3, which is shown in Table 2.20.

Just as the algebraic properties of a local role algebra may be expressed in the formal terms of a role algebra, so the formal algebraic properties of a partial local role algebra may be characterised by a "partial role algebra". The essential properties of a partial local role algebra that have been described lead to the definition:

DEFINITION. A *partial role algebra* is a set W, a partial binary operation on W (that is, a binary operation defined for some ordered pairs of elements (s, t) from W) and a binary relation Q on X satisfying

1 $s(tu) = (st)u$ whenever all expressions are defined;
2 Q is a quasi-order (that is, Q is reflexive and transitive); and

Table 2.20. *Partial local role algebra Q_1^3 for ego in the network* **L**

	Right mult. table			
		Generator[a]		
Element	Class	C	A	Partial order
1	C	3	4	1 0 0 0 0 0 0 0
2	A	5	6	1 1 0 0 0 0 0 0
3	CC	3	6	1 0 1 0 0 0 0 0
4	CA	7	6	1 1 1 1 1 0 1 1
5	AC	8	6	1 0 1 0 1 0 0 0
6	AA	7	6	1 1 1 1 1 1 1 1
7	CAC	*	*	1 1 1 0 1 0 1 1
8	ACC	*	*	1 1 1 0 0 0 0 1

[a] An asterisk in the right multiplication table indicates a product that is not defined.

3 $(s, t) \in Q$ implies $(su, tu) \in Q$ for any u for which su and tu are both defined.

The definition is useful for making the kinds of comparisons among algebras that are described in chapter 3.

The nesting relation for partial role algebras

A nesting relation for partial role algebras can be defined in the same way as for role algebras: we simply compare the quasi-orders of the partial role algebras. We shall assume that each partial role algebra refers to the same set W (that is, to strings in the same set S_k) and to the same binary operation. Then we may define

DEFINITION. A partial role algebra Q is *nested* in a partial role algebra T if for any $s, t \in W$,

$$(s, t) \in T \text{ implies } (s, t) \in Q.$$

We write $Q \leq T$.

We can also extend the definition of partial local role algebras to subsets of network elements in the same way as for local role algebras. We can then show that if Y and Z are subsets of X for a local network **R** on X, with Y a subset of Z, then the partial local role algebra Q_Y^k of the subset Y is nested in the partial local role algebra Q_Z^k of Z.

Analysis of local networks

For any given local network, we have presented a series of algebraic structures that may be used to represent it. We could construct the partial local role algebra of order 1, or of order 2, or of order 3, and so on; we could also construct the local role algebra. A natural question to ask is how are these various structures related to one another and how should we select a particular one in a given situation.

Firstly, we may observe that for some sufficiently high value m, the partial local role algebra Q_1^m of order m is identical to the local role algebra Q_1. Indeed,

THEOREM 2.2. *Let* **R** *be a local network on a set* X *with identified element 1. There is a finite value of* m *for which* Q_1^h *is identical to* Q_1, *for all* $h \geq m$.

The proof follows from the fact that the full relation plane for an element of a local network has only a finite number of distinct relation vectors, so that there is some path length m beyond which no new distinct relation vectors are generated. As a result, the distinct relation vectors are identical for all truncated relation planes of order greater than m and are equal to those of the full relation plane.

The result is helpful in choosing amongst the alternative algebraic structures, because it guarantees that the choice set is finite: namely, $\{Q_1^1, Q_1^2, \ldots, Q_1^m\}$. From there, the choice depends on the kind of methodological restriction on path length in the local network under consideration, as well as on its size. For instance, if a local network gives rise to a large number of distinct relation vectors, then computation of its local role algebra may be difficult. It may then be convenient to deal with representations of bounded sizes, purely on computational grounds. In the examples presented earlier, it is useful to note that though the algebraic precision of the full local role algebra is lost in the partial representation, it is nonetheless the case that many of the relationships captured in the full version are present in the partial one. A suitable order for the partial local role algebra may then be chosen as a value h for which the number of distinct relation vectors in Q_1^h is not too large, and the number of equations and inclusions is not too small.

The effect of methodological restrictions on path length on the choice of an algebra is a little more complex. If the network has no restrictions on path length, then the local role algebra Q_1 is a natural choice, provided that it is not too large. However, in the case of a local network constructed as the kth-order zone of some network member, as in snowball sampling, the partial local role algebras of orders k and $k + 1$ can be argued to have some special significance. Constructions based on the truncated relation plane of order k may be interpreted with some confidence because the truncated plane pertains to all paths of length k or less in relation to all persons reachable by such paths. The truncated relation plane of order $k + 1$ is also a defensible construction because it should permit adequate comparison between paths of length $k + 1$ or less from *ego* to members of his/her *k*th-order zone. For paths of length $k + 2$ or greater, though, there is no guarantee of adequate comparison among paths because there may be paths of length $k + 2$ from *ego* to members of the *k*th order zone which pass through individuals outside the *k*th order zone and which are therefore omitted from the representation.

It is important to note that restrictions both on path length and on the outdegree of each network element in a local network influences the adequacy of the (partial) local role algebra that can be constructed from it. Thus, for example, none of the three partial networks of Figure 2.1 claim completeness in the way that a network with a meaningful boundary does. As a result, a theoretical distinction remains between role algebras defined according to Winship and Mandel and applied to networks with clear boundaries, and the partial role algebras introduced here. Moreover, we can distinguish the three kinds of partial network represented in Figure 2.1 in terms of the quality of the relational information that they afford for primitive and compound ties. From Figure 2.1c, for example, a more complete record of paths of length 2 emanating from *ego* is likely to be obtained than from either Figure 2.1a or 2.1b. Nevertheless, it is an open question whether the difference is a practical one, and one that is best answered by empirical investigation.

Partially ordered semigroups and role algebras: A summary

Before we turn to a more complete discussion of the algebraic means for analysing and comparing these representations of social networks, it is useful to reflect on the relationship between the representations that have been proposed for entire and partial networks.

A complete network and its partially ordered semigroup are "global" entities for a network **R** because both refer to properties of relations on the entire set X. Local networks, role sets and local role algebras are "local" on the other hand, because they pertain only to relations emanating from a single individual. Relation vectors are slices of the relation plane recording the presence of ties of particular types from an identified individual to other individuals. Ordering relations between the relation vectors define the abstract entity of local role algebra. The algebra is abstract in the same sense as the semigroup of a network: distinct collections of relation vectors can give rise to the same local role algebra.

Indeed, as noted earlier, the orderings among relations giving rise to the partially ordered semigroup of a network may also be considered as a role algebra, namely, the role algebra for the entire set X. The orderings among relations for a given network are identical in both structures, that is, any ordering between relations U and V in the partially ordered semigroup is also present in the role algebra, and conversely.

We may therefore summarise the relationship between the partially ordered semigroup of an entire network and the local role algebra of a partial network in the following terms. Both structures are defined by ordering relations among paths in networks, and the orderings are preserved in both cases when the paths are extended by adding a fixed additional path starting at the endpoints of those paths. In particular, if a path of type U from an individual i to any individual j implies a path of type V from i to j, then the presence of a path of type UW from i to some k implies the existence of a path of type VW from i to k also. In the case of the entire network, the path orderings hold for paths beginning with any individual i in the network, whereas for partial networks, the orderings need only apply to paths whose source is the identified *ego* for the network. In the complete network case, moreover, because the orderings hold for paths with any starting point i, they are also preserved when paths are augmented by path components added at the beginning of a path. That is, if the existence of a path of type U from any individual i to some individual j implies the existence of a path of type V from i to j, then whenever there is a path WU from some individual k to j, there is also a path of type WV.

Thus, the algebras representing structure in entire and local networks are directly comparable: the partially ordered semigroup of an entire network has all of the properties of the local role algebra of a partial network and some additional ones. The analogy between these constructions will be pursued in developing the analytic methods introduced in the following chapters.

3

Comparing algebraic representations

For both complete and local social networks a two-level representation of social structure has been defined. The first level constitutes the relational foundation; for entire networks it is a collection of network relations, and for partial networks it is the ego-centred local network. The second level is a derivative algebraic structure, a more abstract representation describing relationships between relational components from the first level. In the case of entire networks, this second-level representation is the partially ordered semigroup of the network; in the case of local or partial networks, it is the local role algebra.

In evaluating this algebraic level of representation, we are not restricted merely to the task of establishing that the definition of the algebraic level from the relational one is meaningful, useful though that is. Rather, some additional mathematical investigation can provide extra information about the usefulness of the representation. This mathematical exploration has two major aspects. The first is a search for an exact account of the way in which properties of the algebraic representation record properties of the relational one. For example, the task of describing the relational implications of equations or orderings in the partially ordered semigroup or local role algebra falls into this class of mathematical problems. The associated empirical problem is that of establishing the empirical significance of relational features made explicit by the representation.

The second aspect is less direct but corresponds to a central issue in the measurement of any phenomenon. It questions the theoretical and substantive value of identifying those configurations whose representations are identical. More specifically, if we have two distinct relational representations with identical algebras, we may ask whether there is an empirical basis for claiming that they possess the same relational structure. The mathematical component of this latter task consists of describing, for any given algebra, the class of all relational configurations that generate it. That is, we need to identify classes of complete networks whose members generate the same partially ordered semigroup and classes of local networks, each of which gives rise to the same local role algebra.

These two approaches to establishing the usefulness (or otherwise) of the algebraic representations will be considered in turn. In this chapter, we shall review some results pertaining to the second problem; that is, we shall consider the description of classes of complete and local networks that have the same algebra. The examination of the first question, which seeks an account of the way in which algebraic properties record relational features of the underlying data, will then be introduced and continued in the following chapter. We shall consider results for complete networks and semigroups first and then turn to local networks and local role algebras.

Isomorphisms of network semigroups

Let $\mathbf{R} = \{R_1, R_2, \ldots, R_p\}$ and $\mathbf{T} = \{T_1, T_2, \ldots, T_p\}$ be two networks consisting of p binary relations (not necessarily distinct). In what follows, we will often assume that relations in **R** and **T** have the same set of relation "labels" and that the relations are listed in the same order. That is, we will assume that the labels for R_1 and T_1 are the same, that those for R_2 and T_2 are the same, and so on. The assumption is reasonable, for example, if two networks have been constructed in different organisations from self-report surveys asking the same questions, such as "Who are your friends in the organisation?", and "To whom in the organisation do you go for help and advice?". The networks possess different sets of members, but the relations, "friend" and "goes to for help and advice", may be assumed to have the same meaning, or relational quality, in each network. Indeed, for some purposes, we may be happy to assert that networks constructed in different ways and possessing different labels are nonetheless sufficiently similar to be comparable. For instance, it may sometimes be useful to compare a network comprising self-report data on "liking" and "antagonism" with a network reporting "positive" and "negative" affect that has been constructed by an observer. In such a case, we would assign the same label to the relations "liking" and "positive affect", and the same label to "antagonism" and "negative affect".

DEFINITION. **R** and **T** are *comparable* networks if there exists a one-to-one mapping (or *bijection*) $f: \mathbf{R} \to \mathbf{T}$, assumed, without loss of generality, to be given by

$$f(R_i) = T_i,$$

where R_i and T_i are relations with the same label, for each i.

We are interested in describing the conditions under which distinct networks give rise to the same semigroup. What, though, do we mean by

Table 3.1. *Two comparable networks* $\mathbf{N}_1 = \{A, B\}$ *and* $\mathbf{N}_2 = \{A, B\}$

		Relation	
Network label	Element	A	B
$\mathbf{N}_1$	1	1 0	1 1
	2	1 0	1 1
$\mathbf{N}_2$	3	1 0	0 1
	4	1 0	0 1

two networks having the same semigroup? Given that the partially ordered semigroup of a network comprises (a) a set of generator labels, (b) a multiplication table and (c) a partial order table, then one reasonable condition for identical semigroups is that all three of these entities must be the same. That is, the partially ordered semigroups of two networks are the same if they have the same labels (that is, if the networks are comparable) and if their multiplication and partial order tables are identical.

DEFINITION. Let $S(\mathbf{R})$ and $S(\mathbf{T})$ be the partially ordered semigroups of comparable networks **R** and **T**. $S(\mathbf{R})$ and $S(\mathbf{T})$ are *isomorphic partially ordered semigroups* if there exists a bijection ϕ from $S(\mathbf{R})$ onto $S(\mathbf{T})$ such that

1 $\phi(R_i) = T_i$ for each $i = 1, 2, \ldots, p$;
2 $\phi(WV) = \phi(W)\phi(V)$ for each $W, V \in S(\mathbf{R})$; and
3 $\phi(W) \leq \phi(V)$ iff $W \leq V$.

The networks **R** and **T** have *isomorphic (abstract) semigroups* if conditions 1 and 2 hold.

It follows from the definition that if the partially ordered semigroups of **R** and **T** are isomorphic, then $S(\mathbf{R})$ and $S(\mathbf{T})$ are also isomorphic as abstract semigroups. The converse does not necessarily hold, however, as the networks in Table 3.1 demonstrate. The networks $\mathbf{N}_1$ and $\mathbf{N}_2$ have isomorphic abstract semigroups, but their partial orders are not isomorphic (Table 3.2). In particular, it can be seen from Table 3.2 that $A < B$ holds in $S(\mathbf{N}_1)$ but not in $S(\mathbf{N}_2)$. Unless stated otherwise, the definition adopted here is that which *includes* the partial order (that is, condition 3 of the preceding definition). For the reasons outlined in chapter 1, the partial order of a network semigroup is considered an important part of the structure of the semigroup. Note, though, that a number of the results derived in this and subsequent chapters apply whether or not part 3 of the preceding definition is included and that a number of earlier researchers have excluded reference to the partial order in their definition of isomorphism (e.g., Bonacich, 1980; Boorman & White, 1976).

Table 3.2. *The partially ordered semigroups* $S(\mathbf{N}_1)$ *and* $S(\mathbf{N}_2)$ *of the networks* $\mathbf{N}_1$ *and* $\mathbf{N}_2$

	Right mult. table				
			Generator		
Semigroup	Element	Word	A	B	Partial order
$S(\mathbf{N}_1)$	1	A	1	2	1 0
	2	B	1	2	1 1
$S(\mathbf{N}_2)$	1	A	1	2	1 0
	2	B	1	2	0 1

Some networks with isomorphic semigroups

A primary result concerning networks with isomorphic partially ordered semigroups identifies the semigroup of a network with the semigroup of its "skeleton" (Lorrain & White, 1971). We observed in chapter 1 that two individuals are structurally equivalent in a network if they possess identical relations to all other network members. We shall show that the partially ordered semigroup of a network is unchanged if (a) we replace a group of structurally equivalent elements by a single "block" having the same relations as elements of the group, or if (b) we add to a group of one or more structurally equivalent individuals, new individuals having the same relations as those in the existing group. The first operation, replacing each group of structurally equivalent individuals by a block representing them, leads to the construction of the skeleton of the network (Lorrain & White, 1971). The second operation, adding individuals who are structurally equivalent to existing network members, defines an "inflation" of the network. The latter construction is formally defined as follows:

DEFINITION. Let $\mathbf{R}$ be a network on a set $X = \{1, 2, \ldots, n\}$ and let $\{B_1, B_2, \ldots, B_n\}$ be a collection of pairwise disjoint nonempty sets. (Thus, each element i of X is associated with a distinct nonempty set B_i.) Let B be the union of the sets $B_1, B_2, \ldots, B_n$, and for each binary relation $R \in \mathbf{R}$, define a binary relation R^* on the set B, by

$$R^* = \cup \{(B_i \times B_j) : (i, j) \in R\}.$$

(In other words, for any elements $k, l \in B$, $(k, l) \in R^*$ if and only if there exist elements $i, j \in X$ such that $k \in B_i$, $l \in B_j$, and $(i, j) \in R$.) The relation R^* is called an *inflation* of R to B, and $\mathbf{R}^* = \{R^* : R^*$ is an inflation of R to B, $R \in \mathbf{R}\}$ is termed the *inflation* of the network $\mathbf{R}$ (Schein, 1970).

Table 3.3. *The network* **B**, *which is an inflation of the network* $\mathbf{N}_1$

Network label	Element	Relation	
		A	*B*
B	*a*	1 1 0 0 0	1 1 1 1 1
	b	1 1 0 0 0	1 1 1 1 1
	c	1 1 0 0 0	1 1 1 1 1
	d	1 1 0 0 0	1 1 1 1 1
	e	1 1 0 0 0	1 1 1 1 1

For example, the network **B** shown in Table 3.3 is an inflation of the network $\mathbf{N}_1$ of Table 3.1. Element 1 in $\mathbf{N}_1$ corresponds to the set $B_1 = \{a, b\}$, and element 2 in $\mathbf{N}_1$ corresponds to $B_2 = \{c, d, e\}$. That is, one obtains an inflation of a network by replacing each element $i \in X$ by a set of elements B_i that are structurally equivalent to one another, in the sense of Lorrain and White, and that possess relations with other elements determined by the relations of i.

The concept of an inflation is the converse of that of a skeleton as defined by Lorrain and White. More specifically, if $\mathbf{R}^*$ is an inflation of $\mathbf{R}$ and there exists no network $\mathbf{R}^\#$ distinct from $\mathbf{R}$ of which $\mathbf{R}$ is an inflation, then $\mathbf{R}$ is the skeleton of $\mathbf{R}^*$. Conversely, if $\mathbf{R}$ is the skeleton of $\mathbf{R}^*$, then $\mathbf{R}^*$ is an inflation of $\mathbf{R}$.

Lorrain and White's theorem (1971, p. 63) may now be stated.

THEOREM 3.1. *Let* $\mathbf{R}^*$ *be an inflation of the network* $\mathbf{R}$. *Then the partially ordered semigroups* $S(\mathbf{R}^*)$ *and* $S(\mathbf{R})$ *are isomorphic.*

Relational structure is therefore invariant under the operation of replacing an element of the network by a collection of elements to which it is structurally equivalent. Inflating a network does not change its structure because no new types of relational interlock are introduced. In fact, it is this property that is at the heart of blockmodel analysis. A blockmodel is an approximation to a skeleton for the network: it is a simpler characterisation of the network which is intended to have approximately the *same* relational structure.

A second simple but important result concerns the disjoint union of networks. Suppose that we observe the same network relations on two distinct groups of individuals and suppose, further, that no links join individuals in the two groups. Then the disjoint union of the two networks is simply the collection of network relations defined on the union of the sets of group members.

Table 3.4. *The network* $\mathbf{N}_3$, *which is the disjoint union of the networks* $\mathbf{N}_1$ *and* $\mathbf{N}_2$ *of Table 3.1*

Network label	Element	Relation A	Relation B
$\mathbf{N}_3$	1	1 0 0 0	1 1 0 0
	2	1 0 0 0	1 1 0 0
	3	0 0 1 0	0 0 0 1
	4	0 0 1 0	0 0 0 1

DEFINITION. Let $\mathbf{R}$ and $\mathbf{T}$ be comparable networks on disjoint sets X and Y, respectively. Suppose, without loss of generality, that relations R_i of $\mathbf{R}$ and T_i of $\mathbf{T}$ are of the same type, for each i. Define the network $\mathbf{R} \cup \mathbf{T}$ on the set $X \cup Y$ with relations $R_i \cup T_i$, given by

$$(u, v) \in R_i \cup T_i \text{ if } (u, v) \in R_i \text{ or } (u, v) \in T_i.$$

$\mathbf{R} \cup \mathbf{T}$ is termed the *disjoint union* of the networks $\mathbf{R}$ and $\mathbf{T}$.

For instance, the network $\mathbf{N}_3$ shown in Table 3.4 is the disjoint union of $\mathbf{N}_1$ and $\mathbf{N}_2$ of Table 3.1.

Now two networks are isomorphic if one is an exact "copy" of the other, that is, if there is a one-to-one mapping of the elements of one network onto the elements of the other such that a relation is present between a pair of elements in the first network if and only if it is also present for the corresponding pair in the second network.

DEFINITION. Let $\mathbf{R} = \{R_1, R_2, \ldots, R_p\}$ and $\mathbf{T} = \{T_1, T_2, \ldots, T_p\}$ be comparable networks defined on sets X and Y, respectively. $\mathbf{R}$ and $\mathbf{T}$ are *isomorphic* if there exists a bijection f from X onto Y such that $(x, y) \in R_i$ if and only if $(f(x), f(y)) \in T_i$; for any $i = 1, 2, \ldots, p$ and for any $x, y \in X$.

It may be established that the disjoint union of isomorphic networks has the same partially ordered semigroup as the constituent networks. Thus, for instance, if two separate divisions of an organisation have isomorphic networks, then the semigroup of the entire structure is the same as the semigroup of each of the constituent networks. That is,

THEOREM 3.2. *Let* $\mathbf{R} \cup \mathbf{T}$ *be the disjoint union of isomorphic networks* $\mathbf{R}$ *and* $\mathbf{T}$. *Then the partially ordered semigroup* $S(\mathbf{R} \cup \mathbf{T})$ *is isomorphic to* $S(\mathbf{R})$, *which is isomorphic to* $S(\mathbf{T})$.

Indeed, Theorem 3.2 may be generalised to show that any finite number of disjoint copies of a network has the same partially ordered semigroup

Table 3.5. *Two comparable networks* $\mathbf{N}_1 = \{A, B\}$ *and* $\mathbf{N}_4 = \{A, B\}$

		Relation	
Network label	Element	*A*	*B*
$\mathbf{N}_1$	1	1 0	1 1
	2	1 0	1 1
$\mathbf{N}_4$	5	1 0 0	1 1 1
	6	1 1 0	1 1 0
	7	1 0 0	1 1 0

as the original network; as for the inflation operation, no novel modes of interlock are imposed.

Theorems 3.1 and 3.2 are almost trivial in a mathematical sense; they are important, however, in evaluating the usefulness of the semigroup representation of relational structure. In both cases, we may argue that the implied changes to network structure, from the operations of inflation and disjoint copying, ought not to change the underlying relational structure. That is, we may argue that the addition of persons to a group in positions identical to some already occupied by existing group members should not alter the pattern of relationships between group roles. Nor should the existence of parallel structures in disjoint groups lead to different patterns of interlock when viewed from a global perspective.

Some other classes of semigroups having identical partially ordered semigroups are described in the later parts of the chapter. Such classes may be argued to comprise networks having the same relational structure because they possess the same orderings and equations among network paths. It is also of interest, though, to make more general kinds of comparisons among networks than those establishing the identity or otherwise of relational structure. In the next section, therefore, we consider the question of how network semigroups may be compared when they are not isomorphic.

Comparing networks: Isotone homomorphisms

An important step towards greater generality in our capacity to compare network semigroups is achieved when we ask whether the equations and orderings among paths in one network are a subset of those in another. If so, the structure of the first network is more "complex" than that of the second, in the sense of making more distinctions among different types of paths. Consider, for example, the networks displayed in Table 3.5,

Table 3.6. *The partially ordered semigroups* $S(\mathbf{N}_1)$ *and* $S(\mathbf{N}_4)$

		Right mult. table			
			Generator		
Semigroup	Element	Word	*A*	*B*	Partial order
$S(\mathbf{N}_1)$	1	*A*	1	2	1 0
	2	*B*	1	2	1 1
$S(\mathbf{N}_4)$	1	*A*	1	3	1 0 0 0
	2	*B*	4	3	1 1 0 1
	3	*AB*	4	3	1 1 1 1
	4	*BA*	4	3	1 0 0 1

whose semigroups are shown in Table 3.6. A set of equations and orderings characterising the semigroup $S(\mathbf{N}_4)$ is

$$AA = A, \quad BB = AB, \quad ABA = BA, \quad BAB = AB$$

and

$$A \le BA \le B \le AB,$$

whereas a set describing the semigroup $S(\mathbf{N}_1)$ is

$$AB = BB = B, \quad BA = AA = A$$

and

$$A \le B.$$

All of the equations and orderings that hold in $S(\mathbf{N}_4)$ also hold in $S(\mathbf{N}_1)$; for instance, since $A = BA$ in $S(\mathbf{N}_1)$, it is also the case that $A \le BA$ in $S(\mathbf{N}_1)$. The equations and orderings for $S(\mathbf{N}_4)$ are therefore a subset of those for $S(\mathbf{N}_1)$, and there are fewer distinct paths in $\mathbf{N}_1$ than $\mathbf{N}_4$. Thus, in the sense just described, $S(\mathbf{N}_1)$ has a simpler structure than $S(\mathbf{N}_4)$.

This type of comparison among network semigroups leads to a partial ordering in terms of complexity among them.

DEFINITION. Let $\mathbf{R} = \{R_1, R_2, \ldots, R_p\}$ and $\mathbf{T} = \{T_1, T_2, \ldots, T_p\}$ be comparable networks with partially ordered semigroups $S(\mathbf{R})$ and $S(\mathbf{T})$, respectively. Let $f: \mathbf{R} \to \mathbf{T}$ be the bijection from the generators of $\mathbf{R}$ onto those of $\mathbf{T}$ that preserves relation labels (that is, R_i and $f(R_i) = T_i$ have the same label). If, whenever $U_1U_2 \cdots U_k \le V_1V_2 \cdots V_m$ in $S(\mathbf{R})$, where $U_i, V_j \in \mathbf{R}$ $(i = 1, 2, \ldots, k;\ j = 1, 2, \ldots, m)$, it follows that $f(U_1)f(U_2) \cdots f(U_k) \le f(V_1)f(V_2) \cdots f(V_m)$ in $S(\mathbf{T})$, then $S(\mathbf{T}) \le S(\mathbf{R})$. In this case, $S(\mathbf{T})$ is termed an *isotone homomorphic image* of $S(\mathbf{R})$.

In the preceding example, for instance, $S(\mathbf{N}_1) \leq S(\mathbf{N}_4)$. It is straightforward to demonstrate that the operation $\leq$ in the definition is both transitive and reflexive and hence is a quasi-order. The equality relation that it induces – that is, $S(\mathbf{R}) = S(\mathbf{T})$ if and only if $S(\mathbf{R}) \leq S(\mathbf{T})$ and $S(\mathbf{T}) \leq S(\mathbf{R})$ – is the relation of isomorphism of partially ordered semigroups. Moreover, the relation may also be expressed in terms of a widely used algebraic construction called a homomorphic mapping.

DEFINITION. Let S and T be two semigroups. A *homomorphism* from S onto T is a mapping $\phi: S \rightarrow T$ satisfying

$$\phi(s_1 s_2) = \phi(s_1)\phi(s_2)$$

for all $s_1, s_2 \in S$. T is said to be a *homomorphic image* of S. For partially ordered semigroups, ϕ is an *isotone*, or *order-preserving*, homomorphism if, in addition,

$$s_1 \leq s_2 \quad \text{implies} \quad \phi(s_1) \leq \phi(s_2)$$

for all $s_1, s_2 \in S$.

The relationship between homomorphic mappings among semigroups and the partial ordering we have defined for the semigroups of comparable networks is summarised in Theorem 3.3.

THEOREM 3.3. *Let* S(**R**) *and* S(**T**) *be the partially ordered semigroups of comparable networks* **R** *and* **T**. *There is an isotone homomorphism from* S(**R**) *onto* S(**T**) *if and only if* S(**T**) $\leq$ S(**R**).

A homomorphism can be described as a structure-preserving mapping; it guarantees that the image under the mapping of the product of two elements of a semigroup is identical to the product of their separate images. It differs from an isomorphism in not requiring that the mapping be a bijection; thus, distinctions between elements in a semigroup S are not necessarily made between their images in any homomorphic image $\phi(S)$. Examples of a homomorphism and an isotone homomorphism are presented in Table 3.7. The semigroup S may be mapped onto the semigroup T by mapping elements A, B, AA, AB and BA in S to elements a, b, a, ab and ba in T, respectively. It may be verified that $\phi(s_1 s_2) = \phi(s_1)\phi(s_2)$, for all $s_1, s_2 \in S$; for instance,

$$\phi(AAB) = \phi(AB) = ab \quad \text{and} \quad \phi(AA)\phi(B) = a{\cdot}b = ab.$$

Similarly, it may be verified that this mapping also preserves the partial order in S: $s_1 \leq s_2$ implies $\phi(s_1) \leq \phi(s_2)$, for all $s_1, s_2 \in S$. Therefore, T is an isotone homomorphic image of S.

The semigroup S of Table 3.7 may also be mapped homomorphically onto the semigroup U by the mapping ϕ. Although products are pre-

Table 3.7. *The partially ordered semigroups S, T and U (T and U are homomorphic images of S, and T is an isotone image of S)*

Semigroup	Full mult. table						Partial order					
S		*A*	*B*	*AA*	*AB*	*BA*		*A*	*B*	*AA*	*AB*	*BA*
	A	*AA*	*AB*	*AA*	*AB*	*AA*	*A*	1	0	0	0	0
	B	*BA*	*B*	*BA*	*B*	*BA*	*B*	0	1	0	1	0
	AA	*AA*	*AB*	*AA*	*AB*	*AA*	*AA*	1	0	1	1	0
	AB	*AA*	*AB*	*AA*	*AB*	*AA*	*AB*	0	0	0	1	0
	BA	*BA*	*B*	*BA*	*B*	*BA*	*BA*	1	1	1	1	1
T		*a*	*b*	*ab*	*ba*			*a*	*b*	*ab*	*ba*	
	a	*a*	*ab*	*ab*	*a*		*a*	1	0	1	0	
	b	*ba*	*b*	*b*	*ba*		*b*	0	1	1	0	
	ab	*a*	*ab*	*ab*	*a*		*ab*	0	0	1	0	
	ba	*ba*	*b*	*b*	*ba*		*ba*	1	1	1	1	
U		*a*	*b*	*ab*	*ba*			*a*	*b*	*ab*	*ba*	
	a	*a*	*ab*	*ab*	*a*		*a*	1	0	0	0	
	b	*ba*	*b*	*b*	*ba*		*b*	0	1	0	0	
	ab	*a*	*ab*	*ab*	*a*		*ab*	0	0	1	0	
	ba	*ba*	*b*	*b*	*ba*		*ba*	0	0	0	1	

served by the mapping, the partial order on S is not; for instance, $BA > B$ in S but $\phi(BA) = ba$, $\phi(B) = b$ and $ba \not> b$ in U. Thus, U is an abstract homomorphic image of S but not an isotone homomorphic image.

The π-relation of an isotone homomorphism

Now each isotone homomorphism ϕ of a partially ordered semigroup S corresponds to a unique binary relation π_ϕ on S. The relation is defined as follows.

DEFINITION. Let ϕ be a homomorphism from the partially ordered semigroup S onto the partially ordered semigroup T. Define the *π-relation* corresponding to ϕ to be the binary relation π_ϕ on S given by

$$(s, t) \in \pi_\phi \text{ iff } \phi(t) \leq \phi(s);\ s, t \in S.$$

It may be established that the relation π_ϕ is transitive and reflexive and hence a quasi-order.

For instance, consider the partially ordered semigroups S and T of Table 3.7. The isotone homomorphism ϕ from S onto T corresponds to the

Table 3.8. *The π-relation corresponding to the isotone homomorphism from S onto T*

	A	*B*	*AA*	*AB*	*BA*
A	1	0	1	1	0
B	0	1	0	1	0
AA	1	0	1	1	0
AB	0	0	0	1	0
BA	1	1	1	1	1

Table 3.9. *The π-relation corresponding to the isotone homomorphism from* $S(\mathbf{N}_4)$ *onto* $S(\mathbf{N}_1)$

	1	2	3	4
1	1	0	0	1
2	1	1	1	1
3	1	1	1	1
4	1	0	0	1

π-relation shown in Table 3.8. It may be verified from Table 3.8 that $\phi(t) \leq \phi(s)$ if and only if $(s, t) \in \pi_\phi$.

Consider also the isotone homomorphism from the partially ordered semigroup $S(\mathbf{N}_4)$ onto $S(\mathbf{N}_1)$ (Table 3.6). The π-relation corresponding to the mapping is presented in Table 3.9. Again it can be seen that $(s, t) \in \pi$ if and only if $\phi(t) \leq \phi(s)$ in the image semigroup $S(\mathbf{N}_1)$.

Now it was observed earlier that the π-relation π_ϕ corresponding to an isotone homomorphism ϕ on S is both reflexive and transitive. It may also be readily established that any ordering present in S is also present in π_ϕ; that is,

$$t \leq s \text{ in } S \text{ implies } (s, t) \in \pi_\phi,$$

for *any* isotone homomorphism ϕ of S. Indeed, π_ϕ is identical to the partial ordering in S precisely when ϕ is an isomorphism.

DEFINITION. Let π_{min} be the relation on the partially ordered semigroup S defined by

$$(s, t) \in \pi_{\text{min}} \text{ iff } t \leq s \text{ in } S.$$

The relation π_{min} is the π-relation associated with an isomorphism of the partially ordered semigroup S and is simply the partial order of S.

For example, the relation $\pi_{\min}$ for the partially ordered semigroup of Table 3.7 is the partial order relation displayed on the right-hand side of the upper panel of Table 3.7.

We have seen, therefore, that the π-relation corresponding to a homomorphism ϕ is a transitive, reflexive relation on S that contains the partial order $\pi_{\min}$ of S. The π-relation of Table 3.8, for instance, contains the partial order of the semigroup of Table 3.7. Under what conditions, though, is a reflexive, transitive relation containing $\pi_{\min}$ the π-relation for some isotone homomorphism ϕ of S? The answer to this question is provided by Theorem 3.4.

THEOREM 3.4. *Let Ω be a reflexive, transitive relation defined on a partially ordered semigroup S that contains the partial order π_{min} of S. Then Ω is a π-relation corresponding to some homomorphism ϕ on S if and only if, for any* s, t $\in$ S,

$$(s, t) \in \Omega \textit{ implies } (su, tu) \in \Omega \textit{ and } (us, ut) \in \Omega \textit{ for each } u \in S.$$

Proof: The proof is contained in Appendix B.

For instance, it may be verified that the π-relation displayed in Table 3.8 satisfies the conditions of the theorem.

A consequence of Theorem 3.4 is the following:

THEOREM 3.5. *There is a one-to-one relationship between isotone homomorphisms of a semigroup* S *and its π-relations.*

Hence we may define:

DEFINITION. The homomorphism corresponding to the π-relation π may be denoted by ϕ_π, and the homomorphic image $\phi_\pi(S)$ of S is termed the *quotient semigroup* of S corresponding to the homomorphism ϕ_π and the π-relation π. We also write $\phi_\pi(S) = S/\pi$.

To construct the homomorphism corresponding to a given π-relation of a partially ordered semigroup S, we need only observe that

1 $\phi(t) = \phi(s)$ if and only if $(s, t) \in \pi$ *and* $(t, s) \in \pi$;
2 $\phi(t) < \phi(s)$ if and only if $(s, t) \in \pi$ but $(t, s) \notin \pi$.

The quotient semigroup S/π may then be derived in two steps. Firstly, we re-write the multiplication and partial order tables of S to include the orderings and equations among images of elements of $\phi(S)$. For the isotone homomorphism ϕ of S onto T of Table 3.7, for instance, we obtain the re-written tables shown in the upper panel of Table 3.10. Secondly, we delete the rows and columns of each table corresponding to redundant elements, such as element AA in Table 3.10 (because

Table 3.10. *Constructing a homomorphic image of the partially ordered semigroup S*

Full mult. table						Partial order				
Step 1										
	A	*B*	*AA*	*AB*	*BA*	*A*	*B*	*AA*	*AB*	*BA*
A	*A*	*AB*	*A*	*AB*	*A*	1	0	1	1	0
B	*BA*	*B*	*BA*	*B*	*BA*	0	1	0	1	0
AA	*A*	*AB*	*A*	*AB*	*A*	1	0	1	1	0
AB	*A*	*AB*	*A*	*AB*	*A*	0	0	0	1	0
BA	*BA*	*B*	*BA*	*B*	*BA*	1	1	1	1	1
Step 2										
	A	*B*	*AB*	*BA*		*A*	*B*	*AB*	*BA*	
A	*A*	*AB*	*AB*	*A*		1	0	1	0	
B	*BA*	*B*	*B*	*BA*		0	1	1	0	
AB	*A*	*AB*	*AB*	*A*		0	0	1	0	
BA	*BA*	*B*	*B*	*BA*		1	1	1	1	

$AA = A$). The outcome is the multiplication table and partial order table of the quotient semigroup, as for example, in the lower panel of Table 3.10. By comparing the lower panel of Table 3.10 with the multiplication and partial order tables for the semigroup T of Table 3.7, it can be seen that $\phi(S)$ is isomorphic to T.

The concept of a homomorphism provides the basis for the comparison of different semigroup structures. If a semigroup S may be homomorphically mapped onto a semigroup T, then one may argue that the structure of the semigroup T is consistent with, but simpler than, the structure of the semigroup S. Equivalently, S is a more complex or structurally articulated version of T, possessing additional structural distinctions to those of T. As we have indicated, most of the time, we shall consider isotone homomorphisms between partially ordered semigroups because these preserve equations *and* orderings under the homomorphic mapping.

Homomorphisms have been used to make comparisons for several types of algebraic representation in the social sciences literature. For instance, Boyd (1969) used the notion of homomorphism in his analysis of marriage class systems, and Boyd, Haehl and Sailer (1972) applied the concept to inverse semigroup models of kinship systems. Friedell (1967) and Pattison and Bartlett (1975, 1982) discussed the application of homomorphisms to Friedell's semilattice model of hierarchical organisation, and Boorman and White (1976), Breiger and Pattison (1978) and Pattison (1982) outlined some applications of homomorphisms to the semigroup representation of relational structure in networks.

Partial orderings among homomorphisms and π-relations

We have established that there is a one-to-one correspondence between isotone homomorphisms of partially ordered semigroups and the class of π-relations on the semigroup having the properties specified in Theorem 3.4. We have also defined a partial ordering on the collection of isotone homomorphic images of a semigroup, with $\phi(S) \leq \tau(S)$ if and only if $\tau(s) \leq \tau(t)$ implies $\phi(s) \leq \phi(t)$, for any $s, t \in S$. This ordering may be extended to the isotone homomorphic mappings themselves:

DEFINITION. Let ϕ and τ be isotone homomorphisms defined on a partially ordered semigroup S. Define $\phi \leq \tau$ if and only if $\phi(S) \leq \tau(S)$.

The π-relations defined on S may also be partially ordered by a *containment* relation:

DEFINITION. Let ϕ and τ be isotone homomorphisms of a partially ordered semigroup S. Define $\pi_\tau \leq \pi_\phi$ if $(s, t) \in \pi_\tau$ implies $(s, t) \in \pi_\phi$, for any $s, t \in S$. We say that π_τ is *contained* in π_ϕ.

Now it follows from the definition that $\pi_\tau \leq \pi_\phi$ if and only if $\phi \leq \tau$. This relationship among the collection of π-relations, on the one hand, and the collection of isotone homomorphisms, on the other, is described as "dual".

DEFINITION. Two partially ordered sets L_1 and L_2 are *dual* if there is a bijection g from L_1 onto L_2 for which

$$x \leq y \text{ in } L_1 \text{ iff } g(y) \leq g(x) \text{ in } L_2.$$

Hence we have established a one-to-one correspondence between the collection of all isotone homomorphisms of a partially ordered semigroup S and the collection of π-relation on S, and the ordering defined among π-relations is dual to that among isotone homomorphisms.

Abstract semigroups. An analogous set of definitions can be introduced for abstract semigroups:

DEFINITION. Let S and T be two finite semigroups with the same set of generator labels. Then we may define

$$T \leq_a S$$

if and only if T is an (abstract) homomorphic image of S. We may also define a relation π_ϕ on S corresponding to each homomorphism ϕ of S by

$$(s, t) \in \pi_\phi \text{ iff } \phi(s) = \phi(t).$$

Clearly, π_ϕ is an equivalence relation on S and has the property that $(s, t) \in \pi_\phi$ implies $(su, tu) \in \pi_\phi$ and $(us, ut) \in \pi_\phi$, for any $u \in S$; $s, t \in S$.

It is also termed a *congruence* relation on S (e.g., Clifford & Preston, 1961).

For example, for the abstract semigroups having multiplication tables of those of S and T of Table 3.7, $T \leq_a S$. The congruence relation on S corresponding to the homomorphism onto T is the relation with equivalence classes (A, AA) (B) (AB) (BA).

Lattices of semigroups and π-relations

The concept of an isotone homomorphism of a semigroup is the fundamental construction for the comparison of different semigroups. If one semigroup is a homomorphic image of the other, then the latter makes structural distinctions not present in the former. More generally, no semigroup is necessarily a homomorphic image of any other, but the two semigroups are located in a partially ordered structure termed a lattice. We may define a lattice as follows:

DEFINITION. Let L be a partially ordered set, that is, a set of elements together with a partial order relation $\leq$. Let J be a subset of L. Then an *upper bound* for J is an element $x \in L$ for which $y \leq x$, for every $y \in J$. A *least upper bound* for J is an element x' such that $x' \leq x$ whenever x is an upper bound for J. If a set J possesses a least upper bound, then it is unique (because $x \leq z$ and $z \leq x$ implies $x = z$). Similarly, a *lower bound* for J is an element z for which $z \leq y$, for every $y \in J$. A *greatest lower bound* for J is an upper bound for the set of lower bounds of J and is unique, if it exists. A *lattice* is a partially ordered set L in which every pair of elements x, y possess a greatest lower bound (*glb*), or *meet*, denoted by $glb(x, y)$, or $meet(x, y)$, or $x \wedge y$, and a least upper bound (*lub*), or *join*, denoted by $lub(x, y)$, or $join(x, y)$ or $x \vee y$.

We may display a finite lattice by drawing its Hasse diagram (see chapter 1). Figure 3.1 presents the Hasse diagram of a finite lattice of six elements. An upper bound of two elements is then any element that is drawn above the two elements and is connected to them, and their least upper bound is the unique upper bound that is below and connected to all other upper bounds. For instance, the upper bounds for the pair S_4 and S_5 in Figure 3.1 are S_1 and S_3, and because S_3 is the minimal element of the set $\{S_1, S_3\}$, S_3 is the least upper bound of S_4 and S_5. A lower bound of two elements in Hasse diagram is any element drawn below both of the elements and connected to them, and a greatest lower bound is the highest element in the set. In Figure 3.1, the element S_6 is the only lower bound for S_4 and S_5 and is therefore the greatest lower bound.

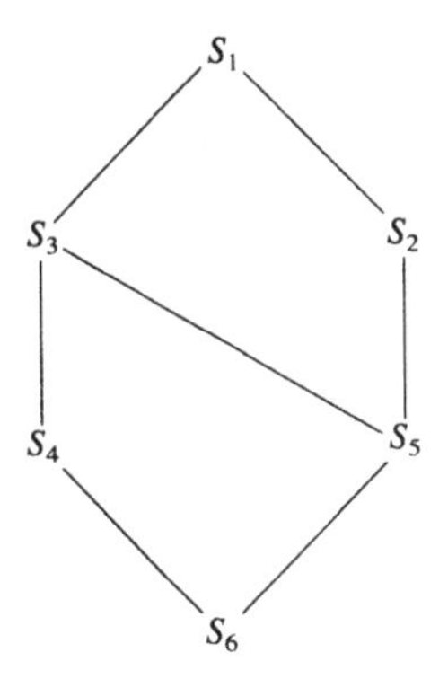

Figure 3.1. The lattice L_S of isotone homomorphic images of $S(\mathbf{N}_4)$

Network semigroups may be seen as elements in at least two different lattices. The first comprises the collection of partially ordered semigroups, partially ordered by isotone homomorphisms. The second is the set of abstract semigroups partially ordered by homomorphic mappings.

DEFINITION. Let **R** be a set of generator labels, and let $L(\mathbf{R})$ be the collection of partially ordered semigroups with generator labels **R**. For semigroups S_1 and S_2 in $L(\mathbf{R})$, define $S_1 \leq S_2$ if and only if there is an isotone homomorphism from S_2 onto S_1. Also, let $A(\mathbf{R})$ be the collection of abstract semigroups having generator labels **R**, partially ordered by $S_1 \leq_a S_2$ if and only if there is an (abstract) homomorphism from S_2 onto S_1. Then:

THEOREM 3.6. *The collection* L(**R**) *of partially ordered semigroups on a set* **R** *of generator labels forms a lattice under the partial ordering* $S_1 \leq S_2$ *if and only if there is an isotone homomorphism from* S_2 *onto* S_1. *Similarly, the collection* A(**R**) *of abstract semigroups with generator labels* **R** *is a lattice under the partial ordering* $S_1 \leq_a S_2$ *if and only if there is an (abstract) homomorphism from* S_2 *onto* S_1. *The set* L_s *of isotone homomorphic images of a finite partially ordered semigroup* S *from* L(**R**) *is a finite sublattice of* L(**R**), *and the set* A_s *of abstract homomorphic images of a semigroup* $S \in A(\mathbf{R})$ *is a finite sublattice of* A(**R**).

Proof: The proof is contained in Appendix B.

For example, the lattice L_s for the semigroup $S = S(\mathbf{N}_4)$ of Table 3.6 is shown in Figure 3.1. The multiplication tables and partial orders for each image of S identified in Figure 3.1 are presented in Table 3.11. The

Table 3.11. *Isotone homomorphic images of* $S(\mathrm{N}_4)$

	Right mult. table				
			Generator		
Image	Element	Word	A	B	Partial order
S_1	1	A	1	3	1 0 0 0
	2	B	4	3	1 1 0 1
	3	AB	4	3	1 1 1 1
	4	BA	4	3	1 0 0 1
S_2	1	A	1	3	1 0 0
	2	B	1	3	1 1 0
	3	AB	1	3	1 1 1
S_3	1	A	1	2	1 0 0
	2	B	3	2	1 1 1
	3	BA	3	2	1 0 1
S_4	1	A	1	2	1 0
	2	B	2	2	1 1
S_5	1	A	1	2	1 0
	2	B	1	2	1 1
S_6	1	$B = A$	1		1

elements in L_s were obtained by systematically generating all possible mappings ϕ from S onto a set T such that ϕ is an isotone homomorphism. A computer program assisting the process is described in chapter 4; but for this small example, the elements can also be constructed by hand. To begin the construction, we observe that any isotone homomorphic image of S must possess all of the equations and orderings of S and, possibly, some additional ones as well. Now the orderings among the four distinct elements of S are presented, in Hasse diagram form, in Figure 3.2. We have already observed that the maximal element of L_s is S itself; it is the image of S having precisely the same orderings and equations as S. Other elements of L_s have additional orderings and/or equations, and they can be generated in two steps. The first is to construct all possible collections of additional orderings to those of S. This step is equivalent to generating all possible transitive and reflexive relations containing the partial order $\pi_{\min}$ of S. The second step is then to check whether each such collection corresponds to an isotone homomorphism of S. This amounts to checking whether the associated transitive and reflexive relation on S satisfies the conditions of Theorem 3.4 and is therefore the π-relation for some isotone homomorphism of S. It can be seen from Figure 3.2, for example, that the following collections of orderings (and

Figure 3.2. Hasse diagram for the partial order of $S(\mathbf{N}_4)$

hence additional orderings and equations induced by the transitivity of the ordering) can be added to those of S:

1 $A \geq BA$ (and hence $A = BA$)
2 $BA \geq B$ (and hence $BA = B$)
3 $B \geq AB$ (and hence $B = AB$)
4 $A \geq BA$, $BA \geq B$ (and hence $A \geq B$ and $A = BA = B$)
5 $BA \geq B$, $B \geq AB$ (and hence $BA \geq AB$ and $BA = B = AB$)
6 $A \geq BA$, $B \geq AB$ (and hence $A = BA$ and $B = AB$)
7 $A \geq BA$, $BA \geq B$, $B \geq AB$ (and hence $A \geq B$, $A \geq AB$, $BA \geq AB$ and $A = BA = B = AB$).

Checking each of these lists in turn, we find that list 1 corresponds to a homomorphism ϕ with $\phi(A) = \phi(BA)$ and other orderings as in S. List 1 leads, in fact, to the homomorphism from S onto S_2. Similarly, lists 3, 5, 6 and 7 correspond to homomorphisms from S onto S_3, S_4, S_5 and S_6, respectively. List 2 does not correspond to an isotone homomorphism of S, because if $\phi(BA) = \phi(B)$, it follows that $\phi(ABA) = \phi(AB)$, that is, $\phi(BA) = \phi(AB)$ as well – as in list 5 but not in list 2.

Figure 3.3 presents the lattice A_s for abstract homomorphic images of the semigroup $S(\mathbf{N}_4)$. The homomorphic images are presented in Table 3.12. They may be constructed in a similar manner to those of Table 3.11 except that we consider all possible additional *equations* to those of S. The collections of possible additional equations are presented in Table 3.13, together with the corresponding abstract homomorphic image of S, where appropriate.

Now recall that the collection L_s of all isotone homomorphisms from a partially ordered semigroup S may be partially ordered by the relation $\phi_1 \leq \phi_2$ if, for all $x, y \in S$, $\phi_2(x) \leq \phi_2(y)$ implies $\phi_1(x) \leq \phi_1(y)$. For the collection of π-relations on S, we have

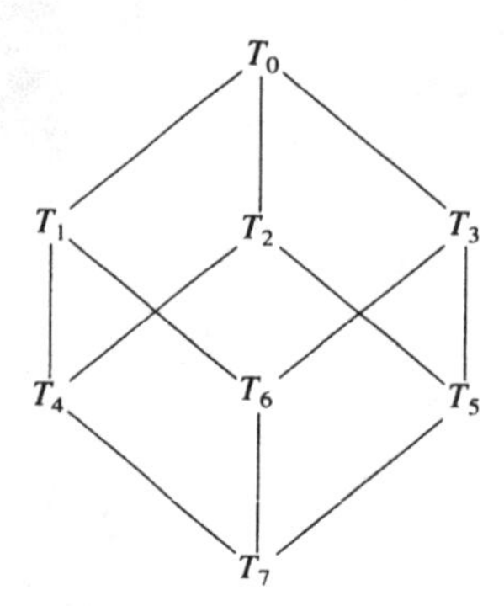

Figure 3.3. The lattice A_S for the abstract semigroup with multiplication table of $S(\mathbf{N}_4)$

DEFINITION. The collection of all π-relations on a partially ordered semigroup S, partially ordered by

$$\pi_r \le \pi_\phi \text{ iff } (s,t) \in \pi_\tau \text{ implies } (s,t) \in \pi_\phi$$

may be denoted $L_\pi(S)$, and termed the *π-relation lattice* of S.

It has already been observed that $\pi_\tau \le \pi_\phi$ in $L_\pi(S)$ if and only if $\phi \le \tau$ in L_s. Thus, because there is a one-to-one relationship between homomorphic images in L_s and π-relations in $L_\pi(S)$, it follows that L_s and $L_\pi(S)$ are dual. We have already established that L_s is a lattice. Consequently, the collection of π-relations $L_\pi(S)$ is also a lattice; indeed

THEOREM 3.7. *The collection* $L_\pi(S)$ *of* π*-relations on the partially ordered semigroup* S *is isomorphic to the dual of the lattice* L_s.

The maximal and minimal elements of $L_\pi(S)$, denoted by π_{max} and π_{min}, respectively, are given by

$$(s,t) \in \pi_{max} \text{ for all } s, t \in S;$$
$$(s,t) \in \pi_{min} \text{ iff } t \le s \text{ in } S;\ s, t \in S.$$

For example, the collection of all isotone homomorphic images and corresponding π-relations for the semigroup $S(\mathbf{N}_4)$ of Table 3.6 are shown in Tables 3.11 and 3.14, respectively. Each homomorphic image is the quotient semigroup associated with exactly one π-relation on S, and the partial ordering among homomorphic images, displayed in Figure 3.1, is clearly the dual of that among corresponding π-relations illustrated in Figure 3.4. (The homomorphic images and π-relations are labelled in Tables 3.11 and 3.14 so that S/π_i is isomorphic to S_i, for $i = 1, 2, \ldots, 6$.)

Table 3.12. *Abstract homomorphic images of* $S(\mathbf{N}_4)$

	Right mult. table					Right mult. table					Right mult. table			
Label	Element	Word	L	A	Label	Element	Word	L	A	Label	Element	Word	L	A
T_0	1	L	1	3	T_1	1	L	1	2	T_2	1	L	1	3
	2	A	4	3		2	A	3	2		2	A	3	3
	3	LA	4	3		3	AL	3	2		3	LA	3	3
	4	AL	4	3										
T_3	1	L	1	3	T_4	1	L	1	2	T_5	1	L	1	1
	2	A	1	3		2	A	2	2		2	A	1	1
	3	LA	1	3										
T_6	1	L	1	2	T_7	1	$A = L$	1						
	2	A	1	2										

Table 3.13. *Finding abstract homomorphic images of* $S(\mathbf{N}_4)$

Possible additional equations	Homomorphism?	Label
$A = B$	No	
$A = AB$	No	
$A = BA$	Yes	T_3
$B = AB$	Yes	T_1
$B = BA$	No	
$AB = BA$	Yes	T_2
$A = B = AB$	No	
$A = B = BA$	No	
$A = AB = BA$	Yes	T_5
$B = AB = BA$	Yes	T_4
$A = B, AB = BA$	No	
$A = AB, B = BA$	No	
$A = BA, B = AB$	Yes	T_6
$A = B = AB = BA$	Yes	T_7

Table 3.14. *π-relations corresponding to isotone homomorphisms of* $S(\mathbf{N}_4)$

π_1	π_2	π_3	π_4	π_5	π_6
1 0 0 0	1 0 0 1	1 0 0 0	1 0 0 0	1 0 0 1	1 1 1 1
1 1 0 1	1 1 0 1	1 1 1 1	1 1 1 1	1 1 1 1	1 1 1 1
1 1 1 1	1 1 1 1	1 1 1 1	1 1 1 1	1 1 1 1	1 1 1 1
1 0 0 1	1 0 0 1	1 0 0 1	1 1 1 1	1 0 0 1	1 1 1 1

The joint homomorphism of two semigroups

Any two semigroups with the same set **R** of generator labels are not necessarily directly comparable in the lattices $L(\mathbf{R})$ and $A(\mathbf{R})$, but they possess both a unique least upper bound and a unique greatest lower bound in each lattice. The greatest lower bound, termed the joint homomorphic image or JNTHOM (Boorman & White, 1976) of the two semigroups, is the largest semigroup that is a homomorphic image of both semigroups. It contains the finest set of equations or structural distinctions consistent with the structure of both of the semigroups and, as such, has been proposed as a precise record of the relational structure shared by their corresponding networks (Boorman & White, 1976; Breiger & Pattison, 1978). The joint homomorphic image of two network semigroups S_1 and S_2 may be defined in either $L(\mathbf{R})$ or $A(\mathbf{R})$.

DEFINITION. Let S_1 and S_2 be two partially ordered semigroups with the same set of generator labels **R**. The *joint isotone homomorphic image*, or

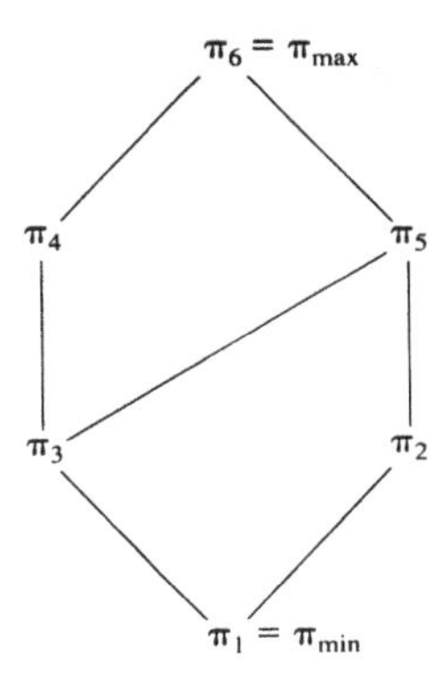

Figure 3.4. The lattice $L_\pi(S(\mathbf{N}_4))$ of π-relations on $S(\mathbf{N}_4)$

JNTIHOM(S_1, S_2), is the maximal semigroup that is an isotone homomorphic image of S_1 and S_2. *JNTIHOM*(S_1, S_2) is the greatest lower bound of S_1 and S_2 in the lattice $L(\mathbf{R})$. If S_1 and S_2 are abstract semigroups, then we may define the *joint homomorphic image* of S_1 and S_2 denoted by *JNTHOM* (S_1, S_2), as the maximal semigroup S that is a homomorphic image of S_1 and S_2 (Boorman & White, 1976). That is, it is the greatest lower bound of S_1 and S_2 in $A(\mathbf{R})$.

The *JNTIHOM* and *JNTHOM* for two partially ordered semigroups S_1 and S_2 may be different, even as abstract semigroups, but each is uniquely defined.

Table 3.15 presents an example of two partially ordered semigroups, their joint isotone homomorphic image (*JNTIHOM*) and their joint homomorphic image (*JNTHOM*). The mappings ϕ_1 and ϕ_2 from V and W, respectively, onto $J = JNTHOM(V, W)$ are given by

$$\phi_1: A, AA, BA \to a \text{ and } B, AB \to b$$

and

$$\phi_2: A \to a \text{ and } B, BB \to b.$$

It may be observed that the homomorphisms ϕ_1 and ϕ_2 partition the semigroup tables of V and W so that (a) the partitioned tables have the same global structure, and (b) the partitioning is consistent within each table, in the sense that any block in the partition contains only elements mapped onto the same element of J by the homomorphism. The mappings ϕ_3 and ϕ_4 from V and W onto $K = JNTIHOM(V, W)$ map all of the elements in each semigroup onto the one element. K is the largest semigroup having all of the equations and orderings of both V and W.

Table 3.15. *The joint homomorphic image J and the joint isotone homomorphic image K of two semigroups V and W*

Semigroup		Full mult. table						Partial order				
V		*A*	*B*	*AA*	*AB*	*BA*		*A*	*B*	*AA*	*AB*	*BA*
	A	*AA*	*AB*	*AA*	*AB*	*AA*	*A*	1	0	0	0	0
	B	*BA*	*B*	*BA*	*B*	*BA*	*B*	0	1	0	1	0
	AA	*AA*	*AB*	*AA*	*AB*	*AA*	*AA*	1	0	1	1	0
	AB	*AA*	*AB*	*AA*	*AB*	*AA*	*AB*	0	0	0	1	0
	BA	*BA*	*B*	*BA*	*B*	*BA*	*BA*	1	1	1	1	1
W				*A*	*B*	*BB*				*A*	*B*	*BB*
			A	*A*	*BB*	*BB*			*A*	1	0	0
			B	*A*	*BB*	*BB*			*B*	0	1	0
			BB	*A*	*BB*	*BB*			*BB*	1	1	1
J					*a*	*b*						
				a	*a*	*b*						
				b	*a*	*b*						
K						*a*						*a*
					a	*a*						1

Algorithms in the *APL* programming language have been developed for the construction of the joint homomorphism of two arbitrary finite semigroups with the same set of generator labels by White and Lorrain (Boorman & White, 1976, note 23). A program for finding the joint homomorphic image of two abstract semigroups is also available in *UCINET* 4.0 (Borgatti, Everett & Freeman, 1991; also Heil, 1983).

For partially ordered semigroups, the *JNTIHOM* may be computed in a similar fashion to the *JNTHOM*, but the partial order table as well as the multiplication table must be taken into account. An algorithm for finding the joint isotone homomorphic image of two partially ordered semigroups S_1 and S_2 having the same set **R** of generator labels may be constructed by adding the equations and orderings in one semigroup, say S_2, to the partial ordering of the other, say S_1. The result is a relation P on the elements of S_1 containing the partial order π_{min} of S_1. The next step is to compute the least π-relation π on S_1 that contains P. The isotone homomorphic image of S_1 corresponding to π is then the joint isotone homomorphic image of S_1 and S_2.

The joint isotone homomorphic image of two network semigroups is the "simplest" structure that the networks share. In many cases, it may also be useful to be able to describe those aspects of structure that are unique to each network. Suppose that S_1 and S_2 are the semigroups of two comparable networks, $\mathbf{R}_1$ and $\mathbf{R}_2$, so that $S_1 = S(\mathbf{R}_1)$ and $S_2 = S(\mathbf{R}_2)$. Then homomorphic images of S_1 other than *JNTHOM* (S_1, S_2) or *JNTIHOM* (S_1, S_2) can provide a different set of simplifications or perspectives on the semigroup S_1 than those recorded by the *JNTHOM* or *JNTIHOM*. If one could find images of S_1 and of S_2 that had no relational structure in common with *JNTHOM* (S_1, S_2) or *JNTIHOM* (S_1, S_2), then one would be able to describe in quite fine detail the ways in which the two networks $\mathbf{R}_1$ and $\mathbf{R}_2$ are different, as well as the ways in which they are the same.

The common structure semigroup

The JNTHOM is not the only choice for a representative of the structure shared by two semigroups, and it is important to recognise that in proposing the JNTHOM as a candidate for the role, we have indeed made a selection. The decision point occurs when we choose to characterise a homomorphic image T of a semigroup S as a simpler, or a more complex, version of S. Do the additional equations in T, and thus the extra constraints on the interrelationships among elements of T, mean that T has a more complex structure or a simpler one? As Gaines (1977) has observed in the context of general systems, the choice is a substantive one and not one that can be made on mathematical grounds. The argument presented earlier, that the algebra making the greater number of relational distinctions is the more complex, is the one that has been adopted by a number of authors, including Boorman and White (1976), Boyd (1969), Lorrain (1975) and Lorrain and White (1971). An alternative argument, however, that the semigroup making fewer relational distinctions (and so possessing more equations) is the more complex, has also been put forward (Bonacich, 1980; Bonacich and McConaghy, 1979; McConaghy, 1981). In fact, Bonacich and McConaghy have argued that the greater number of equations in the homomorphic image of a semigroup means that the semigroup image has a more complex structure rather than a simpler one. The logical consequence of this view is that the structure shared by two semigroups S_1 and S_2 is the "least" semigroup that has all of the equations of both S_1 and S_2, that is, the most "complex" semigroup that is "simpler" than both of them (Pattison, 1981).

DEFINITION. The *common structure semigroup* (Bonacich & McConaghy, 1979) for two semigroups S_1 and S_2 having generator labels in $\mathbf{R}$ is the least upper bound of S_1 and S_2 in the lattice $A(\mathbf{R})$ and is denoted by $CSS(S_1, S_2)$. An analogous construction can be made in the lattice $L(\mathbf{R})$ for partially ordered semigroups: the *common isotone structure semigroup* $CISS(S_1, S_2)$ of S_1 and S_2 is the least upper bound of S_1 and S_2 in $L(\mathbf{R})$.

Table 3.16. *Common structure semigroups for the semigroups V and W*

a. Common structure semigroup CCS(V, W)

Full mult. table

	A	*B*	*AA*	*AB*	*BA*	*BB*
A	*AA*	*AB*	*AA*	*AB*	*AA*	*AB*
B	*BA*	*BB*	*BA*	*BB*	*BA*	*BB*
AA	*AA*	*AB*	*AA*	*AB*	*AA*	*AB*
AB	*AA*	*AB*	*AA*	*AB*	*AA*	*AB*
BA	*BA*	*BB*	*BA*	*BB*	*BA*	*BB*
BB	*BA*	*BB*	*BA*	*BB*	*BA*	*BB*

b. Common isotone structure semigroup CISS(V, W)

	Full mult. table							Partial order					
	A	*B*	*AA*	*AB*	*BA*	*BB*		*A*	*B*	*AA*	*AB*	*BA*	*BB*
A	*AA*	*AB*	*AA*	*AB*	*AA*	*AB*	*A*	1	0	0	0	0	0
B	*BA*	*BB*	*BA*	*BB*	*BA*	*BB*	*B*	0	1	0	0	0	0
AA	*AA*	*AB*	*AA*	*AB*	*AA*	*AB*	*AA*	1	0	1	0	0	0
AB	*AA*	*AB*	*AA*	*AB*	*AA*	*AB*	*AB*	0	0	0	1	0	0
BA	*BA*	*BB*	*BA*	*BB*	*BA*	*BB*	*BA*	1	0	1	0	1	0
BB	*BA*	*BB*	*BA*	*BB*	*BA*	*BB*	*BB*	0	1	0	1	0	1

The common structure semigroup was proposed by Bonacich and McConaghy as an alternative representative of the structure shared by two semigroups. The common structure semigroups for the semigroups *V* and *W* of Table 3.15 are presented in Table 3.16. It can be seen that *CSS*(*V*, *W*) and *CISS*(*V*, *W*) have the same multiplication table, and it can be shown that this result holds for *any* pair of partially ordered semigroups.

The issue of whether the *JNTHOM* (or *JNTIHOM*) or the common structure semigroup is the more useful representative of common structure is discussed in chapter 8, where more information is available on the nature of the relationship between semigroup images and a network.

Lattices of semigroups: A summary

The structures that have been introduced so far in the chapter are presented in summary form in Table 3.17. It can be seen that we have defined three different lattices for partially ordered semigroups and three different lattices for abstract semigroups. For partially ordered semigroups, we have constructed (a) the lattice *L*(**R**) of semigroups with the

Table 3.17. *Lattices of semigroups and π-relations*

Partially ordered semigroups		Abstract semigroups	
Label	Description	Label	Description
$L(\mathbf{R})$	Lattice of partially ordered semigroups with generator labels **R**	$A(\mathbf{R})$	Lattice of abstract semigroups with generator labels **R**
	$S_1 \leq S_2$ if S_1 is an isotone homomorphic image of S_2		$S_1 \leq S_2$ if S_1 is an abstract homomorphic image of S_2
	$JNTIHOM(S_1, S_2) = glb(S_1, S_2)$		$JNTHOM(S_1, S_2) = glb(S_1, S_2)$
	$CISS(S_1, S_2) = lub(S_1, S_2)$		$CSS(S_1, S_2) = lub(S_1, S_2)$
L_S	Lattice of isotone homomorphic images of the partially ordered network semigroup $S(\mathbf{R})$	A_S	Lattice of abstract homomorphic images of the abstract network semigroup $S(\mathbf{R})$
	L_s is a finite sublattice of $L(\mathbf{R})$		A_S is a finite sublattice of $A(\mathbf{R})$
$L_\pi(S)$	Lattice of π-relations on the partially ordered network semigroup $S(\mathbf{R})$	$A_\pi(S)$	Lattice of symmetric π-relations on the abstract network semigroup $A(\mathbf{R})$
	$\pi_1 \leq \pi_2$ if $(s, t) \in \pi_1$ implies $(s, t) \in \pi_2$		
	$L_\pi(S)$ is dual to L_S		$A_\pi(S)$ is dual to A_S

same set **R** of generator labels, partially ordered by isotone homomorphic mappings; (b) the lattice L_S of isotone homomorphic images of a partially ordered semigroup S, also partially ordered by isotone homomorphic mappings; and (c) the lattice $L_\pi(S)$ of π-relations on S, partially ordered by a containment relation. The lattice L_S is a finite sublattice of $L(\mathbf{R})$, and $L_\pi(S)$ is dual to L_S. For two partially ordered semigroups, S and T, with the same set **R** of generator labels, their joint isotone homomorphic image, $JNTIHOM(S, T)$ is their greatest lower bound in $L(\mathbf{R})$, and their common isotone structure semigroup $CISS(S, T)$ is their least upper bound in $L(\mathbf{R})$.

For abstract semigroups, a parallel set of structures has been defined. The lattice $A(\mathbf{R})$ is the collection of (abstract) semigroups with the same set **R** of generator labels, partially ordered by (abstract) homomorphic mappings. The lattice A_S is a finite sublattice of $A(\mathbf{R})$, comprising all (abstract) homomorphic images of S. The collection of all π-relations on the abstract semigroup S constitutes the lattice $A_\pi(S)$ with partial ordering by set containment; $A_\pi(S)$ is the dual of A_S and is sometimes termed the congruence lattice of the abstract semigroup S. For two abstract semigroups, S and T, with the same set **R** of generator labels, their joint homomorphic image $JNTHOM(S, T)$ and common structure semigroup $CSS(S, T)$ are given by their greatest lower bound and least upper bound, respectively, in the lattice $A(\mathbf{R})$.

Local networks with isomorphic local role algebras

In comparison with semigroups of networks, less is known of the relationship between a local network and its local role algebra. One basic and useful result has been established by Mandel (1978), however, and may be expressed in the form of an identical role-set condition. We must first define what we mean by identical role algebras. A natural definition follows:

DEFINITION. Let Q_1 and Q_2 be role algebras having the set $\mathbf{R}$ of generator labels. Then Q_1 and Q_2 are *isomorphic* if $U \leq V$ in Q_1 if and only if $U \leq V$ in Q_2, for any $U, V \in FS(\mathbf{R})$.

Thus, Q_1 and Q_2 are isomorphic whenever Q_1 is nested in Q_2 *and* Q_2 is nested in Q_1, that is, whenever Q_1 and Q_2 have identical right multiplication and partial order tables.

Now, recall that each local network $\mathbf{R}$ defined on a set X of elements gives rise to a relation plane. The columns of the relation plane, termed role-relations, correspond to elements of X, and the rows of the relation plane correspond to elements in the free semigroup $FS(\mathbf{R})$. The role-set for the local network is the set of distinct role-relations in the relation plane. Then:

THEOREM 3.8 *(Mandel, 1978). Two elements whose local networks possess identical role-sets have the same local role algebra.*

The result is established by noting that the role algebra of an element is invariant under alterations to the relation plane that leave the role-set unchanged.

A question that follows from this condition is that of the relational conditions under which two elements have identical role-sets. Winship (1988) discussed two cases in his original formulation of the role-set as a characterisation of local role. The first is the familiar case of structural equivalence: two elements that are structurally equivalent have identical role-sets. The second is a generalisation of structural equivalence to the case in which elements have the same type of relationships with the same types of individuals, but not necessarily with the same individuals. The idea may be captured formally by defining an automorphism on a network. As outlined in chapter 1, an automorphism is a mapping that re-labels the elements of the network in such a way that the relations among the elements appear unchanged. Then two elements may be regarded as the same type if one can be re-labelled by the other in an automorphism. Such elements possess the same "kinds" of relations. For example, the network displayed in Figure 1.5b has an automorphism corresponding to the re-labelling of A by B, B by A, C by D and D by C. The re-labelling is displayed in Figure 1.5c and it can be seen that the relations between elements in Figures 1.5b and 1.5c are identical. Thus, the re-labelling constitutes an automorphism.

DEFINITION. Let $\mathbf{R} = \{R_1, R_2, \ldots, R_p\}$ be a network on X. An *automorphism* α of the network $\mathbf{R}$ is a bijection α of X onto itself such that for any $u, v \in X$ and $i = 1, 2, \ldots, p$,

$$(u, v) \in R_i \text{ iff } (\alpha(u), \alpha(v)) \in R_i.$$

Elements $x, y \in X$ are *automorphically equivalent* if there exists an automorphism α for which

$$\alpha(x) = y.$$

Then:

THEOREM 3.9. *Two elements that are automorphically equivalent have identical role-sets and hence identical local role algebras.*

Proof: The proof follows from the observation that the role-relation R_{xj} in the role-set for element x is identical to the role-relation $R_{\alpha(x)\alpha(j)}$ in the role-set for element $y = \alpha(x)$.

A further generalisation may be derived by considering the mapping from a network $\mathbf{R}$ to an automorphically reduced network $AE(\mathbf{R})$. The mapping is obtained by noting that automorphic equivalence is an equivalence relation and hence that the following mapping is well-defined:

1 $AE(x) = \alpha_x$, where α_x is the equivalence class containing element x, and
2 $(\alpha_x, \alpha_y) \in AE(R_i)$ if and only if $(x, y) \in R_i$ for some $x \in \alpha_x$, $y \in \alpha_y$.

It should be noted that the role-set of the class α_x in $AE(\mathbf{R})$ is not necessarily the same as that of x in $\mathbf{R}$ although the former is a subset of the latter. It is possible, however, that distinct classes x, y of $AE(\mathbf{R})$ will be automorphically equivalent and therefore have identical role-sets, so that one can define a further automorphic reduction of $AE(\mathbf{R})$ to obtain $AE^2(\mathbf{R})$. In general, Pattison (1980; also Borgatti et al., 1989) defined

$$AE^h(\mathbf{R}) = AE(AE^{h-1}(\mathbf{R}))$$

for integers $h > 1$.

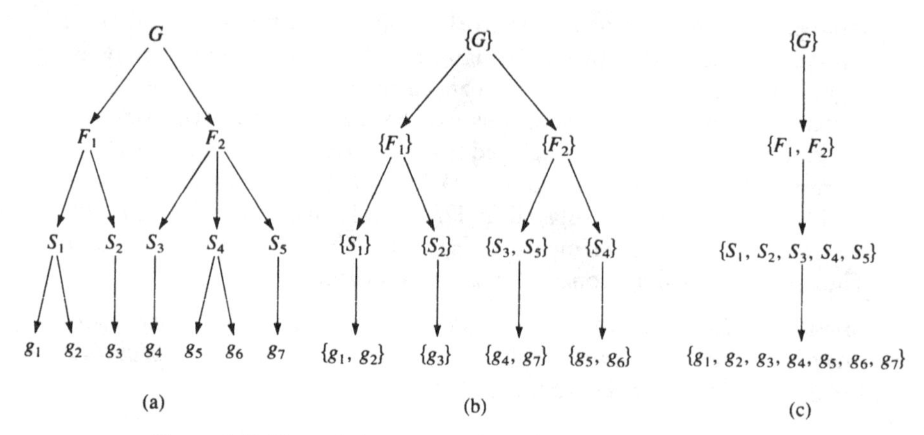

Figure 3.5. Extended automorphic equivalence: (a) the relation R; (b) $AE(R)$; (c) $AE^2(R) = AE^*(R)$

It may be observed that for some minimum integer h, $AE^h(\mathbf{R})$ is automorphically irreducible (in the sense that the only automorphism is the identity mapping).

For such an integer h, define

$$AE^*(\mathbf{R}) = AE^h(\mathbf{R}).$$

Then AE^* induces an equivalence relation on X, termed *extended automorphic equivalence*, and we have established the following relationship among role-sets.

THEOREM 3.10. *Let* x *and* y *be elements in a network* **R** *that are mapped to the same element* x *by a sequence of* AE *mappings (i.e.,* x *and* y *are in the same equivalence class induced by* AE*). *Then the role-sets of* x *and* y *both contain the role-set of the* AE* *class containing them.*

An illustration of the result is provided in Figure 3.5. The relation "is a father of" is shown for four generations of family in Figure 3.5a; in Figures 3.5b and 3.5c the reduced systems $AE(\mathbf{R})$ and $AE^2(\mathbf{R}) = AE^*(\mathbf{R})$ are presented. The example illustrates the manner in which AE^* is a generalisation of AE: elements are equivalent in AE^* if they have the same types of relationships but not necessarily the same number of each.

The problem of characterising those elements that have identical local role algebras but different role-sets is an open one. Some examples of the latter are shown in Table 3.18, which contains a list of the occurrence of identical local role algebras for elements having different role-sets in two-element two-relation networks (Lorrain, 1973). It can be seen in each case that the *egos* of the local networks have the same local role algebra but different role-sets.

Table 3.18. *Some small local networks with identical role-sets*

Local role algebra				Local networks	
	Right mult. table				
Element[a]	Gen. 1	Gen. 2	Partial order	Gen. 1	Gen. 2
1	1	1	1 0	0 0	0 1
2	2	2	1 1	0 1	0 1
				0 0	1 1
				1 1	0 1
1	1	1	1 0	0 1	1 1
2	1	2	1 1	0 1	0 1
				0 0	1 1
				1 1	0 0
1, 2 = 1	1		1	1 0	1 0
				0 1	0 0
				1 1	1 1
				1 0	1 1
				0 0	0 0
				0 1	1 0

[a] *Ego* is element 1.

Comparing local role algebras: The nesting relation

For semigroups of networks, it was suggested that comparisons among partially ordered semigroups having generator labels **R** may take place in the lattice $L(\mathbf{R})$ of partially ordered semigroups on **R**. In particular, it was claimed that if one semigroup is an isotone homomorphic image of another, then the first is a simpler version of the second. This claim led to the characterisation of the structure shared by two semigroups by their joint isotone homomorphic image, *JNTIHOM*. In the case of role algebras, an analogous lattice may be defined using the nesting relation (Mandel, 1983; Pattison, 1989).

THEOREM 3.11. *The collection* $\mathrm{M}(\mathbf{R})$ *of role algebras having the set* **R** *of generator labels forms a lattice whose partial order is given by* $\mathrm{T} \le \mathrm{Q}$ *if and only if* T *is nested in* Q. *For any local role algebra* $\mathrm{Q} \in \mathrm{M}(\mathbf{R})$, *the collection* $\mathrm{L_Q}$ *of role algebras nested in* Q *defines a finite sublattice of* $\mathrm{M}(\mathbf{R})$.

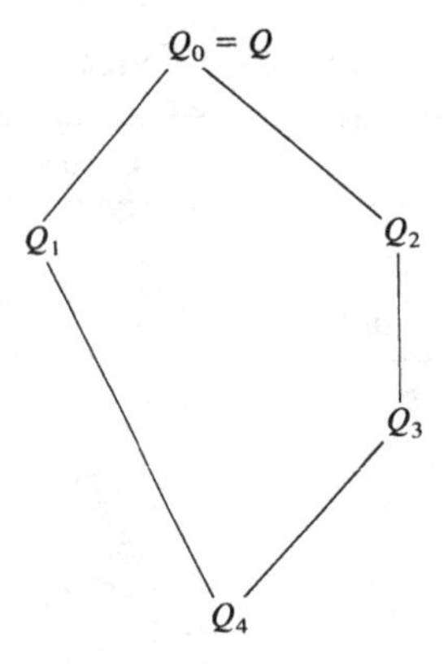

Figure 3.6. The lattice L_Q of role algebras nested in Q

Table 3.19. *Role algebras nested in the role algebra Q*

	Right mult. table				
			Generator		
Label	Element	Class	L	A	Partial order
Q_0	1	L	1	2	1 0 0
	2	A	3	2	0 1 0
	3	AL	3	2	1 1 1
Q_1	1	L	1	2	1 1
	2	A	1	2	0 1
Q_2	1	L	1	2	1 0 0
	2	A	3	2	1 1 0
	3	AL	3	2	1 1 1
Q_3	1	L	1	2	1 0
	2	A	2	2	1 1
Q_4	1	$A = L$	1		1

Proof: The proof is contained in Appendix B.

For instance, the lattice L_Q for the role algebra displayed in Table 2.7 is presented in Figure 3.6. The role algebras nested in Q are presented in Table 3.19.

Just as each isotone homomorphism of a partially ordered semigroup is associated with a binary π-relation on S, so is each nested role algebra of Q associated with a unique binary relation on Q. That is, we may define

Table 3.20. *π-relations in $L_\pi(Q)$ for the role algebra Q*

π_0	π_1	π_2	π_3	π_4
1 0 0	1 1 1	1 0 0	1 0 0	1 1 1
0 1 0	0 1 0	1 1 0	1 1 1	1 1 1
1 1 1	1 1 1	1 1 1	1 1 1	1 1 1

a π-relation π_T on the classes of e_Q corresponding to each role algebra T nested in Q as follows.

DEFINITION. Let Q be a role algebra and let T be a role algebra nested in Q. Define the *π-relation π_T corresponding to T* on the classes of e_Q by: $(s^*, t^*) \in \pi_T$ if there exist relations $s \in s^*$, $t \in t^*$ such that $(s, t) \in T$, where s^*, t^* are classes of e_Q.

For instance, role algebras nested in the role algebra Q of Table 2.7 were presented in Table 3.19. The corresponding π-relations are shown in Table 3.20 and are labelled so that Q_i corresponds to π_i ($i = 0, 1, \ldots, 4$).

In fact, just as for partially ordered semigroups, there is always a one-to-one correspondence between nested role algebras of a given role algebra and its π-relations. The relationship is summarised in the following theorem.

THEOREM 3.12. *If* T *is a role algebra nested in the role algebra* Q, *then* π_T *is a reflexive and transitive relation on* e_Q *with the property that* $(s^*, t^*) \in \pi_T$ *implies* $(su^*, tu^*) \in \pi_T$, *for any* $u \in FS(\mathbf{R})$. *Conversely, if* π *is a transitive and reflexive relation on the classes of* e_Q *with the property that* $(s^*, t^*) \in \pi$ *implies* $(su^*, tu^*) \in \pi$, *for any* $u \in FS(\mathbf{R})$, *then* π *is the π-relation corresponding to some role algebra nested in* Q.

Proof: The proof is given in Appendix B.

As a result of Theorem 3.12 we may define the following:

DEFINITION. Let Q be a role algebra and let π_T be a π-relation on Q corresponding to a role algebra T nested in Q. We may term T the *quotient role algebra* corresponding to the π-relation π_T, and write $T = Q/\pi_T$.

For example, each role algebra displayed in Table 3.19 is nested in the role algebra Q of Table 2.7. The corresponding π-relations are displayed in Table 3.20. Observe that the partial order and multiplication tables for each nested role algebra may be inferred from Table 2.7 and the corresponding π-relation (also see chapter 2).

Now we have already established that the collection of role algebras

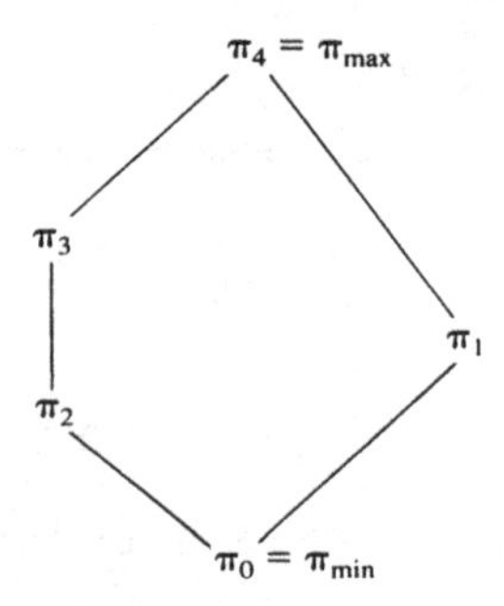

Figure 3.7. The lattice $L_\pi(Q)$ of π-relations on the role algebra Q

nested in a role algebra Q form a lattice L_Q under the nesting relation (with $T \leq Q$ if T is nested in Q). A dual lattice may be constructed using the π-relations corresponding to nested role algebras, under the partial ordering:

DEFINITION. Let π_T and π_P be π-relations corresponding to role algebras T and P nested in role algebra Q. Define

$$\pi_T \leq \pi_P$$

if and only if

$$(s^*, t^*) \in \pi_T \text{ implies } (s^*, t^*) \in \pi_P, \text{ for any classes } s^*, t^* \text{ of } e_Q.$$

Denote by $L_\pi(Q)$ the collection of all π-relations corresponding to role algebras nested in Q, with the preceding partial ordering. $L_\pi(Q)$ is termed the *π-relation lattice* of Q.

For example, Figure 3.7 presents the π-relation lattice of the role algebra Q of Table 2.7. (The π-relations appearing in Figure 3.7 are listed in Table 3.20.)

It may be observed that the minimal element of $L_\pi(Q)$ is the π-relation corresponding to Q itself, which is simply the partial order for Q, presented on the classes of e_Q. The maximal element of $L_\pi(Q)$ is the universal relation π_{max} on e_Q, given by $(s^*, t^*) \in \pi_{max}$ for every pair of classes s^*, t^* of e_Q. Further, the greatest lower bound of two relations, π_T and π_P, in $L_\pi(Q)$ is given by

$$glb(\pi_T, \pi_P) = \pi_T \cap \pi_P;$$

that is, $(s^*, t^*) \in glb(\pi_T, \pi_P)$ if and only if $(s^*, t^*) \in \pi_T$ and $(s^*, t^*) \in \pi_P$.

Now, if one local role algebra is nested in another, then all of the orderings and equations in the second local role algebra are also in the first, and the first contains some additional orderings and/or equations. The presence of additional constraints on the relationships of the first local role algebra led Mandel to argue that the first is simpler (or more constrained) than the second. His argument is consistent with Nadel's (1957) portrayal of "coherent" role-systems as those that are most constrained; it is also a natural complement to that made earlier for semigroups of networks. A natural consequence of this view is to characterise the shared structure of two local role algebras as the most complex role algebra that is simpler than both of them (Breiger & Pattison, 1986).

DEFINITION. Let Q and T be two role algebras defined on the same generator set $\mathbf{R}$. Then

$$JRA(Q, T) = \max\ \{U: U \text{ is nested in } Q \text{ and } U \text{ is nested in } T\}$$

is the *joint nested role algebra* for Q and T. Clearly, $JRA(Q, T)$ is the greatest lower bound of Q and T in the lattice $M(\mathbf{R})$.

The position adopted by Bonacich and others for semigroup algebras (Bonacich, 1980; Bonacich & McConaghy, 1979; McConaghy, 1981) is consistent with the alternative view that a nested role algebra is more complex than the algebra in which it is nested because it possesses more orderings than the latter. Such a view leads to the definition of shared structure in terms of shared inclusions.

DEFINITION. Let Q and T be two role algebras defined on the same set $\mathbf{R}$ of generator relations. Then the *common role algebra* $CRA(Q, T)$ of Q and T is defined as their intersection $Q \cap T$ – that is, $Q \cap T$ is the intersection of the binary relations Q and T on $FS(\mathbf{R})$. It may be shown that $CRA(Q, T)$ is the least upper bound of Q and T in $M(\mathbf{R})$.

The issue of selecting a representative of common structure in local networks is discussed in detail in chapter 8, when the relationship between a network and its algebra has been made clearer. For the moment, it may be observed that more than one position can be argued from the representations as they stand.

Other classes of networks with identical algebras

In the final section of this chapter, some additional conditions under which comparable networks are known to have the same network algebra are presented. The results are relevant to the task of evaluating the algebraic representations of network structure that have been proposed, because they describe classes of networks in which we would expect relational structure to be the same. Such expectations can, in principle at least, be submitted to empirical investigation. The results described in this section are somewhat technical and are not central to the methods developed in the remaining chapters of the book; those developments are taken up again at the start of chapter 4.

Trees

An important class of networks that have been involved in many attempts to model social relationships, especially those referring to influence and power, may be described in graph-theoretical terms as (directed) trees (e.g., Harary, 1959a; Harary et al., 1965; Oeser & Harary, 1964). Boorman (1977), for instance, has constructed a model of structural information about an authority hierarchy that various participants in the hierarchy may infer, and Friedell (1967) has presented a critique of the basic tree model and a suggested alternative.

Figure 3.8a presents a hypothetical network for the relation of being a direct superordinate in a small work group. That is, an arrow is directed from one individual to another if the first is the supervisor of the second. The converse relation, of being a direct subordinate, is shown in Figure 3.8b; the relation is sometimes termed a "reporting" relation because it indicates who reports to whom. It can be seen in Figure 3.8 that each member of the group is the direct subordinate of no more than one other group member. In addition, there is just one individual (labelled A in Figure 3.8) who is directly subordinate to no one. Structures possessing these two properties are termed out-trees.

DEFINITION. A weakly connected graph with edge set R is a *(directed) out-tree* if it has a single *source* (i.e., exactly one element with zero indegree) and no *semicycles* [i.e., no subset $\{i_1, i_2, \ldots, i_f\}$ of distinct elements of X such that $(i_k, i_{k+1}) \in R$ or $(i_{k+1}, i_k) \in R$; $k = 1, 2, \ldots, f-1$ with $i_f = i_1$]. If R is an out-tree, then a path from element i to element j is unique if it exists (e.g., Harary et al., 1965). The *depth* of an element in an out-tree is the length of the path to the element from the source of the out-tree; the depth of the source is defined to be zero. A *minimal element* of an out-tree R is an element with zero outdegree.

Further, we can define the following:

DEFINITION. The class $\mathbf{U}_n$ of out-trees of maximal depth $(n-1)$ comprises all out-trees T for which the maximum length of a path to a minimal element from the source of the tree is $(n-1)$. The class $\mathbf{T}_n$ of out-trees of constant depth $(n-1)$ consists of all out-trees T for which the length of the path to every minimal element from the source is $(n-1)$.

Some members of the classes $\mathbf{U}_3$ and $\mathbf{T}_3$ are illustrated in Figure 3.9. In Boorman's (1977) terms, $\mathbf{T}_n$ consists of all out-trees of constant depth

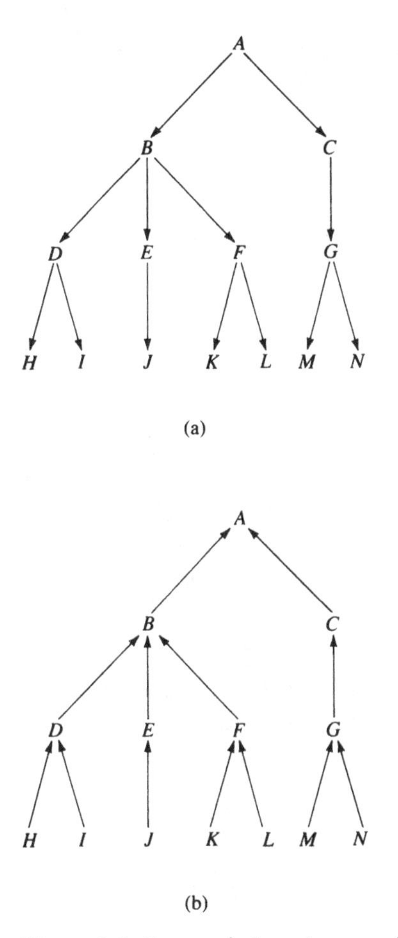

Figure 3.8. Some relations in a small work group: (a) direct superordinate relation; (b) direct subordinate relation

($n-1$) with no constraints on the span of control of any of their elements. (The *span of control* of an individual in a reporting relation is that individual's indegree.) Clearly, the class $\mathbf{T}_n$ is a subset of the class $\mathbf{U}_n$. Theorem 3.13 establishes that each member of the class $\mathbf{U}_n$ generates the same partially ordered semigroup.

THEOREM 3.13. *For any* $\mathrm{T} \in U_n$, $\mathrm{S(T)}$ *has distinct relations* $\mathrm{T}, \mathrm{T}^2, \ldots, \mathrm{T}^n$ *and products in* $\mathrm{S(T)}$ *defined by*

$$\mathrm{T^iT^j = T^{i+j}} \quad \textit{if} \quad \mathrm{i + j < n}$$

and

$$\mathrm{T^iT^j = T^n} \quad \textit{if} \quad \mathrm{i + j \geq n}.$$

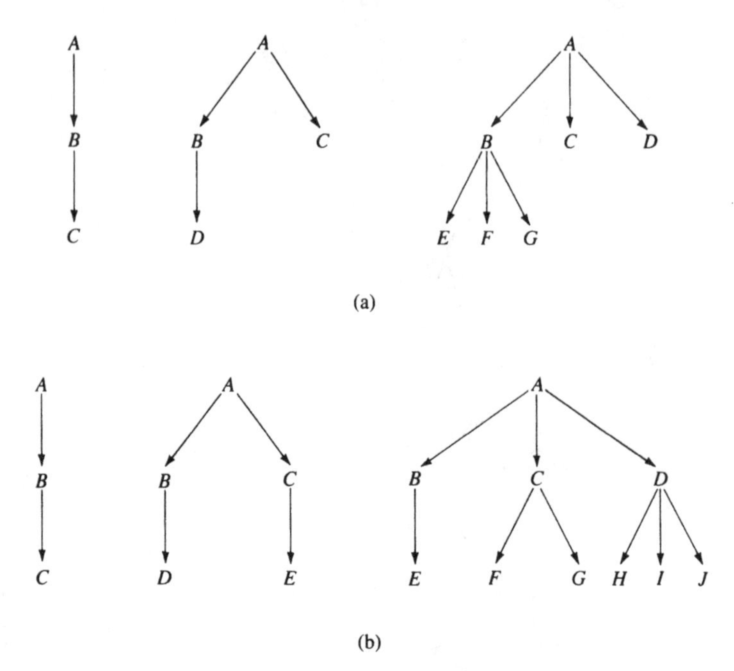

Figure 3.9. Some directed out-trees: (a) some members of the class U_3; (b) some members of the class T_3

The partial order in S(T) *is given by*

$$T^i < T^j \quad \textit{iff} \quad i \geq n \text{ and } j < n.$$

(In fact, T^n *is the null relation; that is, no elements are linked by* T^n.*)*

Proof: The proof is given in Appendix B.

A consequence of Theorem 3.13 is that all members of the class U_n generate the same partially ordered semigroup; that is, they all display the same structure of relations. For the semigroup of an out-tree, the length of the longest path in the tree determines its structure.

Another result on out-trees concerns the semigroup generated by an out-tree of constant depth $(n - 1)$ and its converse, that is, the semigroup generated by networks comprising a direct subordinate relation and a direct superordinate relation.

THEOREM 3.14. *Let* $T_1, T_2 \in \mathbf{T}_n$. *Then the multiplication tables of* $S(\{T_1, T_1'\})$ *and* $S(\{T_2, T_2'\})$ *are isomorphic, where* T′ *denotes the converse of the relation* T *[i.e.,* (i, j) ∈ T′ *if and only if* (j, i) ∈ T.*]*

Proof: The proof is given in Appendix B.

The result does not hold for the multiplication tables generated by members of $\mathbf{U}_n$ and their converses. Nor does it hold for the partially ordered semigroups generated by members of $\mathbf{T}_n$ and their converses. That is, although all members of $\mathbf{T}_n$ generate, with their converses, the same semigroup multiplication table, the partial ordering among the distinct semigroup elements is not always the same. It can easily be shown, however, that the partial order for $S(\{T, T'\})$ generated by any out-tree T from $\mathbf{T}_n$ is contained in the partial order for the semigroup $S(\{C_n, C_n'\})$, where C_n is the unique element of $\mathbf{T}_n$ with exactly n elements and a single maximal path of length $(n - 1)$.

Thus, out-trees of a given constant depth interlock with their converses in the same abstract way; the "boss" and "deputy" relations described by members of $\mathbf{T}_n$ (Boorman, 1977) and their converses give rise to the same distinct compound relations and the same equations among them.

The abstract semigroup table encodes a part of the relational structure claimed to be common to members of $\mathbf{T}_n$. It was noted earlier that the task that naturally followed the identification of collections of binary relations with the same semigroup was that of finding some empirical basis for the identity. In the case of members of $\mathbf{T}_n$, the demand appears to be reasonable. Trees in $\mathbf{T}_n$ are exactly the class of out-trees that lend themselves to unambiguous categorisation in terms of n levels. Thus, whether one proceeds to assign tree elements to levels beginning with the unique maximal element, or source, or whether one starts from any one of the minimal tree elements, the categorisation is the same. The relational structure of the out-tree is equivalent to interlock among the n ranks thus defined, providing a useful simplification of the problem of describing it. Those ranks, in fact, have long been part of our language for describing traditional authority structures as exist, for example, in armies or bureaucracies. The reality of the ranks is suggested by the relationship inferred among members of different rank from different branches of an army tree; the person of higher rank is often presumed to command the respect of the person of lesser rank.

In practice, it is probably the case that few kinds of relations give rise to out-trees of constant depth. Examples include formal reporting relations in organisations such as armies, bureaucracies, and corporations in which policy prescribes chains of command of constant length. In other cases, authority structures may contain branches of various lengths, as in the relations in $\mathbf{U}_n$, and may even contain individuals with more than one immediate superior (Friedell, 1967). Such structures may be approximated by a system of ranks, in a sense to be made precise later. Aberrations from the "ideal" constant-depth tree structures would then be translated into relational complexities in the semigroup structure, but it would be possible to recover the pure form later as a simplification. Thus, provided that the networks considered are essentially tree-like in pattern, some stratification of their elements may provide a reasonable approximation to their relational structure.

Idempotent relations

The simplest semigroup is the trivial one, that is, the semigroup consisting of one element R that satisfies the equation $R^2 = R$. Relations that generate a one-element semigroup are termed *idempotent*: Boyd (1983) proposed that idempotent relations are useful generalisations of equivalence relations on networks. The class of all idempotent relations, that is, the class of all binary relations with the simplest possible relational structure, has been described independently by Schein (1970) and Schwarz (1970a). Schein's formulation is followed here.

DEFINITION. Let Q be a quasi-order relation on a set X, that is, a reflexive and transitive relation on X. Let $e_Q = Q \cap Q'$, where $Q' = \{(j, i): (i, j) \in X\}$ is the converse of Q. Then an element $x \in X$ is called *Q-strict* if the e_Q-class containing x is a singleton. The subset Y of X is *Q-permissible* if each of its elements is Q-strict and there are no pairs of elements $x, y \in Y$ such that x covers y or y covers x (an element i *covers* an element j in Q if $(i, k) \in Q$ and $(k, j) \in Q$ implies $k = i$ or $k = j$). Define the relation W_Y by

$$(x, y) \in W_Y \text{ iff } x = y \text{ and } x \in Y.$$

Then a binary relation P is a *pseudo-order* relation if

$$P = Q \backslash W_Y,$$

where Q is a quasi-order relation and Y is a Q-permissible subset of X.

THEOREM 3.15. *A binary relation is idempotent if and only if it is a pseudo-order relation (Schein, 1970; Schwarz, 1970a).*

Idempotent relations are therefore a class of transitive relations (satisfying $R^2 = R$) that are constructed from inflations of order relations by deleting subsets of reflexive ties according to the preceding definition. The least complicated relations are those defining orderings on the elements of X in the specified way; traversing paths of lengths greater than one provides no additional information to that contained within the given relational ties. Figure 3.10 presents some examples of pseudo-order relations. Pseudo-order relations are very unlikely to define the prevailing structure in actual social networks, except in very small ones. Despite the fact that social ties of groups of young children become more transitive as children grow older (Leinhardt, 1972), strict transitivity is rarely observed (e.g., Johnsen, 1985, 1986), and even then it applies

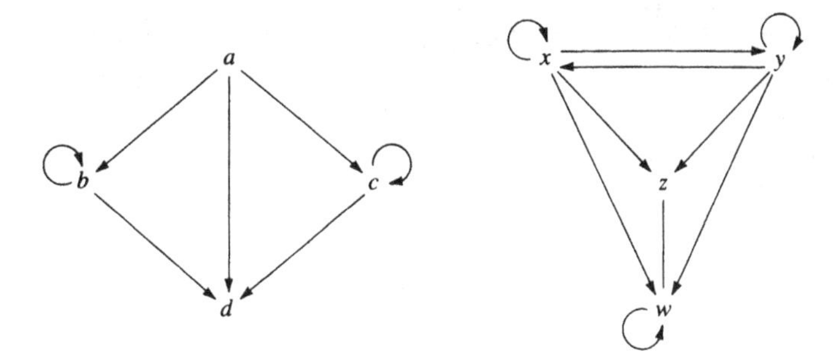

Figure 3.10. Some pseudo-order relations on four elements

in approximate form to only some kinds of relationships (e.g., Granovetter, 1973, argued that only "strong" interpersonal relationships tend to be transitive, and Hallinan and Felmlee, 1975, reported evidence supportive of his proposition).

An interesting subset of the class of idempotent relations are those relations that are the partial orders corresponding to the class of semilattices (Friedell, 1967). All members of that subset generate (trivial) idempotent semigroups but are not unique in doing so, as Lorrain and White (1971) have observed. This observation illustrates the dependence of relational structure in a network on the "content" of its relational constituents. In this instance, the structure of the relation "is superordinate to" is different from the structure of the relation "is *directly* superordinate to", as one might expect.

Monogenic semigroups

Another simple class of semigroups consists of those possessing a single generator; such semigroups are termed monogenic. Monogenic semigroups result from networks possessing a single network relation, that is, from networks of the form $\mathbf{R} = \{R\}$. For instance, the semigroup generated by the friendship network in the hypothetical work group shown in Figure 1.1 is monogenic. The distinct elements in the semigroup generated by the friendship network are F and F^2. The relation F^3 is equal to F, F^4 is equal to F^2, F^5 is equal to F and so on. Indeed the structure of the multiplication table of the semigroup is described completely by the equation

$$F^3 = F,$$

because all other equations can be derived from it.

More generally, the structure of the multiplication table of monogenic semigroups is described as follows.

THEOREM 3.16 *(e.g., Schwarz, 1970b). Powers of a binary relation follow the sequence* $R, R^2, \ldots, R^j, R^{j+1}, \ldots, R^{j+d-1}, R^{j+d} = R^j, R^{j+d+1} = R^{j+1}, \ldots$ *for some minimal pair of natural numbers* j *and* d. *A monogenic semigroup for which* $R, R^2, \ldots, R^{j+d-1}$ *are distinct, but* $R^{j+d} = R^j$ *is said to be a* monogenic semigroup (hereafter, semigroup) of type (j, d). *The natural numbers* j *and* d *are termed the* index *and* period *of the semigroup, respectively.*

For example, the friendship network of Figure 1.1 has index 1 and period 2.

Kim (1982) has reviewed some of the results relating the structure of a network to the index and period of its semigroup. The findings are summarised as follows.

THEOREM 3.17. *Let* $\mathbf{R} = \{R\}$ *be a network on a set* X, *and let* $S = S(\mathbf{R})$ *be its semigroup with index* j *and period* d.

1 *If the network relation* R *contains no cycles, then* $d = 1$ *[and* R^j *is equal to the null relation, that is, the relation* N *for which* $(x, y) \in N$ *for no pair of elements* $x, y \in X$*].*
2 *If* R *is strongly connected, then* d *is the greatest common divisor of the set of lengths of cycles in* R.
3 *The period* d *of* $S(\mathbf{R})$ *is the least common multiple of the periods of the semigroups of the strong components of* R.
4 *The index* $j \leq (n-1)^2 + 1$, *where* n *is the number of elements in* X.

Applying these results to the hypothetical friendship network of Figure 1.1, we may infer that the period of its semigroup is 2 (because the network has two strong components $\{A, B\}$ and $\{C, D\}$ and each of these components has a single cycle of length 2) and that its index is no greater than 10. For the larger network of relations of association in the local network L of Figure 2.1, there are cycles of length 2, 3, 4 and so on, so that the period of the semigroup of the network is 1. The index of the semigroup is no greater than $(9-1)^2 + 1$.

In the case of a particularly simple class of relations termed transition graphs, we can describe the monogenic semigroup of a relation more explicitly. *A transition graph* is a relation in which each individual in the network is linked to no more than one other individual. Such graphs may arise from self-report data if the question eliciting relational information requires that at most one individual is named in response. Examples include questions of the forms, "Who is the person in the group to whom you feel most close?", "Whom do you respect most in this organization?", and "To whom in your family are you least likely to turn for help when you need it?".

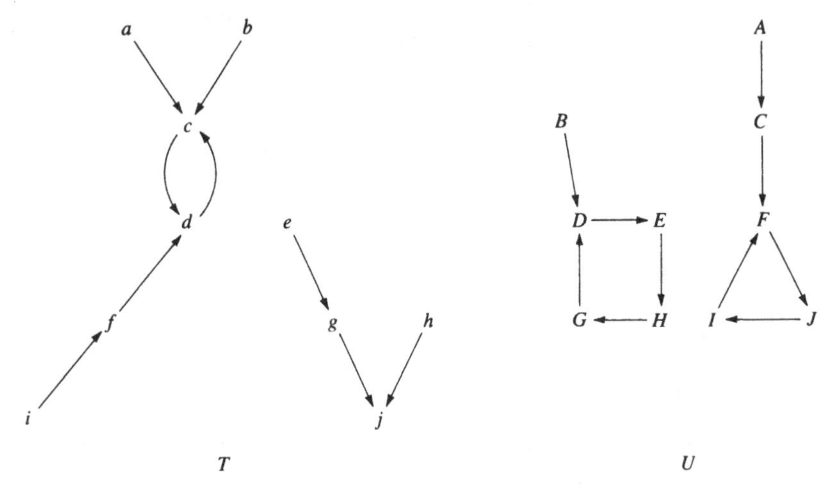

Figure 3.11. Two transition graphs T and U

DEFINITION. A *transition graph* is an irreflexive relation in which the out-degree of each element is either 0 or 1. Transition graphs can take two basic forms. A weakly connected transition graph is

1 a *flower* if every vertex has outdegree one, and
2 an *in-tree (hereafter, tree)* if exactly one vertex has outdegree zero. (The definitions are from Meyer, 1972.)

Figure 3.11 presents two illustrative transition graphs. The second component of the transition graph T is a tree; all other components in Figure 3.11 are flowers.

Because a connected transition graph can have at most one vertex with outdegree zero, it follows that every weak component of a transition graph is either a flower or a tree. If x and y are vertices of the same weak component of a transition graph T, and if there is a path from x to y, then there exists a unique shortest path $[x, y]$ whose length can be denoted by $l[x, y]$.

Now, every component of a transition graph that is a flower has a unique *cycle* (that is, a path from some vertex to itself); a *cycle point* is any vertex that lies on a cycle. A *tree point* of an arbitrary transition graph is any vertex that is not a cycle point. The *period* of a flower T is the number of cycle points of T, that is, the length of its unique cycle.

DEFINITION. Let T be an arbitrary transition graph on n vertices. Define the *period sequence* of T to be

$$\pi(T) = (r_1, r_2, \ldots, r_n)$$

where r_j is the number of components of T that are flowers of period j, and the *depth sequence* of T to be

$$\delta(T) = (d_0, d_1, \ldots, d_{n-1}),$$

where d_j is the number of tree points $x \in T$ such that

$$\max\{l[y, x]: [y, x] \text{ is a path in } T\} = j$$

(Meyer, 1972).

For the graphs presented in Figure 3.4,

$\pi(T) = (0, 1, 0, 0, 0, 0, 0, 0, 0, 0)$ and $\delta(T) = (5, 2, 1, 0, 0, 0, 0, 0, 0, 0)$;

$\pi(U) = (0, 0, 1, 1, 0, 0, 0, 0, 0, 0)$ and $\delta(U) = (2, 1, 0, 0, 0, 0, 0, 0, 0, 0)$.

The monogenic semigroup of a transition graph can now be specified by the following theorem:

THEOREM 3.18. *Let* T *be a transition graph on* n *vertices, with period sequence* $\pi(T) = (r_1, r_2, \ldots, r_n)$ *and depth sequence* $\delta(T) = (d_0, d_1, \ldots, d_{n-1})$. *Then the monogenic semigroup generated by* T *is of type* (h + 1, d), *where*

$$h = \begin{cases} \max\limits_{j=0,1\ldots n-1} \{j: d_j > 0\} \\ 0, \text{ if } d_j = 0 \text{ for all } j \end{cases}$$

and

$$d = \begin{cases} \text{l.c.m.}_{i=1,2,\ldots n} \{i: r_i > 0\} \\ 1, \text{ if } r_i = 0 \text{ for all } i \end{cases}$$

Proof: The proof is contained in Appendix B.

Thus, the monogenic semigroup of a transition graph has a multiplication table determined by its period and depth sequences. For example, the semigroup generated by T of Figure 3.11 is of type (3, 2) and U generates a semigroup of type (2, 12). The partial order of the semigroup is not completely determined by the period and depth sequences; instead, it depends on the presence (or otherwise) of tree points in flowers of the transition graph.

The occurrence of transition graphs in natural social networks is probably rare except where certain methodological restrictions prevail (e.g., where respondents are limited to nominating at most one person as the recipient of a particular kind of network tie, as in the previous examples). Usually, networks are denser than that (e.g., Pool & Kochen,

Table 3.21. *Two-element two-relation networks with identical partially ordered semigroups*

Partially ordered semigroup				Network relations	
	Right mult. table				
Elements	Gen. 1	Gen. 2	Partial order	1	2
1	1	2	1 0	1 0	1 1
2	2	2	1 1	0 1	1 1
				1 0	1 0
				0 1	1 1
				1 0	1 1
				1 1	1 1
1	1	2	1 1	1 0	1 0
2	2	2	0 1	0 1	0 0
				1 0	0 0
				1 1	1 1
				1 1	0 1
				0 1	0 1
1	1	2	1 0	1 0	0 0
2	2	2	0 1	0 1	1 1
				1 0	0 1
				0 1	0 1

1978). Indeed, even if the average density of ties in a network is in the range for transition graphs (less than or equal to 1), ties are usually distributed more unequally among pairs of elements (e.g., Bernard, 1973). Nevertheless, the result is interesting: the interrelations among paths of length 2 or more in transition graphs depend on the length of cycles in the graph and the length of the longest chain in the graph.

Handbooks of small networks

An alternative approach to establishing classes of networks with the same semigroup is computational in nature. Lorrain (1973), for example, has constructed the semigroups generated by networks comprising two elements and two network relations and has thereby classified the set of two-element two-relation networks according to their relational structure. The pairs of binary relations in the set giving rise to the same partially ordered semigroup structure are presented in Table 3.21. Classes of networks generating the same semigroup may be examined for features leading to their identical relational structure and so assist the evaluation of the semigroup representation. Further, the techniques developed in later chapters will demonstrate how such lists have a wider application in the assessment process.

Summary

In this chapter, we have described some constructions that may be used to compare network algebras. We have observed that the algebras of comparable networks lie in a partially ordered structure termed a lattice, with the partial ordering defined by either isotone homomorphic mappings or the nesting relation. The greatest lower bound of two algebras in the lattice is the largest algebra having all of the equations and orderings present in either of the algebras, and the least upper bound of two algebras is the smallest algebra whose orderings and equations hold simultaneously in both of the algebras.

We have also seen that certain classes of networks can be shown to give rise to the same algebra. In each case, it can argued that the members of the class possess some salient features suggesting that they may be associated with similar social implications. Whether they do, of course, is a matter for empirical verification but, at this stage, the algebraic constructions that have been proposed appear to hold some promise as representations of social relational structure.

4

Decompositions of network algebras

The algebraic constructions that have been introduced to represent relational structure in complete and local social networks make few structural assumptions in the hope of preserving the faithfulness with which they represent the structure of network paths. A cost is associated with this approach, however, in that the mathematical structures generated have relatively weak and, generally, poorly understood mathematical and numerical properties. Clyde Coombs summarised the trade-off between the faithfulness of a model and its mathematical power when he asked of measurement, "Do we know what we want, or do we want to know?" (Linzell, 1975). Do we use what might be a poor representation of the data but which nonetheless has strong numerical properties or do we sacrifice mathematical power for a model whose mathematical relationships more adequately reflect known or, at least, plausible relationships among the social phenomena in question?

It is implicit in the structural representations that have been proposed that the latter course is believed to be the wiser one. Detailed analyses of social network data indicate that at this stage it is difficult to add mathematical properties to those already assumed without seriously misrepresenting at least some types of social network data. We are left, therefore, with structures that are often complex and for which we have no convenient methods of analysis. It is the aim of this chapter to describe some useful analytic procedures for structures of this kind. Indeed, the methods presented have been devised so as to apply to a wide range of finite mathematical structures.

Most types of structural analysis are, in essence, methods for structural decomposition. They attempt to identify simpler structural components, or building blocks, in terms of which a structure can be described. For instance, one well-known example of a structural analytic method is principal components analysis. The analysis is a means of decomposing a finite set of variables defining a multidimensional real space into a set of orthogonal components. Calculations may be performed for each component separately in the knowledge that the component is independent of all other such components. The analysis thus breaks down a complex system into simpler parts, each of which may be dealt with in turn. The problem of the complexity of the set of variables is thereby addressed, and subsequent analysis is simplified as a consequence.

A second example of a useful analytic strategy comes from the decomposition theory for finite-state machines (Krohn & Rhodes, 1965; Krohn, Rhodes & Tilson, 1968). The theory deals with finite algebraic structures, termed finite-state machines, which bear some similarity to network algebras. A finite-state machine is conceptualised as a finite set of memory states with a finite input alphabet and a finite output alphabet, and two functions relating them. The *state-transition* function determines the next memory state from the combination of present state and current input, and the *output* function specifies the next output symbol from present state and current input. Now each finite-state machine may be represented at a more abstract level by a finite semigroup whose elements are the input sequences to the machine, regarded as state-transition functions on the set of memory states of the machine. The composition of two semigroup elements corresponds to the concatenation of the input sequences that they represent, and two distinct input sequences are considered equal if they induce the same state-transition function.

Krohn and Rhodes' (1965) fundamental theory is based on a Wreath-product decomposition of the semigroup of a machine into simple components. The decomposition is literal in the sense that the semigroup decomposition induces a corresponding decomposition of the machine itself into simple machines and gives a procedure for connecting these simple machines to form a larger machine that can simulate the behaviour of the original machine. In this way, the complexity of the original machine is resolved: its simple components of known structure are identified and, although the components are not independent of one another, the nature of their interaction is well specified and well understood.

The analysis to be described draws some guidance from these examples. From both, it draws the principle that the components should be as simple as possible and that they should not be further reducible into simpler components. From the second, it extracts the property that the higher-level algebraic representation (semigroup level) should dictate the decomposition of the lower-level relational representation (machine level). Finally, from the first example, it derives the notion that the independence of the components should be emphasised, although instead of insisting on strict independence, we shall argue that it is more useful to maximise the degree of independence subject to some other constraints.

In general terms, therefore, we present a means of analysing a mathematical structure into simple and maximally independent parts. By using such a technique, we hope to overcome the problem of large and complex mathematical structures with which our insistence on representational faithfulness has endowed us. In so doing, we are also concerned with "the fundamental structure problem of algebra", namely, "analysing a given [algebraic] system into simpler components, from which the given system can be reconstructed by synthesis" (Birkhoff, 1967, p. 55).

There are many ways of approaching the problem of decomposing any type of mathematical structure. In particular, the nature of a decomposition procedure depends on the synthesis rule implicit in it (Birkhoff, 1967) or, equivalently, on the nature of its implicit definition of independence (Naylor, 1981, 1983). In the case of such structures as network semigroups or local role algebras, moreover, the definition of independence may be made at the relational level, for example, in terms of specific network links or at the level of the algebra that the network links define. Given that distinct networks may give rise to the same abstract semigroup, there will not necessarily be a one-to-one correspondence between definitions operating at the two levels.

In choosing the strategy to be described, we have followed two main guidelines. Firstly, because the algebraic representations of the two types of network structure are taken as the main focus of the analysis, only decomposition procedures operating at the algebraic level have been considered. Secondly, generality has been deemed to be an important characteristic of the resulting strategy, so that the method selected is based on several quite general synthesis rules from the universal algebra literature. From these rules, a decomposition strategy has been developed that is applicable to finite algebras, in general, and network semigroups and local role algebras, in particular. The technique we use is based on a more general procedure described in Pattison and Bartlett (1982) and is modified here to apply to partially ordered structures, of which partially ordered semigroups and role algebras are examples. The technique is presented first for finite partially ordered semigroup algebras and then is generalised to role algebras. Mathematical definitions are adapted from those of Birkhoff (1967), Fuchs (1963) and Kurosh (1963).

Decompositions of finite semigroups

The decomposition technique developed by Pattison and Bartlett (1982) is applicable to a variety of algebraic structures. In general terms, an algebra is a partially ordered set of elements and one or more operations defined on that set. An operation is a mapping of an ordered sequence of some fixed number of elements onto a single element of the set. For example, a binary operation f maps an ordered *pair* of elements (x, y) onto a single element denoted, say, by $f(x, y)$. A partially ordered semigroup is an algebra comprising a partially ordered set and a single

Table 4.1. *A partially ordered semigroup T*

	Right mult. table			
		Generator		
Element	Word	*a*	*b*	Partial order
1	*a*	1	3	1 0 0 0 0 0 1 0
2	*b*	4	5	0 1 1 0 0 0 0 0
3	*ab*	6	7	0 0 1 0 0 0 0 0
4	*ba*	4	5	0 1 1 1 0 1 0 0
5	*bb*	8	2	0 0 0 0 1 0 1 0
6	*aba*	6	7	0 0 1 0 0 1 0 0
7	*abb*	1	3	0 0 0 0 0 0 1 0
8	*bba*	8	2	1 0 0 0 1 0 1 1

binary operation. The collection of all algebras of a specified type defines a *family* of algebras, that is, a collection of algebras having the same set of operations and satisfying a specified set of postulates. In the case of partially ordered semigroups, members of the family may be characterised as follows:

DEFINITION. A *partially ordered semigroup* comprises a partially ordered set S and a single binary operation f that

1 satisfies the *associative law*

$$f(x, f(y, z)) = f(f(x, y), z) \text{ and}$$

2 is *isotone*:

$$x \leq y \text{ implies } f(x, z) \leq f(y, z) \text{ and } f(z, x) \leq f(z, y),$$

for any $x, y, z \in S$.

The operation $f(x, y)$ is usually written more simply as xy, and each finite partially ordered semigroup may be represented in the familiar form of a multiplication table reporting the binary operation and a binary relation recording the partial order. For instance, Table 4.1 reports a partially ordered semigroup in the form introduced in chapter 1.

Direct representations

As we observed earlier, the nature of a decomposition procedure for an algebraic structure depends on its implicit synthesis rule. That is, the way in which the structure is broken down into smaller pieces depends on the

Table 4.2. *Two partially ordered semigroups S_1 and S_2*

Label	Element	Multiplication table: m	n			Partial order
S_1	m	m	m			1 0
	n	n	n			1 1
Label	Element	Multiplication table: x	y	z	w	Partial order
S_2	x	x	y	z	w	1 0 0 1
	y	z	w	x	y	0 1 0 0
	z	z	w	x	y	0 1 1 0
	w	x	y	z	w	0 0 0 1

rules by which those pieces may be recombined to produce the original structure. Different "re-synthesis" rules are generally associated with different component pieces. One widely used synthesis rule is that of the *direct product*. The direct product of two algebras is defined on the set of all ordered pairs of elements from the two algebras. The operations in the direct product are defined as the conjunction of operations in the constituent algebras. For example, consider the partially ordered semigroups S_1 and S_2 whose full multiplication tables and partial orders are presented in Table 4.2. The direct product $S_1 \times S_2$ of S_1 and S_2 is presented in Table 4.3. The elements of $S_1 \times S_2$ are elements of the Cartesian product of the sets of elements for S_1 and S_2; that is, they are formed by pairing each element of S_1 with each element of S_2. The partial order in the direct product $S_1 \times S_2$ is defined so that an element $(s_1, s_2) \leq (t_1, t_2)$ if and only if the ordering holds for each of the constituents, that is, if and only if $s_1 \leq t_1$ *and* $s_2 \leq t_2$. The binary operation is similarly defined in terms of the constituent operations in S_1 and S_2:

$$(s_1, s_2)\ (t_1, t_2) = (s_1 t_1, s_2 t_2),$$

for any $s_1, t_1 \in S_1$; $s_2, t_2 \in S_2$. More generally:

DEFINITION. Let $S_1, S_2, \ldots, S_r$ be a collection of partially ordered semigroups. The *direct product* $S_1 \times S_2 \times \cdots \times S_r$ of $S_1, S_2, \ldots, S_r$ is the partially ordered semigroup consisting of the set $S_1 \times S_2 \times \cdots \times S_r$ and the binary operation

$$(s_1, s_2, \ldots, s_r)\ (t_1, t_2, \ldots, t_r) = (s_1 t_1, s_2 t_2, \ldots, s_r t_r).$$

The partial order for $S_1 \times S_2 \times \cdots \times S_r$ is given by

Table 4.3. *The direct product* $S_1 \times S_2$ *of the semigroups* S_1 *and* S_2

	Multiplication table								
Element	(m, x)	(m, y)	(m, z)	(m, w)	(n, x)	(n, y)	(n, z)	(n, w)	Partial order
(m, x)	(m, x)	(m, y)	(m, z)	(m, w)	(m, x)	(m, y)	(m, z)	(m, w)	1 0 0 0 0 0 1 0
(m, y)	(m, z)	(m, w)	(m, x)	(m, y)	(m, z)	(m, w)	(m, x)	(m, y)	0 1 1 0 0 0 0 0
(m, z)	(m, z)	(m, w)	(m, x)	(m, y)	(m, z)	(m, w)	(m, x)	(m, y)	0 0 1 0 0 0 0 0
(m, w)	(m, x)	(m, y)	(m, z)	(m, w)	(m, x)	(m, y)	(m, z)	(m, w)	0 1 1 1 0 1 0 0
(n, x)	(n, x)	(n, y)	(n, z)	(n, w)	(n, x)	(n, y)	(n, z)	(n, w)	0 0 0 0 1 0 1 0
(n, y)	(n, z)	(n, w)	(n, x)	(n, y)	(n, z)	(n, w)	(n, x)	(n, y)	0 0 1 0 0 1 0 0
(n, z)	(n, z)	(n, w)	(n, x)	(n, y)	(n, z)	(n, w)	(n, x)	(n, y)	0 0 0 0 0 0 1 0
(n, w)	(n, x)	(n, y)	(n, z)	(n, w)	(n, x)	(n, y)	(n, z)	(n, w)	1 0 0 0 1 0 1 1

$$(s_1, s_2, \ldots, s_r) \le (t_1, t_2, \ldots, t_r)$$

if and only if

$$s_i \le t_i$$

for each $i = 1, 2, \ldots, r$; $s_i, t_i \in S_i$ $(i = 1, 2, \ldots, r)$.

Some of the classical algebraic decomposition theorems (e.g., those for finite Abelian groups and finite dimensional vector spaces) depend on the direct product as their implied construction rule. That is, a structure is decomposed in such a way that it can be represented as a direct product of simpler structures. The direct product encodes a strong principle of independence or orthogonality: the elements of the compound structure are the members of the Cartesian product of the elements of the components, and operations in the compound structure are performed as the conjunction of their independent operations in component structures. If a partially ordered semigroup is isomorphic to the direct product of nontrivial partially ordered semigroups (i.e., of semigroups whose sets contain more than one element), then we term it directly reducible. For instance, the partially ordered semigroup T presented in Table 4.1 is isomorphic to $S_1 \times S_2$ in Table 4.3, as can be seen by mapping elements a, b, ab, ba, bb, aba, abb and bba of T onto (m, x), (n, y), (m, y), (n, z), (n, w), (m, z), (m, w) and (n, x) of $S_1 \times S_2$, respectively. Thus, T is directly reducible. The task of understanding the structure of T may now be replaced with that of understanding the structure of the simpler components of T: the two components operate independently of one another, and their conjunction is isomorphic to T.

The notion of direct reducibility can be captured with the following definition.

DEFINITION. A partially ordered semigroup S is termed *directly reducible* if it is isomorphic to a nontrivial direct product of partially ordered semigroups (i.e., as the direct product of semigroups, at least two of which are not one-element semigroups). Otherwise, S is *directly irreducible*. The representation of S as the direct product $S_1 \times S_2 \times \cdots \times S_r$ is termed a *direct reduction* or *direct representation* of S, and the semigroups $S_1, S_2, \ldots, S_r$ are termed its *direct components*.

Existence of direct representations

The direct decomposition of the partially ordered semigroup of Table 4.1 illustrates the useful descriptive role that direct decompositions may play. We may ask, though, for which semigroups do such decompositions exist and, where they exist, how can they be found? In answer to these questions, some results from universal algebra may be invoked.

Table 4.4. *The π-relation corresponding to the isotone homomorphism from T onto* S_1

	π-relation							
	1	2	3	4	5	6	7	8
1	1	0	1	0	0	1	1	0
2	1	1	1	1	1	1	1	1
3	1	0	1	0	0	1	1	0
4	1	1	1	1	1	1	1	1
5	1	1	1	1	1	1	1	1
6	1	0	1	0	0	1	1	0
7	1	0	1	0	0	1	1	0
8	1	1	1	1	1	1	1	1

For algebras in general, it has been shown that the form and existence of direct product representations may be characterized by the lattice of homomorphisms of the algebra (e.g., Birkhoff, 1967) or, equivalently, by the lattice of its π-relations. Recall that an isotone homomorphism from a partially ordered semigroup S onto a partially ordered semigroup T is a mapping ϕ from S onto T such that, for all $s_1, s_2 \in S$,

1 $\phi(s_1 s_2) = \phi(s_1)\phi(s_2)$, and
2 $s_1 \leq s_2$ in S implies $\phi(s_1) \leq \phi(s_2)$ in T.

The semigroup T is termed an isotone (homomorphic) image of S, and we write $T = \phi(S)$.

Now each isotone homomorphism ϕ defined on a partially ordered semigroup S corresponds to a unique π-relation π_ϕ on S:

$$(s, t) \in \pi_\phi \text{ iff } \phi(t) \leq \phi(s).$$

For instance, consider the partially ordered semigroups T and S_1 of Tables 4.1 and 4.2, respectively. There is an isotone homomorphism from T onto S_1 in which

$$\phi(a) = \phi(ab) = \phi(aba) = \phi(abb) = m$$

and

$$\phi(b) = \phi(ba) = \phi(bb) = \phi(bba) = n.$$

The corresponding π-relation is shown in Table 4.4, and it may be verified that $\phi(t) \leq \phi(s)$ if and only if $(s, t) \in \pi_\phi$.

It was established in chapter 3 that the collection L_s of isotone homomorphic images of a partially ordered semigroup S defines a lattice, as does the collection $L_\pi(S)$ of π-relations on S. Further, the lattice $L_\pi(S)$ is dual to the lattice L_s.

For example, the collection of all homomorphic images and corresponding π-relations for the semigroup T of Table 4.1 are shown in Tables 4.5 and 4.6, respectively. Each homomorphic image is the quotient semigroup associated with exactly one π-relation on S, and the partial ordering among π-relations, displayed in Figure 4.1, is clearly the dual of that among corresponding homomorphic images illustrated in Figure 4.2. (The homomorphic images and π-relations are labelled in Tables 4.5 and 4.6 so that T/π_i is isomorphic to T_i, for $i = 1, 2, \ldots, 7$.)

The manner in which the π-relation lattice $L_\pi(S)$ – or, equivalently, the homomorphism lattice L_s – of a partially ordered semigroup S determines the existence and nature of its direct product representations may now be expressed as follows:

THEOREM 4.1. *The partially ordered semigroup* S *is the direct product of partially ordered semigroups* $S_1, S_2, \ldots, S_r$ *if and only if there exist* π-*relations* $\pi_1, \pi_2, \ldots, \pi_r \in L_\pi(S)$ *such that*

1 S/π_i *is isomorphic to* S_i, *for each* i;
2 $glb(\pi_1, \pi_2, \ldots, \pi_r) = \pi_{min}$;
3 $lub(\pi_i, glb(\pi_1, \pi_2, \ldots, \pi_{i-1})) = \pi_{max}$, *for each* i; *and*
4 *the* π-*relations are* permutable, *that is,*

$$\pi_i\pi_j = \pi_j\pi_i$$

for all $i = 1, 2, \ldots, r$; $j = 1, 2, \ldots, r$.

Proof: The proof is given in Appendix B (also Birkhoff, 1948, p. 87).

Thus, to find a direct product representation of a partially ordered semigroup, it is necessary and sufficient to identify permutable π-relations in the lattice $L_\pi(S)$ that satisfy conditions 2 and 3 of Theorem 4.1. If such a collection can be found, then the partially ordered semigroup is isomorphic to the direct product of the corresponding quotient semigroups. The identified quotient semigroups become the components of the partially ordered semigroup S, that is, the collection of semigroups of simpler structure from which it can be exactly reproduced.

For example, we can illustrate the procedures and theorem just described by considering the partially ordered semigroup T presented in Table 4.1. The π-relations π_2 and π_5 displayed in Table 4.6 satisfy (a) $glb(\pi_2, \pi_5) = \pi_{min}$, (b) $lub(\pi_2, \pi_5) = \pi_{max}$, and (c) $\pi_2\pi_5 = \pi_5\pi_2 = \pi_{max}$. Theorem 4.1 thus establishes that the semigroup T of Table 4.1 can be expressed as the direct product of the quotient semigroups T/π_2 and T/π_5. It may be seen that the quotient semigroups T/π_2 and T/π_5 are isomorphic

Table 4.5. *Isotone homomorphic images of the partially ordered semigroup T*

	Right mult. table						Right mult. table				
			Generator						Generator		
Label	Element	Word	*a*	*b*	Partial order	Label	Element	Word	*a*	*b*	Partial order
T_0	1	*a*	1	3	10000010	T_1	1	*a*	1	3	1010
	2	*b*	4	5	01100000		2	*b*	4	2	0110
	3	*ab*	6	7	00100000		3	*ab*	1	3	0010
	4	*ba*	4	5	01110100		4	*ba*	4	2	1111
	5	*bb*	8	2	00001010						
	6	*aba*	6	7	00100100						
	7	*abb*	1	3	00000010						
	8	*bba*	8	2	10001011						
T_2	1	*a*	1	2	1001	T_3	1	*a*	1	3	1000
	2	*b*	3	4	0100		2	*b*	2	4	0110
	3	*ba*	3	4	0110		3	*ab*	3	1	0010
	4	*bb*	1	2	0001		4	*bb*	4	2	1001
T_4	1	*a*	1	2	11	T_5	1	*a*	1	1	10
	2	*b*	1	2	01		2	*b*	2	2	11
T_6	1	*a*	1	2	10	T_7	1	$a = b$	1		1
	2	*b*	2	1	01						

Table 4.6. *π-relations in $L_\pi(T)$*

Label	π-relation	Label	π-relation
π_0	10000010 01100000 00100000 01110100 00001010 00100100 00000010 10001011	π_1	10100110 01101010 00100010 11111111 01101010 10100110 00100010 11111111
π_2	10001011 01100000 01100000 01110100 00001010 01110100 00001010 10001011	π_3	10000010 01110100 00100100 01110100 10001011 00100100 10000010 10001011
π_4	11111111 01101010 01101010 11111111 01101010 11111111 01101010 11111111	π_5	10100110 11111111 10100110 11111111 11111111 10100110 10100110 11111111
π_6	10001011 01110100 01110100 01110100 10001011 01110100 10001011 10001011	π_7	11111111 11111111 11111111 11111111 11111111 11111111 11111111 11111111

to S_2 and S_1 in Table 4.2, respectively, and that T is isomorphic to the direct product displayed in Table 4.3.

Where they exist, direct product representations are clearly of great use. Few partially ordered semigroups admit direct product representations, however. For instance, no partially ordered semigroup having an odd number of distinct elements is directly reducible. For semigroups with an even number of distinct semigroups, the proportion of semigroups which are directly reducible is not known, but it is probably small. For instance, of the 126 distinct semigroups with four elements and a partial order equal to the identity relation (Forsythe, 1955) only nine possess

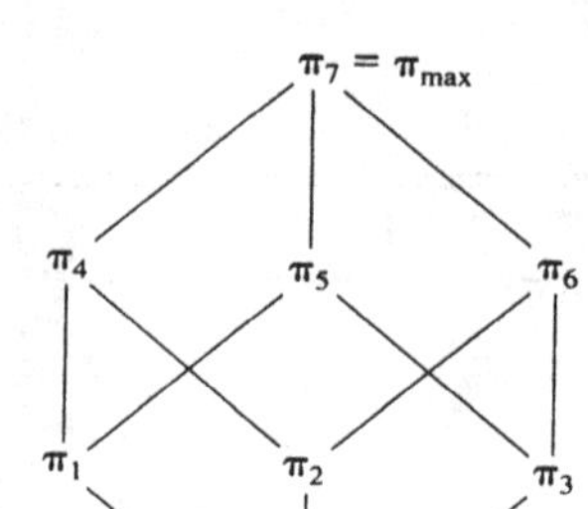

Figure 4.1. The π-relation lattice $L_\pi(T)$ of the partially ordered semigroup T

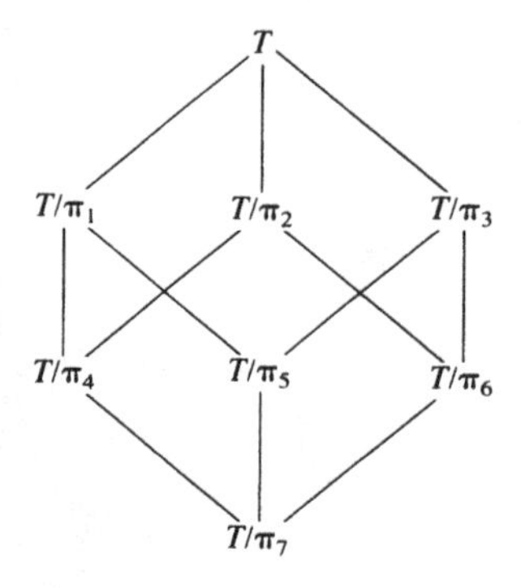

Figure 4.2. The lattice L_T of isotone homomorphic images of T

nontrivial direct reductions. There is some value, therefore, in also considering other forms of decomposition.

Subdirect representations

A universal algebraic construction that generalises the direct product to allow some overlap in the structural content of components of associated decompositions is that of the *subdirect* product. A subdirect product of semigroups is defined as any subsemigroup of their direct product for which the projection mapping into each component semigroup is a surjection.

DEFINITION. A partially ordered semigroup T is a *subsemigroup* of the partially ordered semigroup S if

1 T is a subset (possibly empty) of S;
2 for any $s_1, s_2 \in T$, $s_1 \le s_2$ in T if and only if $s_1 \le s_2$ in S (i.e., T has the partial order induced from S); and
3 T is a partially ordered semigroup (so that $s_1, s_2 \in T$ implies $s_1 s_2 \in T$).

DEFINITION. A partially ordered subsemigroup S of a direct product $S_1 \times S_2 \times \cdots \times S_r$ of partially ordered semigroups $S_1, S_2, \ldots, S_r$ is called a *subdirect product* of the S_k ($k = 1, 2, \ldots, r$) if, given $s_k \in S_k$, there exists an element $s \in S$ having s_k as its component in S_k. If S is isomorphic to a subdirect product of two or more nontrivial partially ordered semigroups (i.e., partially ordered semigroups having more than one element), then S is termed *subdirectly reducible* and the collection $\{S_1, S_2, \ldots, S_r\}$ defines a *subdirect reduction* or *subdirect representation* of S; otherwise, S is *subdirectly irreducible*. The semigroups S_k in a subdirect reduction of S are termed its *subdirect components*.

For example, the semigroups U_1 and U_2 displayed in Table 4.7 have the direct product also shown in Table 4.7. The subsemigroup U of the direct product presented in Table 4.8 is a subdirect product of U_1 and U_2 because each element of U_1 and U_2 appears in some element of the subsemigroup U. That is, elements x and y of U_1 each appear at least once as the first component of elements in U; similarly, elements a, b, c, d, e, f and g of U_2 each appear at least once as the second component of an element in U. The semigroup V appearing in Table 4.9 can be seen to be isomorphic to the semigroup U of Table 4.8 by mapping elements 1 to 8 of V onto (x, a), (x, b), (y, c), (y, d), (y, e), (y, f), (x, c) and (x, g) of U. The semigroup V of Table 4.9 is therefore subdirectly reducible, with subdirect components isomorphic to U_1 and U_2.

The definition of subdirect product guarantees that the projection mapping

$$\phi_k: S \rightarrow S_k$$

from a semigroup onto any of its subdirect components given by

$$\phi_k(s_1, s_2, \ldots, s_r) = s_k$$

is an isotone homomorphism, for each $k = 1, 2, \ldots, r$. For example, the π-relations π_1 and π_2 appearing in Table 4.10 are associated with homomorphisms from the semigroup V onto U_1 and U_2. (Table 4.10 presents all π-relations in $L_\pi(V)$.)

It may be observed that the operations in a subdirect product of semigroups are performed as the conjunction of their independent operations in the subdirect components, just as for the direct product. The sense in which the components of a subdirect product fail to be independent is not in

Table 4.7. *Two partially ordered semigroups U_1 and U_2 and their direct product $U_1 \times U_2$*

Label	Element	Right mult. table: Generator x	y	Partial order
U_1	x	y	x	1 0
	y	x	y	0 1

Label	Element	Right mult. table: Generator a	b	Partial order
U_2	a	c	d	1 1 0 0 0 0 0
	b	e	f	0 1 0 0 0 0 0
	$aa = c$	c	c	1 1 1 1 1 1 1
	$ab = d$	a	a	0 0 0 1 0 1 0
	$ba = e$	c	g	0 0 0 0 1 1 0
	$bb = f$	a	b	0 0 0 0 0 1 0
	$bab = g$	e	e	0 1 0 0 0 0 1

Label	Element	Right mult. table: Generator (x, a)	(x, b)	Partial order
$U_1 \times U_2$	(x, a)	(y, c)	(y, d)	11000000000000
	(x, b)	(y, e)	(y, f)	01000000000000
	(x, c)	(y, c)	(y, c)	11111110000000
	(x, d)	(y, a)	(y, a)	00010100000000
	(x, e)	(y, c)	(y, g)	00001100000000
	(x, f)	(y, a)	(y, b)	00000100000000
	(x, g)	(y, e)	(y, e)	01000010000000
	(y, a)	(x, c)	(x, d)	00000001100000
	(y, b)	(x, e)	(x, f)	00000000100000
	(y, c)	(x, c)	(x, c)	00000001111111
	(y, d)	(x, a)	(x, a)	00000000001010
	(y, e)	(x, c)	(x, g)	00000000000110
	(y, f)	(x, a)	(x, b)	00000000000010
	(y, g)	(x, e)	(x, e)	00000000100001

terms of the operations in the component structures. Rather, it is the inability of the components to be independently realisable in the compound structure, that is, the failure of the compound algebra to exhibit the full Cartesian product structure from the components, which renders the components nonindependent. (For further discussion of these two

Table 4.8. *A subsemigroup U of $U_1 \times U_2$ that defines a subdirect product of U_1 and U_2*

	Right mult. table		
	Generator		
Element	(x, a)	(x, b)	Partial order
(x, a)	(y, c)	(y, d)	1 1 0 0 0 0 0 0
(x, b)	(y, e)	(y, f)	0 1 0 0 0 0 0 0
(x, c)	(y, c)	(y, c)	1 1 1 1 0 0 0 0
(x, g)	(y, e)	(y, e)	0 1 0 1 0 0 0 0
(y, c)	(x, c)	(x, c)	0 0 0 0 1 1 1 1
(y, d)	(x, a)	(x, a)	0 0 0 0 0 1 0 1
(y, e)	(x, c)	(x, g)	0 0 0 0 0 0 1 1
(y, f)	(x, a)	(x, b)	0 0 0 0 0 0 0 1

Table 4.9. *A partially ordered semigroup V isomorphic to the semigroup U*

	Right mult. table			
		Generator		
Element	Word	F	H	Partial order
1	F	3	4	1 1 0 0 0 0 0 0
2	H	5	6	0 1 0 0 0 0 0 0
3	FF	7	7	0 0 1 1 1 1 0 0
4	FH	1	1	0 0 0 1 0 1 0 0
5	HF	7	8	0 0 0 0 1 1 0 0
6	HH	1	2	0 0 0 0 0 1 0 0
7	FFF	3	3	1 1 0 0 0 0 1 1
8	HFH	5	5	0 1 0 0 0 0 0 1

aspects of independence, both of which are usually required, see Krantz, Luce, Suppes & Tversky, 1971, pp. 246–7.)

Existence of subdirect representations

As for direct product representations, the existence and nature of subdirect representations of a given partially ordered semigroup S are also determined by its π-relation lattice, $L_\pi(S)$ (e.g., Birkhoff, 1967).

THEOREM 4.2. *A partially ordered semigroup S is isomorphic to a subdirect product of partially ordered semigroups $S_1, S_2, \ldots, S_r$ if and only if there exist π-relations $\pi_1, \pi_2, \ldots, \pi_r \in L_\pi(S)$, such that*

Table 4.10. *π-relations in* $L_\pi(V)$

Label	π-relation	Label	π-relation	Label	π-relation	Label	π-relation
π_0	11000000	π_1	11000011	π_2	11000000	π_3	11000011
	01000000		11000011		01000000		01000000
	00111100		00111100		11111111		00111100
	00010100		00111100		00010100		00111100
	00001100		00111100		00001100		00111100
	00000100		00111100		00000100		00000100
	11000011		11000011		11111111		11000011
	01000001		11000011		01000001		11000011
π_4	11111111	π_5	11111111	π_6	11111111		
	01000000		01000100		11111111		
	11111111		11111111		11111111		
	11111111		11111111		11111111		
	11111111		11111111		11111111		
	00000100		01000100		11111111		
	11111111		11111111		11111111		
	11111111		11111111		11111111		

1 S/π_i *is isomorphic to* S_i, *and*
2 $\text{glb}(\pi_1, \pi_2, \ldots, \pi_r) = \pi_{min}$.

Proof: See, for example, Birkhoff (1967, p. 193).

From Theorem 4.2, it follows that a partially ordered semigroup S is subdirectly irreducible if, and only if, its lattice $L_\pi(S)$ of π-relations has a unique element covering its minimal element.

DEFINITION. An *atom* of a lattice L with minimal element z is an element $a > z$ satisfying the condition:

$$a \geq b \text{ and } b > z \text{ implies } a = b, \text{ for any } b \in L;$$

that is, it is an element that covers the minimal lattice element.

Hence:

THEOREM 4.3. *A partially ordered semigroup* S *is subdirectly irreducible if and only if its π-relation lattice* $L_\pi(S)$ *has exactly one atom.*

As a result of Theorem 4.3, inspection of $L_\pi(S)$ determines whether S is subdirectly irreducible or not. In many cases, it also reveals that there exist a large number of possible choices for the components of the subdirect reduction, through the many possible combinations of choices for the π_k satisfying condition 2 of Theorem 4.2.

For example, the partially ordered semigroup V of Table 4.9 is subdirectly reducible. Its π-relations are listed in Table 4.10 and the diagram of its

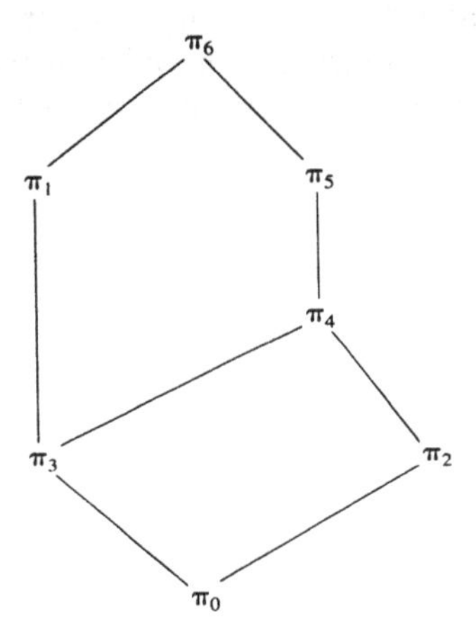

Figure 4.3. The π-relation lattice $L_\pi(V)$ of the partially ordered semigroup V

Figure 4.4. The π-relation lattice $L_\pi(U_2)$ of the semigroup U_2 (π-relations of U_2 are presented as π-relations on V; see Table 4.10)

π-relation lattice is shown in Figure 4.3. The π-relation lattice clearly contains more than one atom. The π-relations π_1 and π_2 satisfy the conditions of Theorem 4.2, and so $\{S/\pi_1, S/\pi_2\}$ defines a subdirect representation of V, as we have observed. Consider now, however, the partially ordered semigroup U_2 whose π-relation lattice $L_\pi(U_2)$ is shown in Figure 4.4. (Because U_2 is an isotone homomorphic image of V, the π-relations can be presented as π-relations on V, as in Figure 4.4.) The lattice contains just one atom, and as a result, U_2 is subdirectly irreducible (Theorem 4.3).

If a partially ordered semigroup is subdirectly reducible, it usually admits a number of different subdirect representations, and the problem is raised of how to select *one* subdirect representation from the often large set of possibilities. For example, the semigroup V may be written as a subdirect product of quotient semigroups associated with each of the following sets of π-relations:

1 $\{\pi_2, \pi_3\}$;
2 $\{\pi_2, \pi_1\}$;
3 $\{\pi_2, \pi_3, \pi_5\}$;
4 $\{\pi_1, \pi_2, \pi_4\}$;
5 $\{\pi_2, \pi_3, \pi_4, \pi_5\}$;
6 $\{\pi_1, \pi_2, \pi_4, \pi_5\}$;
7 $\{\pi_1, \pi_2, \pi_3, \pi_4, \pi_5\}$.

Some means of ordering the various subdirect representations in terms of efficiency would clearly be useful, a claim that is illustrated by comparing the subdirect representation of V in terms of π_1 and π_2 with that in terms of π_2, π_3, π_4 and π_5. The representation based on $\{\pi_1, \pi_2\}$ is in terms of a subsemigroup of the direct product of two subdirectly irreducible semigroups. The subsemigroup has eight elements from the direct product of order 14. In the case of the representation based on $\{\pi_2, \pi_3, \pi_4, \pi_5\}$, V is represented as a subdirect product of four semigroups, not all of which are subdirectly irreducible, and the size of the direct product semigroup of which V is a subsemigroup is 168. Indeed, of the subdirect representations of V just listed, only the second, fourth and sixth are representations in terms of subdirectly irreducible semigroups. The example suggests that one useful way of restricting candidate subdirect representations is to exclude those whose components are subdirectly reducible.

Factorisation

We define a special form of subdirect decomposition of a partially ordered semigroup, termed a factorisation, in two steps. In the first step, we eliminate subdirect representations with reducible components. In the second, we define a partial ordering on the remaining subdirect representations in terms of efficiency and eliminate all those representations that are not of maximal efficiency.

For convenience of expression, the subdirect representation corresponding to the equation

$$\pi_{\min} = glb(\pi_1, \pi_2, \ldots, \pi_r)$$

is identified with the collection of π-relations $\{\pi_1, \pi_2, \ldots, \pi_r\}$.

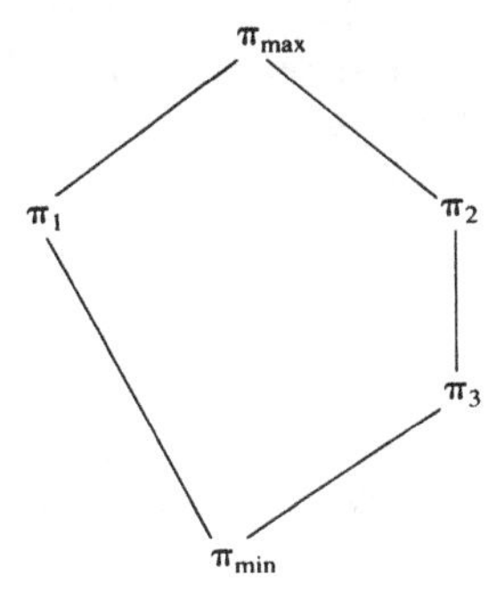

Figure 4.5. A π-relation lattice admitting two irredundant subdirect decompositions

DEFINITION. An element π of a lattice L is *meet-irreducible* (hereafter, *irreducible*) if

$$\pi = glb(\pi_1, \pi_2) \text{ implies } \pi = \pi_1 \text{ or } \pi = \pi_2.$$

A subdirect decomposition $\{\pi_1, \pi_2, \ldots, \pi_r\}$ is *irredundant* if

1 π_i is irreducible, for each $i = 1, 2, \ldots, r$, and
2 $glb(\pi_1, \pi_2, \ldots, \pi_{i-1}, \pi_{i+1}, \ldots, \pi_r) > \pi_{min}$, for each $i = 1, 2, \ldots, r$.

For instance, in Figure 4.3 the elements π_1, π_2, π_4 and π_5 are irreducible, but π_3 is not because $\pi_3 = glb(\pi_1, \pi_4)$. The representation $\{\pi_1, \pi_2\}$ is irredundant because it comprises two irreducible π-relations, each lying above π_{min}. The representations $\{\pi_2, \pi_3\}$ and $\{\pi_1, \pi_2, \pi_5\}$ are not irredundant, however, because the first contains a reducible element and the second contains an element π_5 whose omission leaves a subdirect representation. Thus, in an irredundant subdirect decomposition of a semigroup, no component is reducible (condition 1) and no component can be removed from the collection to leave a set still serving as a subdirect representation (condition 2). In the interests of efficiency, therefore, it is useful to restrict attention to irredundant subdirect representations.

In general, however, irredundant subdirect representations are not necessarily as parsimonious as possible. The π-relation lattice diagrammed in Figure 4.5, for example, indicates the existence of two irredundant subdirect reductions, $\{\pi_1, \pi_2\}$ and $\{\pi_1, \pi_3\}$. In such a case, the quotient semigroup S/π_2 is a homomorphic image of S/π_3, indicating that $\{\pi_1, \pi_2\}$ is the more concise irredundant subdirect representation of S. Relative parsimonies of this kind may be captured by the following definition.

DEFINITION. Let $F_1 = \{\pi_1, \pi_2, \ldots, \pi_r\}$ and $F_2 = \{\delta_1, \delta_2, \ldots, \delta_q\}$ be irredundant subdirect representations of a partially ordered semigroup S. Define a *partial ordering on irredundant subdirect representations* according to $F_1 \le F_2$ iff, for each $j = 1, 2, \ldots, q$, there exists some i ($i = 1, 2, \ldots, r$) such that

$$\delta_j \le \pi_i$$

in $L_\pi(S)$. Further, given a collection F of irredundant subdirect representations of S, define $F \in$ F to be *minimal* if $G \le F$ and $G \in$ F implies $G = F$.

Then the following definition is proposed as characterising a useful class of subdirect decompositions.

DEFINITION. A *factorisation* of a partially ordered semigroup S is any minimal irredundant subdirect representation of S. Each component of a factorisation of S is termed a *factor* of S; the factors may be said to constitute a *factorising set.*

For example, a semigroup with Figure 4.5 as its π-relation lattice has a unique factorisation, corresponding to the factorising set $F = \{\pi_1, \pi_2\}$. For the subdirect representations of the semigroup V listed earlier, only $\{\pi_1, \pi_2\}$ is irredundant, so that V also has a unique factorisation.

In general, the collection of factorisations of a partially ordered semigroup is claimed to define a particularly useful class of subdirect representations of the semigroup. Any factorising set of a semigroup is sufficient to describe all of its structural content; moreover, the algebraic conditions of the factorisation definition provide an operationalisation for the intuitively useful requirement that the factors be maximally independent of one another. Hence, the structural complexity implicit in a full algebraic representation, recorded in the multiplication table and partial order of a semigroup, may be resolved by identifying collections of structurally simpler components of the semigroup that are known to combine in reasonably independent ways to produce the original semigroup. Such collections permit the structural content of the semigroup to be manipulated in a manageable way. Thus, the identification of all factorisations of a partially ordered semigroup may be seen as a first step in reducing the possible complexity of an algebraic description.

Some more examples of factorisation are presented later in the chapter after an algorithm for constructing factorisations is described. Before describing the algorithm, though, the question of whether a semigroup has just one or a number of possible factorisations is briefly discussed.

Uniqueness of factorisations

Some partially ordered semigroups have unique factorisations, as in all the examples shown so far, but for others there is more than one possible factorisation. In part, the question of the uniqueness of irredundant reductions of a partially ordered semigroup is resolved by knowing the type of structure that the π-relation lattice of the semigroup possesses.

DEFINITION. A *distributive* lattice L is one in which

$$glb(x, lub(y, z)) = lub(glb(x, y), glb(x, z))$$

for all $x, y, z \in L$. A *modular* lattice L is one in which

$$x \leq z \text{ implies } lub(x, glb(y, z)) = glb(lub(x, y), z)$$

for all $y \in L$; for any $x, z \in L$.

Note that any lattice that is distributive is necessarily modular, but the converse does not hold. The lattices displayed in Figures 4.1 and 4.4 are both distributive and therefore modular; the lattices of Figures 4.3 and 4.5 are nonmodular. In fact, the lattices displayed in Figures 4.3 and 4.5 are the only nondistributive lattices of five elements (e.g., Birkhoff, 1967).

THEOREM 4.4. *If the π-relation lattice $L_\pi(S)$ of a partially ordered semigroup S is distributive, then S possesses a unique irredundant subdirect representation.*

Proof: For example, see Birkhoff (1967, p. 58); the theorem is implied by the dual of the corollary to Lemma 1.

For example, because the lattice of Figure 4.1 is distributive, Theorem 4.4 guarantees that the factorisation of any partially ordered semigroup whose π-relation lattice is isomorphic to Figure 4.1 is unique. In particular, the semigroup T necessarily possesses a unique factorisation.

If the π-relation lattice is modular but not distributive, then irredundant subdirect representations are not necessarily unique, but it is at least guaranteed that all irredundant reductions have the same number of components. Specifically:

THEOREM 4.5 *(Kurosh–Ore Theorem). If the π-relation lattice $L_\pi(S)$ of a partially ordered semigroup S is modular, then the number of components in any irredundant subdirect representation of S is independent of the representation.*

Proof: See, for example, Birkhoff (1967, pp. 75–6).

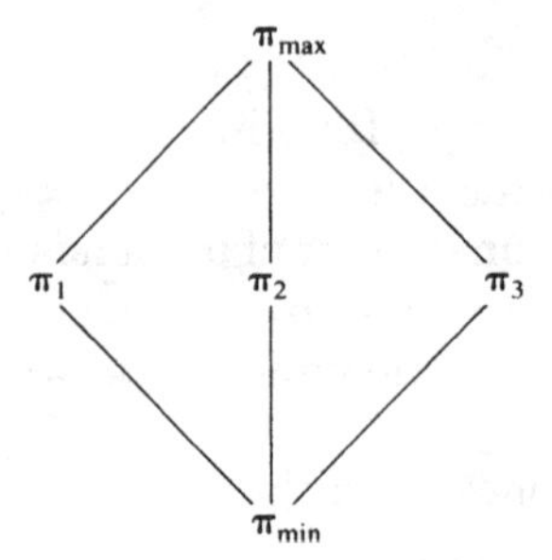

Figure 4.6. A nondistributive, modular lattice

An example of a modular lattice that is not distributive is presented in Figure 4.6. A π-relation lattice isomorphic to Figure 4.6 has three irredundant subdirect representations, namely, $\{\pi_1, \pi_2\}$, $\{\pi_1, \pi_3\}$ and $\{\pi_2, \pi_3\}$. On grounds of parsimony, there appears no reason to choose one of the three representations as optimal.

Lattices that are not even modular give rise to nonunique irredundant subdirect representations in a different way, as Figure 4.5 illustrates. The π-relation lattice presented in Figure 4.5 is actually the critical sublattice for nonmodularity. That is, a lattice is nonmodular if, and only if, it contains the lattice of Figure 4.5 as a sublattice (where a subset M of a lattice L is a *sublattice* if, for any $x, y \in M$, $glb(x, y) \in M$ and $lub(x, y) \in M$). The nonmodular lattice of Figure 4.5 possesses two irredundant subdirect representations, corresponding to $\{\pi_1, \pi_2\}$ and $\{\pi_1, \pi_3\}$. The factorisation definition selects the unique representation in terms of $\{\pi_1, \pi_2\}$, because $\{\pi_1, \pi_2\} < \{\pi_1, \pi_3\}$. In this case, we have argued that the representation using $\{\pi_1, \pi_2\}$ is more efficient than that in terms of $\{\pi_1, \pi_3\}$.

An algorithm for factorisation

Like direct and subdirect representations of a semigroup, the existence and form of factorisations is determined by its π-relation lattice. However, as for direct and subdirect representations also, the definition of a factorisation is of the existence type. That is, the discovery of the complete set of factorisations of the semigroup involves (a) examination of the π-relation lattice to ensure that all irredundant subdirect decompositions have been considered as potential factorisations and (b) verification of whether any proposed irredundant subdirect decomposition is minimal. For many partially ordered semigroups of interest, however, the π-relation lattice is quite large, and the implied search of the π-relation lattice is a demanding task. For a partially ordered semigroup S of n

elements, the number of elements in the π-relation lattice of S can be as large as the number of distinct quasi-orders on n elements. Thus, even for a semigroup with a small number of elements, the π-relation lattice may have a large number of π-relations. As a result, an algorithm that identifies precisely the collection of all factorisations of a given finite partially ordered semigroup is of considerable use and will be described.

Let $M = \{z_1, z_2, \ldots, z_a\}$ be the set of atoms of the π-relation lattice $L_\pi(S)$ of a partially ordered semigroup S.

DEFINITION. Let $\pi \in L_\pi(S)$. Define a *meet-complement* (hereafter, *complement*) of π to be any element $\delta \in L_\pi(S)$ such that

1 $\delta > \pi_{\min}$, and
2 $glb(\delta, \pi) = \pi_{\min}$.

Further, for any atom $z \in L_\pi(S)$, define the *set of maximal complements* of z by

$$C(z) = \{z^* \in L_\pi(S) : \text{(a) } z^* \text{ is a complement of } z \text{ and (b) } glb(w, z) = \pi_{\min} \text{ and } w \geq z^* \text{ implies } w = z^*\}.$$

For instance, the lattice displayed in Figure 4.3 has atoms π_2 and π_3, and their corresponding sets of maximal complements are $\{\pi_1\}$ and $\{\pi_2\}$, respectively.

Now consider sets of the form

$$F^{**} = \{y_i \in C(z_i);\ i = 1, 2, \ldots, a\}. \tag{4.1}$$

Each set F^{**} comprises a collection of complements, exactly one from the set of maximal complements for each atom in the set M. We can show that each set F^{**} corresponds to a subdirect representation of S.

LEMMA 4.1. *Each* F^{**} *defines a subdirect representation of* S *in terms of irreducible components.*

Proof: The proof is given in Appendix B.

Further, the collection of factorisations of a partially ordered semigroup may be constructed from the collection of subdirect representations of the form F^{**}. First, we define $\mathbf{F}^*$ to be the family of irredundant subdirect representations derived from the F^{**}; that is, let

$$\mathbf{F}^* = \{F^* : F^* \text{ is a subset of } F^{**} \text{ defined by equation (4.1) and } F^* \text{ is an irredundant subdirect representation}\}.$$

Second, let $\mathbf{F}$ be the collection of minimal members of $\mathbf{F}^*$; that is, let

$$\mathbf{F} = \{F^* \in \mathbf{F}^* : F \leq F^* \text{ and } F \in \mathbf{F}^* \text{ implies } F = F^*\}.$$

That is, the set $\mathbf{F}$ is constructed from the set of maximal complements of the form F^{**} in two stages. In the first stage, all possible subsets of

F^{**} that are irredundant subdirect representations are constructed. In the second, minimal members of that collection are identified as members of the set F.

THEOREM 4.6. *F is the set of factorisations of* S.

Proof: The proof is given in Appendix B.

Thus, to construct the collection of all factorisations of a partially ordered semigroup S, one may proceed by

1 finding all atoms of its π-relation lattice $L_\pi(S)$; then
2 finding the set of maximal meet-complements of each atom; then
3 selecting irredundant sets of elements, obtained by deleting redundant members of sets of the form defined by (4.1); and finally,
4 deleting any nonminimal members of that collection.

For example, consider first the partially ordered semigroup V of Table 4.9. The π-relation lattice $L_\pi(V)$ of V (Figure 4.3) possesses two atoms, π_3 and π_2. Each of these atoms has a unique maximal complement, so that $\{\pi_1, \pi_2\}$ is the only set of the form F^{**}. It may quickly be seen that $\{\pi_1, \pi_2\}$ is irredundant, so that $\{\pi_1, \pi_2\}$ constitutes a unique factorisation of the semigroup V.

Consider also the semigroup $S(\mathrm{N}_4)$ of Table 3.6. The π-relation lattice of $S(\mathrm{N}_4)$, shown in Figure 3.4, indicates that $S(\mathrm{N}_4)$ has two atoms, π_3 and π_2. Each of the atoms has a unique maximal complement: π_4 is a unique maximal complement for π_2, and π_2 is a unique maximal complement for π_3. Hence $\{\pi_2, \pi_4\}$ is a unique factorisation of $S(\mathrm{N}_4)$.

In fact, if the maximal complement of an atom of the π-relation lattice of a partially ordered semigroup happens to be unique, then an explicit expression for it may be given without constructing $L_\pi(S)$. Rather, it may be expressed in terms of readily generated π-relations.

DEFINITION. Let π_{st} be the least π-relation including the partial ordering

$$s \geq t$$

for $s, t \in S$. π_{st} is termed the π-relation *generated by* the ordering $s \geq t$. Let z be an atom of the π-relation lattice $L_\pi(S)$. Define a relation $\pi(z)$ by

$$\pi(z) = \{(s, t): glb(\pi_{st}, z) = \pi_{\min}; s, t \in S\}.$$

For example, consider the partially ordered semigroup V of Table 4.9. The semigroup does not contain the ordering $1 > 6$, and the π-relation π_{16} has the form shown in Table 4.11. Table 4.12 records the

Table 4.11. *The π-relation π_{16} generated by the ordering 1 > 6 on the semigroup V*

π_{16}
1 1 1 1 1 1 1 1
0 1 0 0 0 0 0 0
1 1 1 1 1 1 1 1
1 1 1 1 1 1 1 1
1 1 1 1 1 1 1 1
0 0 0 0 0 1 0 0
1 1 1 1 1 1 1 1
1 1 1 1 1 1 1 1

Table 4.12. *π-relations generated by each possible additional ordering i > j on the semigroup V*

i	*j*	Relation	*i*	*j*	Relation	*i*	*j*	Relation	*i*	*j*	Relation	*i*	*j*	Relation
1	3	π_4	1	4	π_4	1	5	π_4	1	6	π_4	1	7	π_3
1	8	π_3	2	1	π_1	2	3	π_6	2	4	π_6	2	5	π_6
2	6	π_5	2	7	π_1	2	8	π_1	3	1	π_2	3	2	π_2
3	7	π_2	3	8	π_2	4	1	π_4	4	2	π_4	4	3	π_3
4	5	π_3	4	7	π_4	4	8	π_4	5	1	π_4	5	2	π_4
5	3	π_3	5	4	π_3	5	7	π_4	5	8	π_4	6	1	π_6
6	2	π_5	6	3	π_1	6	4	π_1	6	5	π_1	6	7	π_6
6	8	π_6	7	3	π_2	7	3	π_2	7	4	π_2	7	5	π_2
7	6	π_2	8	1	π_3	8	3	π_4	8	4	π_4	8	5	π_4
8	6	π_4	8	7	π_3									

π-relations generated by each potential additional ordering on *V*, using the labelling of π-relations appearing in Table 4.10. It can be seen from Table 4.12 that the relation $\pi(\pi_1)$ constructed according to the preceding definition is π_2, whereas $\pi(\pi_3) = \pi_1$. In fact, it can be established more generally that:

THEOREM 4.7. *If* z *has a unique maximal complement, then* π(z) *is a* π*-relation and* C(z) = {π(z)}. *Conversely, if* π(z) *is a* π*-relation, then it is the unique maximal complement of* z.

Proof: The proof is given in Appendix B.

Theorem 4.7 establishes that a check on whether $\pi(z)$ is a π-relation is equivalent to a check on whether *z* has a unique maximal meet-complement; if so, it identifies that maximal meet-complement as $\pi(z)$. Furthermore:

THEOREM 4.8. *Let* $M = \{z_1, z_2, \ldots, z_a\}$ *be the set of atoms of the* π*-relation lattice of a partially ordered semigroup* S. *If, for each* $z_i \in M$, z_i *has a unique maximal complement* $\pi(z_i)$*, then the factorisation of* S *is unique and* $\{\pi(z_1), \pi(z_2), \ldots, \pi(z_a)\}$ *is the factorising set.*

Proof: The proof is given in Appendix B.

For example, Theorem 4.8 establishes that $\{\pi_1, \pi_2\}$ is a unique factorisation for the semigroup V of Table 4.9. We may also attempt to apply Theorems 4.7 and 4.8 to the π-relation lattice $L_\pi(T)$ of the partially ordered semigroup T of Table 4.1. The lattice $L_\pi(T)$ possesses three atoms, π_1, π_2 and π_3. The relations $\pi(\pi_1)$, $\pi(\pi_2)$ and $\pi(\pi_3)$ corresponding to the atoms are π-relations and are equal to π_6, π_5 and π_4, respectively. Thus, each atom has a unique maximal complement, and $\{\pi_6, \pi_5, \pi_4\}$ defines a unique factorisation of T. The quotient semigroups for the factorisation are presented in Table 4.5.

Each of these steps is potentially programmable, but at this stage, only a partial implementation has proved necessary. The latter takes the form of a program that (a) generates the π-relation π_{st} induced by adding each possible additional ordering $s \geq t$ not already present in S, (b) identifies from the list the atoms of the π-relation lattice of the semigroup and (c) determines which of the π-relations generated in step (a) are complements of each atom. In the case of an atom possessing a unique maximal complement, that complement is constructed according to Theorem 4.7. For semigroups whose atoms all possess unique maximal complements, the factorisation of the semigroup is found explicitly. The C program *PSFACT* performing these steps is available from the author on request.

As another example of the application of the program, consider the partially ordered semigroup S(N) generated by the blockmodel N = $\{L, A\}$ of Table 2.6. The semigroup is displayed in Table 1.13 (also Table 2.10). *PSFACT* identifies that its π-relation lattice $L_\pi(S(\mathrm{N}))$ has the atoms shown in Table 4.13. It also establishes that each of the atoms has a unique maximal complement and hence constructs a unique factorisation of S(N) in terms of those maximal complements. The maximal complements are also shown in Table 4.13 together with their corresponding isotone homomorphic images.

Using factorisation to analyse network semigroups

So far in this chapter, it has been proposed that factorisation is a useful procedure for breaking down a partially ordered semigroup into simpler structural components. It has been shown how the lattice of π-relations of the semigroup determines the existence of factorisations, and an algorithm for finding all factorisations has been presented.

Table 4.13. *Atoms in $L_\pi(S(\mathrm{N}))$ and their unique maximal complements and corresponding factors*

	Atom		Maximal complement	Isotone homomorphic image corresponding to maximal complement				
				Right mult. table				Partial order
				Element	Word	L	A	
π_1	1 0 0 0 0 0	π_4	1 1 1 1 1 1	1	L	1	2	1 1
	0 1 0 0 0 0		0 1 1 0 1 0	2	A	1	2	0 1
	0 1 1 0 0 0		0 1 1 0 1 0					
	0 1 0 1 0 0		1 1 1 1 1 1					
	1 1 1 1 1 1		0 1 1 0 1 0					
	1 1 1 1 1 1		1 1 1 1 1 1					
π_2	1 0 0 0 0 0	π_5	1 0 0 0 0 0	1	L	1	3	1 0 0
	0 1 0 0 0 0		0 1 0 1 0 0	2	A	2	3	0 1 0
	0 1 1 0 0 0		1 1 1 1 1 1	3	LA	3	3	1 1 1
	1 1 0 1 0 0		0 1 0 1 0 0					
	0 1 1 0 1 0		1 1 1 1 1 1					
	1 1 1 1 1 1		1 1 1 1 1 1					
π_3	1 0 0 0 0 0	π_6	1 0 0 0 0 0	1	L	1	2	1 0 0
	0 1 0 0 0 0		1 1 1 0 0 0	2	A	3	3	1 1 0
	0 1 1 0 1 0		1 1 1 0 0 0	3	AL	3	3	1 1 1
	0 1 0 1 0 0		1 1 1 1 1 1					
	0 1 1 0 1 0		1 1 1 1 1 1					
	1 1 1 1 1 1		1 1 1 1 1 1					

The factorisation procedure is presented as a first step in the analysis of the partially ordered semigroup of a network. By determining the factors of a number of different but comparable network semigroups, we can begin to identify the algebraic components that arise commonly in network semigroups. We can therefore begin to describe commonly occurring simple structural forms for networks comprising the given set of relation types. The second step in the proposed analysis of a network semigroup is that of determining which features of network relations give rise to any factor of its semigroup. This step is discussed in detail in the next chapter and is followed by a general procedure for describing, or "interpreting", network semigroups using the factorisation procedure.

The reduction diagram

The factors of any factorisation of a partially ordered semigroup correspond to π-relations that are irreducible. That is, each factor of a factorisation is subdirectly irreducible and possesses a unique (nontrivial)

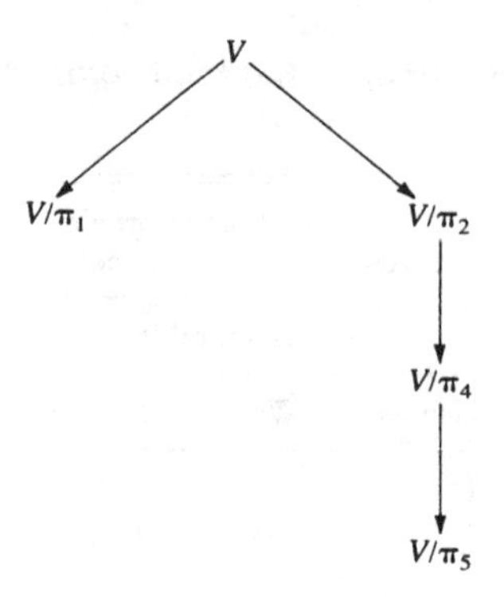

Figure 4.7. Reduction diagram for the factorisation of V

maximal isotone homomorphic image. That maximal image is either factorisable or is itself subdirectly irreducible, and in the latter case, it has a unique maximal homomorphic image.

One may systematically find the unique images of all factors appearing in a factorisation, their factorisations or unique images and so on. The procedure results in a subset of the lattice L_s of homomorphic images of S and, together with the partial ordering among its elements, it summarises the structure of the partially ordered semigroup. The diagram illustrating this summary statement of the structure of L_s and, hence, of S is called a *reduction diagram*; the following conventions are adopted in its presentation:

1 The elements of the reduction diagram of a partially ordered semigroup S are selected members of L_s and include S itself.
2 A single arrow directed from an element of the reduction diagram represents a mapping of that element onto a unique maximal (nontrivial) homomorphic image.
3 A collection of arrows directed from an element of the reduction diagram connect that element to the homomorphic images appearing in its factorisation.
4 If no arrow is directed from an element of the reduction diagram, then that element has only trivial homomorphic images.

For example, the reduction diagram for the factorisation of the partially ordered semigroup V of Table 4.9 is presented in Figure 4.7.

Co-ordination of a partially ordered semigroup

Direct product decompositions may be constructed to provide a means of finding a set of "co-ordinates" of the structure in question. A system

Table 4.14. *Co-ordinates for elements of the partially ordered semigroup V in the subdirect representation corresponding to* $\{\pi_1, \pi_2\}$

Element of V	Co-ordinates
1	(x, a)
2	(x, b)
3	(y, c)
4	(y, d)
5	(y, e)
6	(y, f)
7	(x, c)
8	(x, g)

of co-ordinates may be obtained from the set of projections of a partially ordered semigroup onto its components: the set of images of an element under the projection set are its co-ordinates. For instance, in the direct decomposition of the semigroup T of Table 4.1, the element a is mapped onto the element (m, x) in the representation of T by $S_1 \times S_2$ so that (m, x) is a pair of co-ordinates for the element a of T.

The same procedure for co-ordination may be applied to subdirect representations of a partially ordered semigroup. The co-ordinates for an element are given by its images under projection mappings onto the components, and operations on a set of elements can be determined from operations on their co-ordinates. An example of a system of co-ordinates from a factorisation is given in Table 4.14. The co-ordinates are for the elements of the partially ordered semigroup V of Table 4.9 and are obtained by the mapping of each element onto elements of the semigroups U_1 and U_2, respectively. For example, element 1 of the semigroup V is mapped onto element x of U_1 and element a of U_2 and is therefore assigned the co-ordinates (x, a). Element 2 of V is mapped onto x in U_1 and b in U_2 and so has co-ordinates (x, b). Note that multiplication of elements can be computed from multiplication in the component structures; the product of elements 1 and 2 from V, for example, is given by

$$(x, a)(x, b) = (xx, ab) = (y, d),$$

and (y, d) corresponds to element 4 in V.

Relationships between factors

A factorisation of a partially ordered semigroup was observed earlier to give rise to a set of maximally independent irreducible components of the

Table 4.15. *Association indices for factors of the semigroups T and V*

Semigroup T				Semigroup V		
Factor	T/π_4	T/π_5	T/π_6	Factor	V/π_1	V/π_2
T/π_4	1.00	0.00	0.00	V/π_1	1.00	0.46
T/π_5	0.00	1.00	0.00	V/π_2	0.46	1.00
T/π_6	0.00	0.00	1.00			

semigroup. The extent of the dependence among components can, in fact, be quantified by an association index such as the one defined here:

DEFINITION. Let S/π_1 and S/π_2 be quotient semigroups of a partially ordered semigroup S. The *association* between the quotient semigroups may be indexed by

$$r(\pi_1, \pi_2) = \frac{\left|\frac{S}{\pi_1}\right|\left|\frac{S}{\pi_2}\right| - \left|\frac{S}{glb(\pi_i, \pi_2)}\right|}{\left|\frac{S}{\pi_1}\right|\left|\frac{S}{\pi_2}\right| - \left|\frac{S}{\text{lub}(\pi_1, \pi_2)}\right|}$$

where $|S/\pi|$ is the number of distinct classes in the congruence relation σ_π corresponding to π – that is, the relation given by $(s, t) \in \sigma_\pi$ if and only if $(s, t) \in \pi$ and $(t, s) \in \pi$. The index may also be written as $r(S/\pi_1, S/\pi_2)$, understood to have the same meaning as $r(\pi_1, \pi_2)$.

The properties of the index are summarised in Theorem 4.9.

THEOREM 4.9. *The association index just defined for quotient semigroups of a partially ordered semigroup satisfies*

1 $0 \leq \mathrm{r}(\pi_1, \pi_2) \leq 1$, *and* $\mathrm{r}(\pi_1, \pi_2) = \mathrm{r}(\pi_2, \pi_1)$, *for all* $\pi_1, \pi_2 \in \mathrm{L}_\pi(\mathrm{S})$;
2 $\mathrm{r}(\pi_1, \pi_2) = 0$ *if and only if* $\mathrm{S/glb}(\pi_1, \pi_2)$ *is isomorphic to* $\mathrm{S}/\pi_1 \times \mathrm{S}/\pi_2$; *and*
3 *if* $\pi_1 = \pi_2$, *then* $\mathrm{r}(\pi_1, \pi_2) = 1$.

Proof: The proof is given in Appendix B.

Thus, the index takes values in the range 0 to 1 and achieves its lower bound precisely when the quotient semigroups are independent in the full sense (Krantz et al., 1971). The upper bound of 1 is obtained when the π-relations coincide, as well as under the condition of nonequal π-relations having associated congruence relations that are equal.

The association indices for every pair of factors of the semigroups T (Table 4.1) and V (Table 4.9) are presented in Table 4.15. For instance,

Table 4.16. *Multiplication table for a semigroup S*

	1	2	3	4	5	6
1	1	3	3	6	5	6
2	4	5	5	6	5	6
3	6	5	5	6	5	6
4	4	5	5	6	5	6
5	6	5	5	6	5	6
6	6	5	5	6	5	6

to calculate the value of the index for V/π_1 and V/π_2, we determine that $glb(V/\pi_1, V/\pi_2)$ is V/π_0 (i.e., V) and $lub(V/\pi_1, V/\pi_2)$ is V/π_6 (see Figure 4.3), and that the semigroups V/π_1, V/π_2, $glb(V/\pi_1, V/\pi_2)$ and $lub(V/\pi_1, V/\pi_2)$ have 2, 7, 8 and 1 elements, respectively. Consequently, the value of the index of association between the factors V/π_1 and V/π_2 of V is $(2 \times 7 - 8)/(2 \times 7 - 1)$, or 0.46.

Factorisation of finite abstract semigroups

It was observed in chapter 2 that some authors have confined their attention to the multiplication table of the semigroup of a network and have ignored the partial order of the semigroup. In such a case, the scheme of analysis previously outlined may be readily modified to yield an analysis of an abstract semigroup into abstract semigroup components. The modification takes the form of a condition on the partial order of a semigroup, namely, that it is symmetric (so that if $s \geq t$ in a semigroup S, then it must be the case that $t \geq s$ and thus that $s = t$). We then construct π-relations on the semigroup S corresponding to homomorphisms, as before, but we add the requirement that π-relations must be symmetric. For instance, consider the abstract semigroup S of Table 4.16. (S is the abstract semigroup generated by the blockmodel N of Table 2.6.) To construct relations on S corresponding to homomorphisms of S(N), we first replace the partial order of the semigroup S(N) with the identity relation (so that $i \leq j$ in S if and only if $i = j$; $i, j \in S$). We then generate π-relations containing the identity relation, which not only are reflexive and transitive and satisfy the condition

$$(s, t) \in \pi \text{ implies } (us, ut) \in \pi \text{ and } (su, tu) \in \pi, \text{ for any } u \in S,$$

but also are symmetric. Each such π-relation generated corresponds to an (abstract) homomorphism of the semigroup S, and any homomorphism of S is associated with such a π-relation.

THEOREM 4.10. *Let* ϕ *be a homomorphism on a semigroup* S. *Then the relation* π_ϕ *on* S *given by*

$$(s, t) \in \pi_\phi \text{ iff } \phi(s) = \phi(t);\ s, t \in S$$

is reflexive, symmetric and transitive and has the property:

$$(s, t) \in \pi_\phi \text{ implies } (us, ut) \in \pi_\phi \text{ and } (su, tu) \in \pi_\phi, \text{ for any } u \in S.$$

The property is equivalent to the substitution *property, namely:*

$$(s_1, t_1) \in \pi_\phi \text{ and } (s_2, t_2) \in \pi_\phi$$

implies

$$(s_1 s_2, t_1 t_2) \in \pi_\phi.$$

The relation π_ϕ *is termed the* congruence relation *corresponding to the homomorphism* ϕ. *Conversely, any reflexive, symmetric and transitive relation having the substitution property corresponds to a homomorphism of* S.

Proof: The proof is contained in Appendix B.

The collection of all such π-relations generated for a semigroup S form a lattice $A_\pi(S)$ that is the dual of the lattice A_s of abstract homomorphic images of S (see chapter 3). Direct representations, subdirect representations and factorisations may all be constructed from the lattice $A_\pi(S)$. For example, the list of all such π-relations on the semigroup $S(\mathbf{N})$ is presented in Table 4.17. Each π-relation is presented in the form of a partition on the elements of S because each π-relation is reflexive, symmetric and transitive, and therefore is an equivalence relation (or partition) on S. The atoms of the lattice $A_\pi(S)$ are the relations π_{14}, π_{15} and π_{16}. Each atom has a unique maximal complement, which are, respectively, π_8, π_3 and π_9. The latter relations, therefore, give rise to a factorisation of the abstract semigroup S.

A FORTRAN program MINLATT, which is the abstract semigroup analogue of the program *PSFACT* for partially ordered semigroups, is available from the author on request.

A decomposition procedure for role algebras

A similar procedure to that just outlined for partially ordered semigroups also yields a useful decomposition procedure for structures such as role algebras.

Recall, firstly, that a role algebra comprises a set **R** of generator relations, a free composition operation on **R** giving rise to a free semigroup

Table 4.17. *π-relations in $A_\pi(S)$, presented as partitions on S*

π-relation	Partition on S
π_1	(123456)
π_2	(1) (23456)
π_3	(146) (235)
π_4	(13456) (2)
π_5	(146) (2) (35)
π_6	(1) (2) (3456)
π_7	(1) (235) (46)
π_8	(1) (24) (356)
π_9	(1) (23) (456)
π_{10}	(1) (23) (46) (5)
π_{11}	(1) (2) (3) (456)
π_{12}	(1) (2) (35) (46)
π_{13}	(1) (2) (356) (4)
π_{14}	(1) (2) (3) (46) (5)
π_{15}	(1) (2) (3) (4) (56)
π_{16}	(1) (2) (35) (4) (6)
π_{17}	(1) (2) (3) (4) (5) (6)

$FS(\mathbf{R})$ and a binary relation Q on $FS(\mathbf{R})$ satisfying the conditions (a) Q is reflexive and transitive (i.e., a quasi-order) and (b) $(s, t) \in Q$ implies $(su, tu) \in Q$, for any $u \in FS(\mathbf{R})$, $s, t \in FS(\mathbf{R})$. Each role algebra may be presented in a finite form as a partial order and right multiplication table on the equivalence classes of the relation e_Q defined by

$$(s, t) \in e_Q \text{ iff } (s, t) \in Q \text{ and } (t, s) \in Q.$$

Tables 2.7, 2.8 and 2.10, for instance, present role algebras in such a form.

Secondly, we may recall that the nesting relation for role algebras has been introduced as a means of comparing role algebras. A role algebra T is nested in a role algebra Q if whenever $(s, t) \in Q$, then also $(s, t) \in T$ $(s, t \in FS(\mathbf{R}))$. That is, if T is nested in Q, T contains all of the orderings present in Q plus some additional ones. Thus, the nesting relation is an analogue for role algebras of the homomorphic mapping for partially ordered semigroups: one partially ordered semigroup is a homomorphic image of another if it contains all of its orderings and, possibly, some additional ones.

Thirdly, we observed in chapter 3 that a unique π-relation π_T on the classes of e_Q is associated with each role algebra T nested in Q. The relation π_T is given by:

> $(s^*, t^*) \in \pi_T$ if there exist relations $s \in s^*$, $t \in t^*$ such that $(s, t) \in T$, where s^*, t^* are classes of e_Q.

For instance, the role algebras nested in the role algebra Q of Table 2.7 were presented in Table 3.19, and their corresponding π-relations were given in Table 3.20.

Fourthly, we established in chapter 3 that the collection L_Q of role algebras nested in Q is a lattice under the nesting relation ($T \leq Q$ if T is nested in Q). The lattice $L_\pi(Q)$ of π-relations of Q is dual to L_Q and has the partial ordering: $\pi_T \leq \pi_P$ if and only if $(s^*, t^*) \in \pi_T$ implies $(s^*, t^*) \in \pi_P$, for any classes s^*, t^* of e_Q. For example, Figure 3.7 presents the π-relation lattice of the role algebra Q of Table 2.7.

Finally, we noted that the minimal element of $L_\pi(Q)$ is simply the partial order for Q itself and that the greatest lower bound of two π-relations in $L_\pi(Q)$ is their intersection.

The factorisation procedure of this chapter establishes that the minimal element a_{min} of a lattice L may be expressed as the meet of lattice elements

$$a_{min} = glb(a_1, a_2, \ldots, a_r),$$

where the set $F = \{a_1, a_2, \ldots, a_r\}$ has the following properties:

1 it is irredundant – that is, (a) each a_i is irreducible (so that $a_i = glb(d_1, d_2)$ implies $a_i = d_1$ or $a_i = d_2$), and (b) for each i, $glb(a_1, \ldots, a_{i-1}, a_{i+1}, \ldots, a_r) > a_{min}$ – and
2 it is minimal – that is, $F \leq G$ where $G = \{b_1, b_2, \ldots, b_s\}$ is any other irredundant representation.

The set F has been termed a factorising set and may be identified by the algorithm described earlier.

Because the minimal element of $L_\pi(Q)$ is Q itself, and because the greatest lower bound of elements in Q is their intersection, the factorisation technique can be applied to $L_\pi(Q)$ to give an expression for the π-relation π_{min} corresponding to the role algebra Q as the intersection of π-relations corresponding to role algebras nested in Q:

$$\pi_{min} = \pi_T \cap \pi_U \cap \cdots \cap \pi_V.$$

The collection $\{T, U, \ldots, V\}$ of nested role algebras, which is not necessarily unique, is a set of maximally different role algebras nested in Q; as such, it defines a decomposition of the role algebra Q into a collection of "simpler" (Mandel, 1983) role algebra components.

Potential factorisations of a role algebra may be identified using the algorithm described earlier. Where maximal meet-complements of atoms of $L_\pi(Q)$ are unique, a unique factorisation of Q may be constructed explicitly. As for partially ordered semigroups, we may define

Table 4.18. *The π-relations π_{st} for each possible additional ordering s > t on Q*

s	t	π_{st}	Label in Table 3.20
L	A	1 1 1 0 1 0 1 1 1	π_1
L	AL	1 1 1 0 1 0 1 1 1	π_1
A	L	1 0 0 1 1 0 1 1 1	π_2
A	AL	1 0 0 1 1 1 1 1 1	π_3

DEFINITION. Let π_{st} be the least π-relation on the classes of Q for which $(s, t) \in \pi_{st}$; it is termed the *π-relation generated by the ordering* $s \geq t$. Let z be an atom of $L_\pi(Q)$ and define $\pi(z)$ by

$$\pi(z) = \{(s, t): glb(z, \pi_{st}) = \pi_{\min};\ s, t \text{ are classes of } Q\}.$$

A C program *RAFACT*, which constructs the relations π_{st} for all pairs of classes s, t for which $s \not\geq t$, is available from the author on request. The list of relations π_{st} for the role algebra of Table 2.7 is presented in Table 4.18.

It may be established that if $\pi(z)$ is a π-relation for each atom $z \in L_\pi(Q)$, then the factorisation of Q is unique. That is,

THEOREM 4.11. *Let* $M = \{z_1, z_2, \ldots, z_a\}$ *be the set of atoms of* $L_\pi(Q)$*. If* z_i *has a unique maximal complement, then that maximal complement is* $\pi(z_i)$*. If each* $z_i \in M$ *has a unique maximal complement, then* $\{\pi(z_1), \pi(z_2), \ldots, \pi(z_a)\}$ *defines a unique factorisation of* Q.

Proof: The proof follows the same steps as the proofs for Theorems 4.7 and 4.8, and is omitted.

As an illustration of Theorem 4.11, consider the role algebra Q of Table 2.7 whose relations π_{st} for classes s, t of Q, with $s \not\geq t$ are shown in Table 4.18. The atoms of the lattice are $z_1 = \pi_1$ and $z_2 = \pi_2$. The relations $\pi(z_1)$ and $\pi(z_2)$ constructed from the preceding definition are equal to π_3 and π_1, respectively (and so, clearly, are π-relations). Hence $\{\pi(z_1), \pi(z_2)\}$ corresponds to a unique factorisation of Q, a conclusion that can be verified by examination of the π-relation lattice $L_\pi(Q)$ of Q

Table 4.19. *Atoms z of the π-relation lattice of the local role algebra of the network* L *and their unique maximal complements* $\pi(z)$

Atoms			
z_1	z_2	z_3	z_4
10000000	10000000	10000000	10000000
11100000	11000000	11000000	11000000
10100000	10100000	10100000	10100000
11111011	11111111	11111011	11111011
10101000	10101000	10101000	11101001
11111111	11111111	11111111	11111111
11101011	11101011	11111011	11101011
11100001	11100001	11100001	11101001
Maximal complements			
$\pi(z_1)$	$\pi(z_2)$	$\pi(z_3)$	$\pi(z_4)$
11000000	10000000	11101011	10100000
11000000	11111011	11101011	11100001
11111111	11111011	11101011	10100000
11111111	11111011	11111111	11111111
11111111	11111011	11101011	10101000
11111111	11111111	11111111	11111111
11111111	11111011	11101011	11111111
11111111	11111011	11101011	11100001

in Figure 3.7. The right multiplication tables and partial orders corresponding to the factors Q/π_1 and Q/π_3 of Q are presented in Table 3.19.

Theorem 4.11 may also be used to establish that the local role algebra presented in Table 2.4 and Figure 2.2 for the network L (Table 2.2) possesses a unique factorisation. The program *RAFACT* was used to find the atoms z_1, z_2, z_3 and z_4 of the π-relation lattice of the role algebra and the corresponding relations $\pi(z_1)$, $\pi(z_2)$, $\pi(z_3)$ and $\pi(z_4)$; all of the relations are presented in Table 4.19. The factors of L are described in detail in chapter 5.

The correspondence between the factorisation procedure for role algebras and partially ordered semigroups establishes the link between the proposed analyses of the algebraic representatives of network structure in entire and partial networks. The only difference between the two analyses is that, for entire networks, the definitions require relational consistency for compound relations obtained by multiplying other relations on the left, as well as on the right. At the network level, this distinction amounts to the difference between comparisons among paths with a fixed source, *ego*, and comparisons among paths with all possible elements as starting points. Apart from this distinction, though, the analyses are the same.

Summary

The outcome of this chapter is a decomposition procedure that yields an efficient collection of maximally independent and simple components of both partially ordered semigroups and local role algebras. We have seen in detail how the procedure may be applied to some simple structures. More examples of the application of the factorisation technique to semigroup and role algebra structures are given in subsequent chapters. Next we describe how it can be used to develop a detailed scheme of analysis for both of these representations of network structure.

5

An analysis for complete and local networks

In chapter 4, a method for analysing the algebra of a complete or local network into simple components was described. In this chapter we develop a means for relating that analysis of the algebraic representation of a network to decompositions at the relational or network level. In particular, we explore the question of whether each component, or factor, of the algebraic representation corresponds to any component of the associated network.

The importance of such a question is both mathematical and substantive. Not only is it relevant to a more complete mathematical understanding of the relationship between a network and the algebra that it defines (and hence of the theoretical value of the representation), but it also plays a significant interpretative role. More specifically, it enables us to determine whether each algebraic component has concrete relational referents and, if so, to identify them explicitly. These referents, if they exist, provide data for the assessment of the representation in terms of the original network data and so provide useful complementary evidence for the value, or otherwise, of the representation.

In particular, the factorisation procedure decomposes the algebra of a network into isotone homomorphic images in the case of partially ordered semigroups, or nested role algebras in the case of local role algebras. In each case, the components of the factorisation are "simpler" algebraic representations than the algebra of the network, in the sense of making fewer distinctions among labelled paths in the corresponding network. It is natural to ask, therefore, whether there is any simplification of the network that corresponds to this algebraic simplification. That is, can we identify any means of reducing a complete or local network so that the resulting structure has an algebraic representation equal to some component of the factorisation of the original network? Because if we can, then we are more likely to be able to link algebraic components of the factorisation to specific features of the network in question. Such a facility adds to the descriptive account of network structure made possible by the factorisation technique, and thereby assists in the evaluation of the algebraic representations that have been proposed.

We consider these questions for complete and local networks in turn.

An analysis for complete networks

There are two general approaches to the problem of understanding the relationship between reductions of a network and homomorphic images of its semigroup. Firstly, when and how can we obtain a reduction of a complete network corresponding to a given isotone homomorphic image of its partially ordered semigroup? Secondly, under what conditions does a given abstraction of a network induce an isotone homomorphism of the network semigroup?

These two questions are examined in two stages. In the first, we review known conditions under which simplifying a network leads to a homomorphism of its semigroup. In the second, we develop an empirical solution to the problem of identifying simplifications of the system consistent with a given homomorphic image of its semigroup. Finally, we show how the latter technique may be used in conjunction with the factorisation procedure of chapter 4 to produce a systematic and efficient analysis of the network.

Relational conditions for semigroup homomorphisms

The graph-theoretical operation in terms of which correspondences between the semigroup and relational levels have generally been sought is that of a homomorphism of the network, in the sense of Heil and White (1976) and White (1977); Boyd (1991) also discusses a number of other types of homomorphism. A network homomorphism of the type considered here is illustrated in Figure 5.1. The network displayed in Figure 5.1a may be mapped homomorphically onto that in Figure 5.1b. Each node in the image network is the image of one or more nodes in the original network. For instance, node *a* in the network **T** is the image of nodes 1 and 2 in **R**, node *b* is the image of node 3, and node *c* is the image of node 4. There is a link of a particular type between two nodes *a* and *b* in the image network whenever a link of that type existed between some node that is mapped onto *a* and some node that is mapped onto *b*. The link need not exist for every pair of nodes in the original network mapped onto nodes *a* and *b*, respectively; rather, it need only have been present for one or more such pairs. Thus, for instance, node 1 is connected by the relation *B* to node 4 in **R**; hence the image *a* of node 1

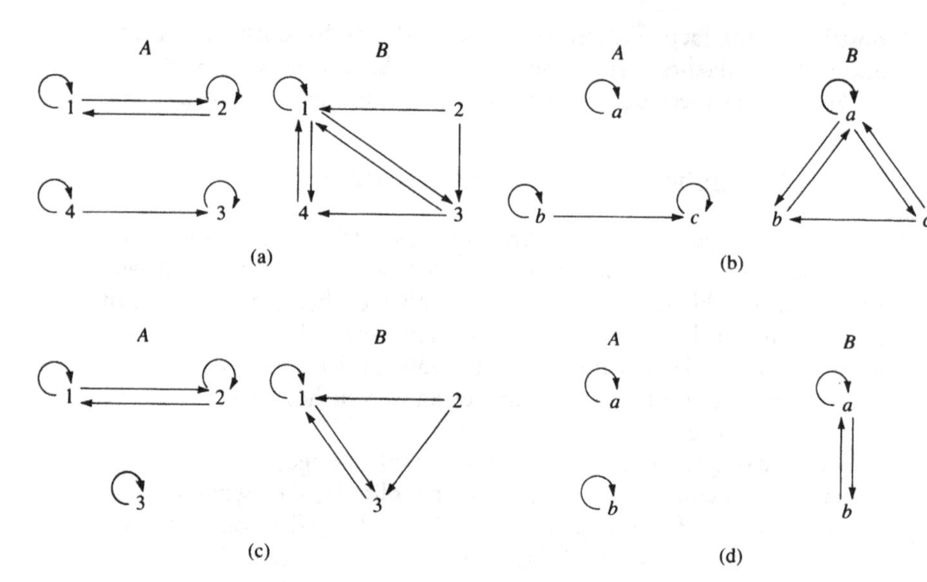

Figure 5.1. Some network mappings: (a) a network **R** on a set $X = \{1, 2, 3, 4\}$; (b) a network **T** on $Y = \{a, b, c\}$ [T is a homomorphic image of **R** under the mapping $\mu(1) = \mu(2) = a$, $\mu(3) = b$, $\mu(4) = c$]; (c) a network **U** on $Z = \{1, 2, 3\}$ (**U** is a subnetwork of **R**); (d) a network **V** on $W = \{a, b\}$ [**V** is a derived network of **R** under the mapping $\mu(1) = \mu(2) = a$, $\mu(3) = b$, $\mu(4)$ is undefined]

is linked by B to the image c of node 4. Note, however, that node 2, which also has the image a, is *not* linked to node 4 by B even though their images are connected by B.

Formally, a network homomorphism can be defined in the following way.

DEFINITION. The network $\mathbf{T} = \{T_1, T_2, \ldots, T_p\}$ on the set Y is a *homomorphic image* of the network $\mathbf{R} = \{R_1, R_2, \ldots, R_p\}$ on the set X if there exists a surjection μ from X onto Y such that

$$(y_1, y_2) \in T_i \text{ iff there exist nodes } x_1, x_2 \in X \text{ such that } \mu(x_1) = y_1,\ \mu(x_2) = y_2 \text{ and } (x_1, x_2) \in R_i;\ i = 1, 2, \ldots, p,\ y_1, y_2 \in Y.$$

The mapping μ is termed *a network homomorphism.*

It may be observed that the preceding definition also contains an implicit mapping between relations in the two networks. In particular, it is assumed that the networks are comparable in the sense described in chapter 3, that is, there is a one-to-one mapping from relations in the first network to relations in the second. The relation R_1 is mapped onto T_1, R_2 is mapped onto T_2 and so on. The mapping is left implicit in the definition to avoid notational complexity; indeed, it is usually also the case that corresponding relations in the two networks have the same label. For instance, in Figure 5.1, the relation A has the same meaning in each network displayed, and so does the relation B.

It may also be observed that each partition of the node set X of a network may be associated with a homomorphism of the network. If P is a partition of X and $[x] = \mu(x)$ denotes the equivalence class of P containing x, then a homomorphism from $\mathbf{R} = \{R_1, R_2, \ldots, R_P\}$ is induced onto $\mathbf{T} = \{T_1, T_2, \ldots, T_P\}$ on the equivalence classes of P with $([x], [y]) \in T_i$ if and only if there exists some $x \in [x]$ and $y \in [y]$ such that $(x, y) \in R_i$. We write $T_i = R_i/P$ and call the derived network $\mathbf{T}$, the *induced* derived network under the partition P. For instance, the homomorphism from $\mathbf{R}$ onto $\mathbf{T}$ represented in Figure 5.1 is associated with the partition (12) (3) (4) on the node set X of $\mathbf{R}$.

Another kind of simplification of a network is a "subnetwork", also illustrated in Figure 5.1. The network $\mathbf{U}$ on the node set $\{1, 2, 3\}$ shown in Figure 5.1c is a subnetwork of the network $\mathbf{R}$ of Figure 5.1a. The node set $\{1, 2, 3\}$ of $\mathbf{U}$ is a subset of the node set $\{1, 2, 3, 4\}$ of $\mathbf{R}$, and links are present in $\mathbf{U}$ whenever links between the nodes $\{1, 2, 3\}$ are present in $\mathbf{R}$.

DEFINITION. The network $\mathbf{T} = \{T_1, T_2, \ldots, T_p\}$ on the set Y is a *subnetwork* of the network $\mathbf{R} = \{R_1, R_2, \ldots, R_p\}$ on the set X if there exists a partial one-to-one function μ from X onto Y such that

$$(y_1, y_2) \in T_i \text{ iff there exist nodes } x_1, x_2 \in X \text{ such that}$$
$$\mu(x_1) = y_1,\ \mu(x_2) = y_2 \text{ and } (x_1, x_2) \in R_i;\ i = 1, 2, \ldots, p,\ y_1, y_2 \in Y.$$

Clearly, any subset Y of a node set X possesses a corresponding partial one-to-one function μ_y from X to Y and so defines a subnetwork of the network on X. The function is given by

$$\mu(x) = x \text{ if } x \in Y; \text{ otherwise } \mu(x) \text{ is not defined.}$$

For instance, the mapping μ for the network $\mathbf{U}$ of Figure 5.1c maps node 1 in $\mathbf{R}$ onto node 1 in $\mathbf{U}$, node 2 in $\mathbf{R}$ onto node 2 in $\mathbf{U}$, node 3 in $\mathbf{R}$ onto node 3 in $\mathbf{U}$ and is undefined for node 4.

More generally, one can define simplifications of a network that are combinations of homomorphic images and subnetworks. An example is presented in Figure 5.1d. The network $\mathbf{V}$ is a subnetwork of the homomorphic image $\mathbf{T}$ of $\mathbf{R}$; it can also be seen as a homomorphic image of the subnetwork $\mathbf{U}$ of $\mathbf{R}$. We shall refer to the network $\mathbf{V}$ as a *derived network* of $\mathbf{R}$, and it may be defined formally as follows.

DEFINITION. The network $\mathbf{T} = \{T_1, T_2, \ldots, T_p\}$ on the set Y is a *derived network* of the network $\mathbf{R} = \{R_1, R_2, \ldots, R_p\}$ on the set X if there exists a partial function μ from X onto Y such that

$(y_1, y_2) \in T_i$ iff there exist nodes $x_1, x_2, \in X$ such that $\mu(x_1) = y_1$, $\mu(x_2) = y_2$ and $(x_1, x_2) \in R_i$; $i = 1, 2, \ldots, p$; $y_1, y_2 \in Y$.

Y is referred to as a *derived set* of X.

If the function μ is a surjection, then **T** is a homomorphic image of **R**, as previously defined. If μ is a partial one-to-one function, then **T** is a subnetwork of **R**. For example, in Figure 5.1, **V** is a derived network of **R** under the partial function μ on X, where

$$\mu(1) = a = \mu(2); \qquad \mu(3) = b; \qquad \mu(4) \text{ is undefined.}$$

The mapping from **R** onto **T** identified in Figure 5.1 is associated with the surjection μ given by

$$\mu(1) = a = \mu(2); \qquad \mu(3) = b; \qquad \mu(4) = c;$$

whereas the mapping from **R** onto **U** is induced by the partial one-to-one function:

$$\mu(1) = 1; \qquad \mu(2) = 2; \qquad \mu(3) = 3; \qquad \mu(4) \text{ is undefined.}$$

Now let S' be an isotone homomorphic image of the partially ordered semigroup $S(\mathbf{R})$ of the network **R** on the node set X. Then the correspondence sought between the homomorphic image S' of $S(\mathbf{R})$ and some derived network **T** of **R** may be characterised as being of the form

$$S(\mathbf{T}) = S'. \tag{5.1}$$

That is, given a homomorphic image S', can one find a set Y derived from X whose induced derived network **T** has a semigroup isomorphic to S'? The questions posed earlier take the form:

1 Given a homomorphic image S' of $S(\mathbf{R})$, under what conditions does there exist a derived network **T** with S' isomorphic to $S(\mathbf{T})$, and how can such derived networks be found?
2 Given a derived network **T** of **R**, under what conditions is $S(\mathbf{T})$ a homomorphic image of $S(\mathbf{R})$?

In general, the answer to the question of the existence of correspondences of the kind represented by equation (5.1) is that only some homomorphisms of the semigroup of a network correspond to derived networks and only some derivations of the network have counterparts in the collection of images of its semigroup (e.g., Bonacich, 1983, 1989). Some of the relational conditions under which the correspondence is known to exist are now presented. The first was described in chapter 3; it is a condition guaranteeing the iosomorphism of the semigroup of a network and the semigroup of a derived network.

A. Structural equivalence

THEOREM 5.1 *(Lorrain & White, 1971). Let* **R** *be a network on a node set* X, *and let* μ: X → Y *where* Y *is a set of classes of a partition* SE *on* X *satisfying the structural equivalence condition (Lorrain & White, 1971; see chapters 1 and 3). Let T be the network derived from* **R** *by the mapping* μ. *Then the partially ordered semigroups* S(**R**) *and* S(*T*) *are isomorphic.*

Proof. Lorrain and White (1971).

White and Reitz (1983, 1989) termed the relation *SE* a *strong equivalence.*

The next nine results describe conditions under which $S(\mathbf{T})$ is not necessarily isomorphic to $S(\mathbf{R})$. The first, that of automorphic equivalence, was described by Winship (1988), and it has been introduced in chapter 3 in relation to local role algebras.

B. Automorphic equivalence. Recall that an automorphism α of a network **R** on X is a bijection α from X onto itself for which $(x, y) \in R_i$ if and only if $(\alpha(x), \alpha(y)) \in R_i$. Two nodes x and y are automorphically equivalent if and only if there exists some automorphism α of the network for which $\alpha(x) = y$. The relation AE on X defined by $(x, y) \in AE$ if and only if x and y are automorphically equivalent is an equivalence relation on X, and it has been argued to define a partition of nodes into classes possessing the same abstract social "role" (Borgatti et al., 1989; Winship, 1988). If a network has no automorphically equivalent nodes, then AE is the identity relation. Two leaders of distinct small groups with the same structure are automorphically equivalent, for instance, because one may be mapped onto the other by an automorphism of the network comprising the two groups. Consider, also, the network displayed in Figure 5.2a. Nodes 5 and 6 are automorphically equivalent, as are nodes 3 and 4, and nodes 7 and 8. Each of these pairs of nodes relates to other types of nodes in the same way and with the same frequency and so may be said to play the same network role. They do not necessarily relate to the *same* other nodes, though, as structural equivalence requires. Hence automorphic equivalence is a generalisation of structural equivalence.

Now we consider the derived network induced by the automorphic equivalence relation.

THEOREM 5.2 *(Winship, 1988). Let* μ *be the surjection on the node set* X *of a network* **R** *associated with the partition of* X *into automorphic equivalence classes. That is,* μ: x → [x], *where* [x] *is the class of*

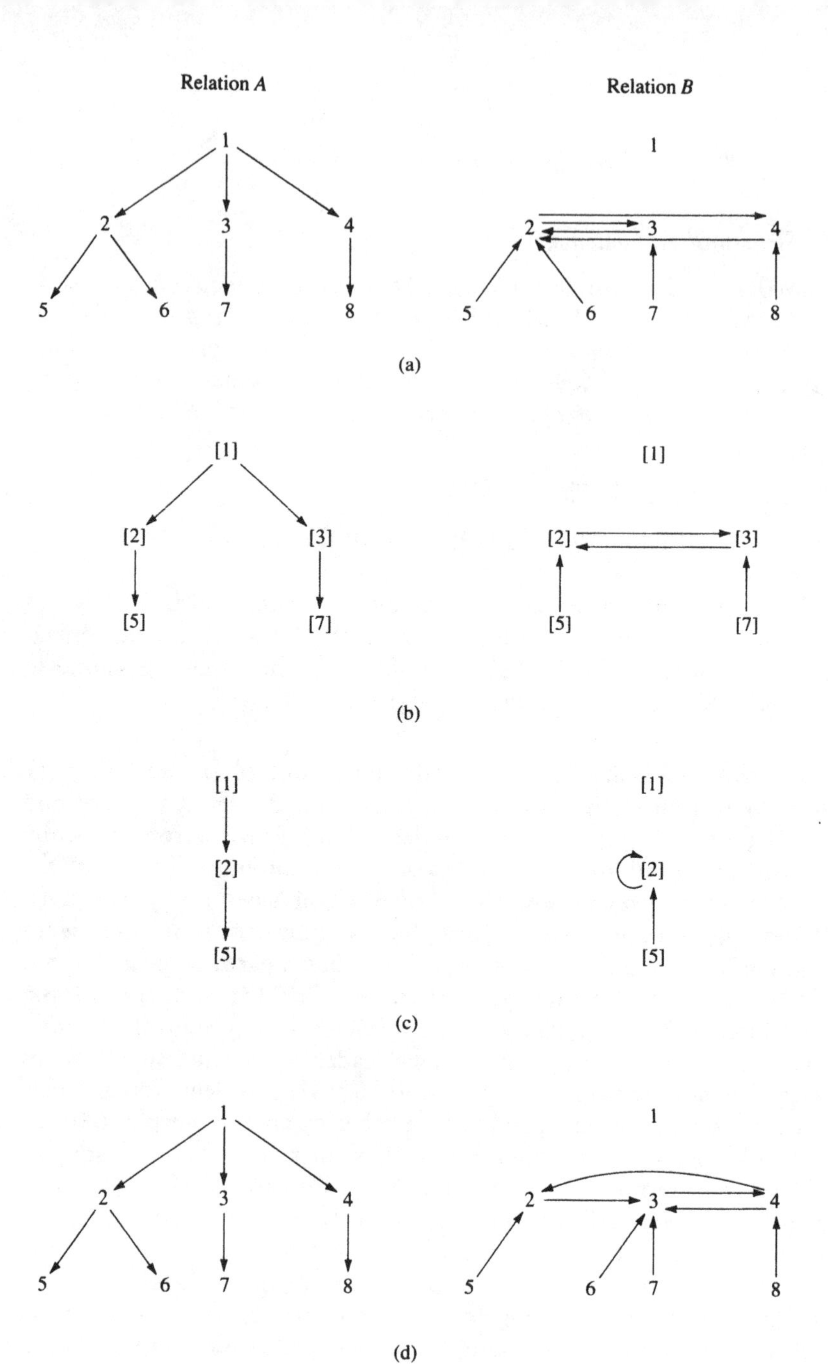

Figure 5.2. Automorphic, extended automorphic and regular equivalences in a network: (a) a network $\mathbf{R} = \{A, B\}$ on a set $X = \{1, 2, 3, 4, 5, 6, 7, 8\}$ [the partition AE = (1) (2) (34) (56) (78) is an automorphic equivalence]; (b) the derived network **T** induced by the automorphic equivalence partition AE; (c) the derived network **U** induced by the extended automorphic equivalence AE^* on X, where AE^* = (1) (234) (5678); (d) the network $\mathbf{V} = \{A, B\}$ on $Y = \{1, 2, 3, 4, 5, 6, 7, 8\}$ [the partition (1) (234) (5678) is a regular equivalence on **V**]

elements automorphically equivalent to x, *and* $([x], [y]) \in \bar{R}_i$ *if and only if* $(x, y) \in R_i$, *for some* $x \in [x]$ *and some* $y \in [y]$. *Then the derived network* **T** *induced on the classes of* X *has a partially ordered semigroup* S(**T**) *which is an isotone homomorphic image of* S(**R**).

For example, the network displayed in Figure 5.2b is derived from the network of Figure 5.1a by the automorphic equivalence relation and represents relations among the classes of nodes possessing the same role. Theorem 5.2 establishes that the relational structure of the derived network is an isotone homomorphic image of that of the original network.

It can be seen from Figure 5.2 that the structure of the derived network **T** is simpler than that of the original network **R**. For instance, in **R**, the paths *BBB* and *B* are distinct, whereas in **T** they are not.

C. Extended automorphic equivalence, or iterated roles. Pattison (1980) and Borgatti et al. (1989) extended the notion of automorphic equivalence to a more general type of equivalence. They observed that the derived network induced by the *AE* relation sometimes possessed a nontrivial *AE* relation and so could define a further derivation of the network. For example, the network in Figure 5.2c is derived by the *AE* relation from the network in Figure 5.2b, and so from Figure 5.2a. It possesses no nontrivial *AE* relation, and the partition of nodes of *X* to which it corresponds is the coarsest partition of *X* that may be obtained by a succession of *AE* relations. In chapter 3, such a partition was termed extended automorphic equivalence and denoted by the relation *AE**. Nodes that belong to the same class of the relation *AE** may also be argued to possess similar social roles: they possess the same kinds of relations to the same other types of nodes, and these relations may vary only in their frequency. The *AE* and *AE** relations depicted in Figure 5.2 illustrate the difference between the two. Nodes 3 and 4 are automorphically equivalent, because they have the same number and types of relations to nodes 7 and 8, which are also automorphically equivalent. Nodes 2, 3 and 4 are combined by the extended automorphic equivalence, however, even though they have different *numbers* of relations to nodes 5, 6, 7 and 8. Thus, the distinction between the relations *AE* and *AE** is in terms of the "quantity" of relations, and both may be said to indicate nodes whose relations possess the same "quality".

THEOREM 5.3. *Let* μ *be the mapping from the node set* X *of a network* **R** *onto the classes of the relation* AE* *on* X:

$$\mu: x \rightarrow [x],$$

where [x] *is the* AE* *class containing* x. *Then the induced derived network* **T** *on the classes of* X *has a partially ordered semigroup that is an isotone homomorphic image of the semigroup of* **R**.

The network **U** in Figure 5.2c has a simpler structure again than that of **R** or **T** in Figure 5.2. In **U**, paths labelled *BB* are the same as those labelled *B*, whereas in both **R** and **T**, they are distinct.

D. Regular equivalence. An even more general operationalisation of the notion of similar social roles has been described by White and Reitz (1983, 1989). They defined a relation termed regular equivalence, which directly partitions nodes into classes having similar relations with other classes. It may be defined formally in the following way.

DEFINITION. Let $\mathbf{R} = \{R_1, R_2, \ldots, R_p\}$ be a network on a set X. Let RE be a partition on X satisfying the condition:

$(x, y) \in RE$ implies

1 $(x, z) \in R_i$ implies that there exists some $w \in X$ such that $(y, w) \in R_i$ and $(z, w) \in RE$; for any $z \in X$, and any $R_i \in \mathbf{R}$; and
2 $(z, x) \in R_i$ implies that there exists some $w \in X$ such that $(w, y) \in R_i$ and $(z, w) \in RE$; for any $z \in X$, and any $R_i \in \mathbf{R}$.

RE is termed a *regular equivalence.*

For example, consider the networks displayed in Figures 5.2a and 5.2d. In each case, the partition (1) (234) (5678) defines a regular equivalence on the node set X. Nodes 2, 3 and 4 are regularly equivalent because (a) each receives an A relation from the class [1]; (b) each expresses an A relation to the class [5]; (c) each expresses and receives a B relation from its own class; and (d) each receives a B tie from class [5].

Borgatti and Everett (1989) showed that there is a unique maximal regular equivalence relation for any network. Further, White and Reitz (1983, 1989) established that any regular equivalence on a network is associated with a homomorphism of its semigroup.

THEOREM 5.4. *Let* RE *be a regular equivalence relation on a node set* X *of a network* **R**. *Then if* **T** *is the derived network induced by the relation* RE, S(**T**) *is an isotone homomorphic image of* S(**R**).

Regular equivalence may be seen as a generalisation of extended automorphic equivalence because if two elements are in the same class of an extended automorphic equivalence, then they are also regularly equivalent. The relationship is illustrated in Figure 5.2. In Figure 5.2a, the partition (1) (234) (5678) is both an extended automorphic equivalence and a regular equivalence, whereas in Figure 5.2d, the same partition is only a regular equivalence.

White and Reitz (1989) describe an algorithm REGE that identifies the maximal regular equivalence of a network. It may also be used to identify approximations to regular equivalence. Exact regular equivalences other than the maximal one may also be found using algorithms described by Borgatti and Everett (1989) and Everett and Borgatti (1988). A number of these algorithms are implemented in *UCINET* IV (Borgatti, Everett & Freeman, 1991).

E. Outdegree and indegree equivalence conditions. Two immediate further generalisations of regular equivalence may be derived by considering the two defining conditions for regular equivalence separately. The first condition ensures that regularly equivalent nodes have similar outgoing or "expressed" network ties. The second requires that they have similar incoming or "received" network relations. By considering these two conditions separately, we can define partitions of nodes whose incoming or outgoing ties are similar, so that they have similar roles with respect to either expressed or received relations. As an example, consider the network of Figure 5.3a. Node 2 has a pattern of outgoing relations similar to that of node 3; hence nodes 2 and 3 can be argued to have a similar role in terms of outgoing relations. Nodes 2 and 3, however, do not have similar patterns of incoming ties because only node 3 receives B ties from nodes 4 and 5. Nodes 4 and 5, though, are the *recipients* of similar types of relations and therefore have similar roles with respect to incoming relations.

DEFINITION. Let $\mathbf{R} = \{R_1, R_2, \ldots, R_p\}$ be a network on X, and let OE and IE be partitions on X satisfying

1. $(x, y) \in OE$ implies that if $(x, z) \in R_i$, then there exists some $w \in X$ such that $(y, w) \in R_i$ and $(z, w) \in OE$, for any $z \in X$, and any $R_i \in \mathbf{R}$; and
2. $(x, y) \in IE$ implies that if $(z, x) \in R_i$, then there exists some $w \in X$ such that $(w, y) \in R_i$ and $(z, w) \in IE$, for any $z \in X$, and any $R_i \in \mathbf{R}$.

The partition OE is termed an *outdegree equivalence* and IE an *indegree equivalence*.

For instance, the partition (1) (23) (4) (5) is an outdegree equivalence in the network of Figure 5.3a and (1) (2) (3) (45) is an indegree equivalence. The derived networks induced by the two partitions are displayed in Figures 5.3b and 5.3c.

The indegree and outdegree equivalence conditions were originally described by J. Q. Johnson (Seiyama, 1977), who also established the following result.

THEOREM 5.5. *Let* T *be a derived network of the network* $\mathbf{R}$ *induced by either an outdegree or an indegree equivalence on the node set* X *of* $\mathbf{R}$. *Then* $S(T)$ *is an isotone homomorphic image of* $S(\mathbf{R})$.

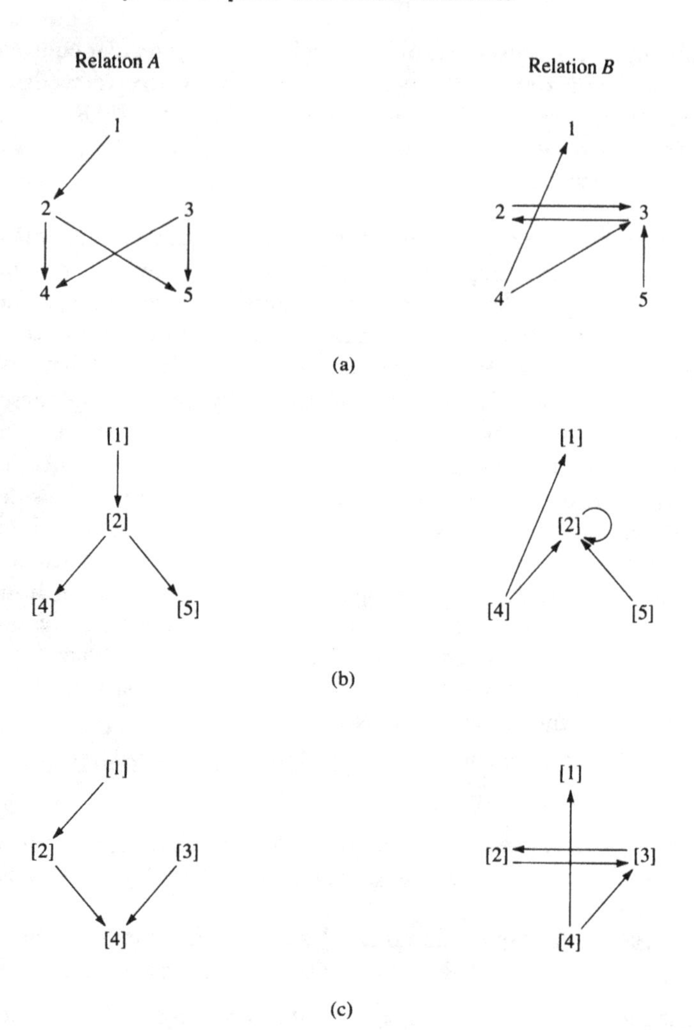

Figure 5.3. Indegree and outdegree equivalences in a network: (a) the network $\mathbf{R} = \{A, B\}$, in which the partition $OE = (1)\,(23)\,(4)\,(5)$ is an outdegree equivalence and the partition $IE = (1)\,(2)\,(3)\,(45)$ is an indegree equivalence; (b) the network induced by the OE relation; (c) the network induced by the IE relation

Of course, if all relations in a network are symmetric, then the collections of indegree, outdegree and regular equivalences on the network coincide.

F. Kim and Roush conditions. An even more general set of conditions, of which all of the preceding equivalences are special cases, has been developed by Kim and Roush (1984). They have defined a quite general condition on relations between classes of a partition of the node set of a network that guarantees that the semigroup of the derived network associated with the partition is an isotone image of the network semigroup. The condition may be described as follows.

DEFINITION. Let **R** be a network on X, and let KR_i be an equivalence relation on X satisfying *Kim and Roush's condition* G_i:

> Let C_1 and C_2 be any two equivalence classes in KR_i, and let the number of their elements be n_1 and n_2, respectively. Let D be any subset of C_1 consisting of i elements, or if $i > n_1$, let $D = C_1$. Then KR_i satisfies the condition G_i if, for any such equivalence classes C_1 and C_2, and for any $j = 1, 2, \ldots, p$, the set $\{z: z \in C_2 \text{ and } (y, z) \in R_j \text{ for some } y \in D\}$ has at least $\min(i, n_2)$ elements.

The condition G_1 is equivalent to the outdegree condition, and the condition G_n (where n is the number of elements in X) is equivalent to the indegree condition. The relationships between the requirements of these and some other conditions are illustrated schematically in Figure 5.4. The condition G_i specifies that if there are links of a particular kind between one class and another, then any i elements in the first class must be linked to at least i distinct elements in the second (or all of them if the second class contains fewer than i elements). As such, it ensures that the "range" of relations expressed from one class to another is sufficient to guarantee that any compound relations between classes of KR_i are also observed for at least some of the individual members of those classes.

Then Kim and Roush (1984) proved the following theorem.

THEOREM 5.6. *Let* **T** *be the derived network of a network* **R** *on a node set X induced by a partition* KR_i *of* X *that satisfies Kim and Roush's condition* G_i. *Then* $S(\mathbf{T})$ *is an isotone homomorphic image of* $S(\mathbf{R})$.

G. Central representatives condition. A different kind of partition inducing a semigroup homomorphism was described by Pattison (1982) and is illustrated in Figure 5.5. In Figure 5.5, the incoming and outgoing relations of node 3 are a subset of those of node 2, and we can think of node 2 as the more "central" or "prominent" of nodes 2 and 3. Indeed, compared with node 3, node 2 is as close or closer to other nodes in the network; it has indegrees and outdegrees in each relation at least as high as those of node 3; and for any geodesic path connecting other network nodes and passing through node 3, there is a corresponding path between the same nodes that passes through node 2. Thus, node 2 is more central than node 3 in each of the senses of centrality described

Structural equivalence (*SE*)

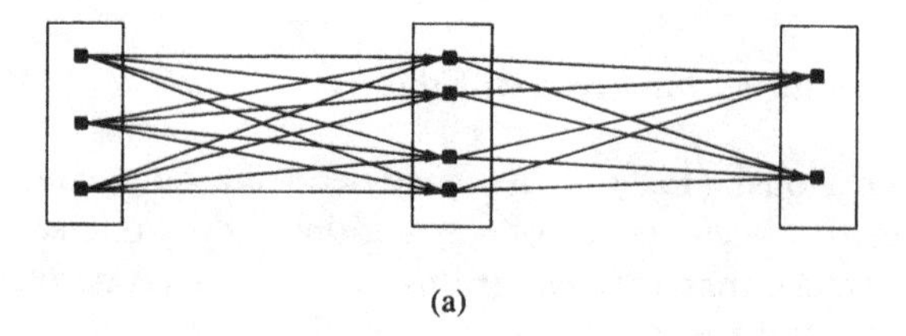

(a)

Automorphic equivalence (*AE*)

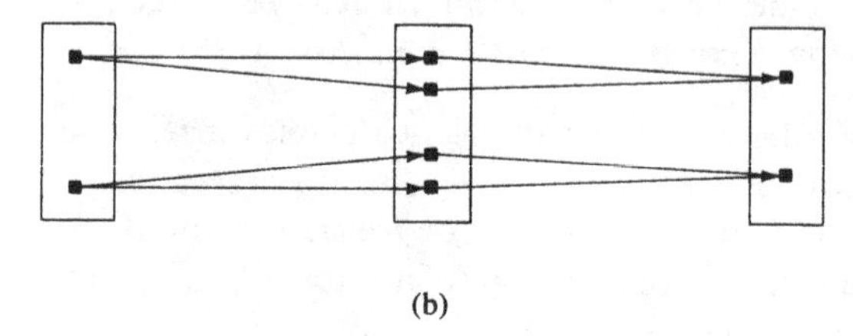

(b)

Regular equivalence (*RE*)

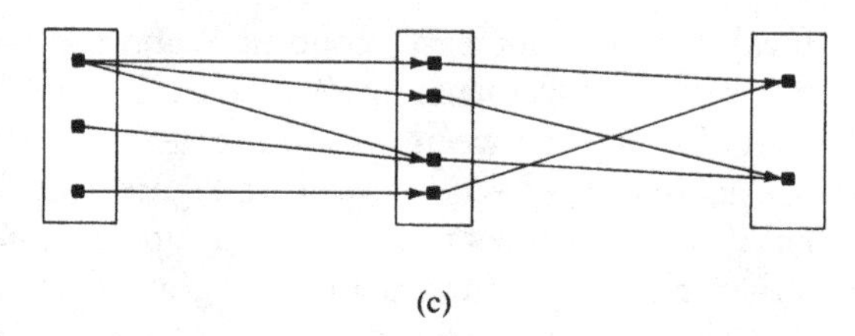

(c)

Indegree equivalence (*IE*)

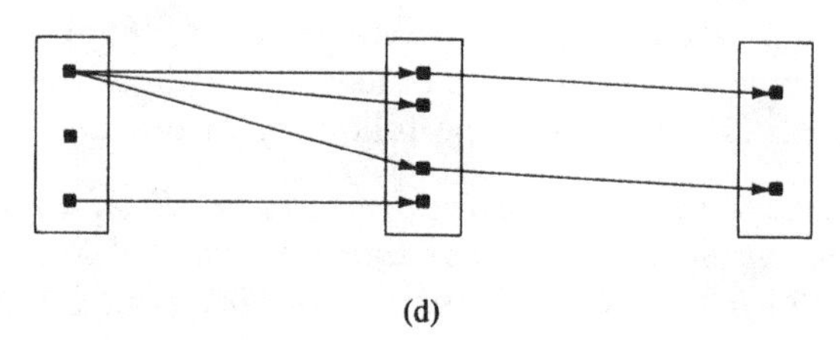

(d)

Figure 5.4. Some conditions for semigroup homomorphisms (elements are shown as small squares, blocks as boxes and relations as directed arrows): (a) structural equivalence (*SE*): *i* and *j* are *SE* if *i* and *j* have identical relations with any other element *k*; (b) automorphic equivalence (*AE*): *i* and *j* are *AE* if they belong to "parallel" positions in the same or different structures; (c) regular equivalence (*RE*): *i* and *j* are RE if, whenever the block containing *i* and *j* is connected to another block, each *i* and *j* is connected to some element in the latter block (true for both incoming and outgoing connections); (d) indegree equivalence (*IE*): *i* and *j* are *IE* if, whenever the block containing *i* and *j* *receives* a tie from another block, then each element in the block receives a tie from that block; (e) outdegree equivalence (*OE*): *i* and *j* are *OE* if, whenever the block containing *i* and *j* *sends* a tie to another block,

Outdegree equivalence (OE)

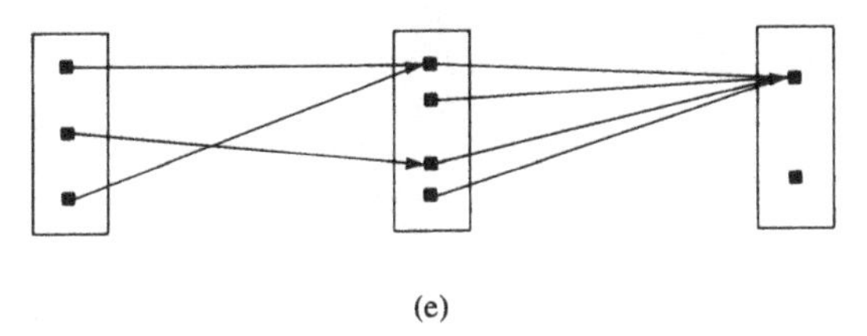

(e)

Kim and Roush conditions G_i

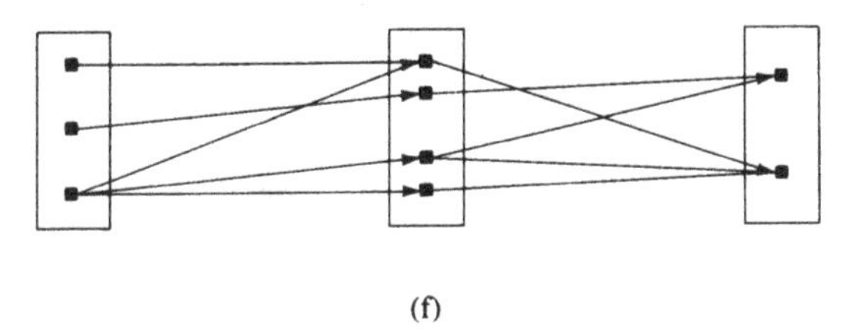

(f)

Central representatives condition

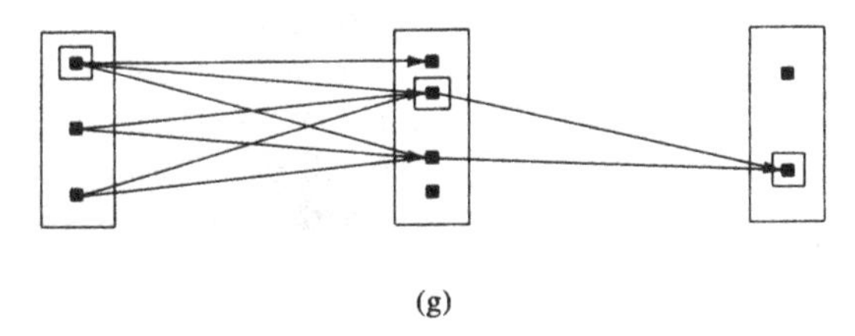

(g)

Kim and Roush condition G_{im}

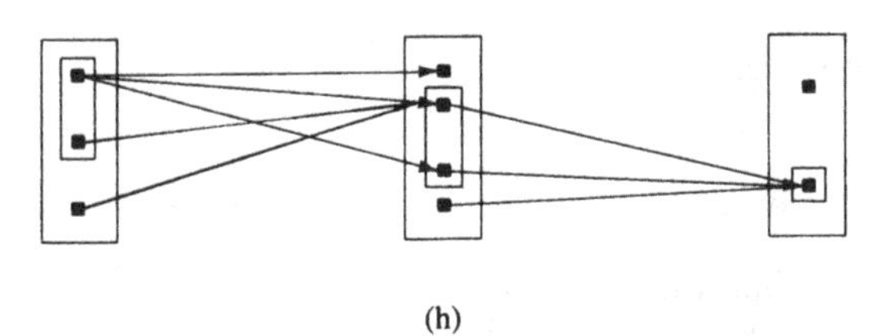

(h)

then each element in the block sends a tie to that block; (f) Kim and Roush condition G_i (KR_i): a partition of X satisfies the condition KR_i if, whenever there is a connection from one block to another, then any i elements in the first block are connected to i distinct elements (or all elements) in the second block; (g) central representatives condition: a partition of X satisfies the central representatives condition if each block in the partition has a central representative whose ties include the ties of all other block members; (h) Kim and Roush condition G_{im}: a partition of X satisfies the condition G_{im} if each block of the partition possesses a central subset whose ties include the ties of all other block members and relations between central subsets satisfy the condition G_i

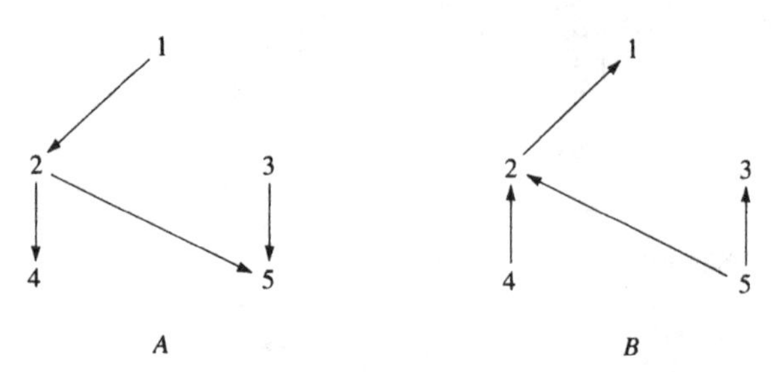

Figure 5.5. The central representatives condition (the partition (1) (23) (45) satisfies the *CR* condition, with nodes 1, 2 and 5 the central representatives of the three classes)

by Freeman (1979). (Freeman actually defined centrality for nondirected graphs, but the properties he described may also be defined for directed relations.) Similarly, node 5 is the more central of nodes 4 and 5, and the relations of node 4 are a subset of those of node 5. Now consider a partition of nodes such that each class of nodes possesses a node whose relations are a superset of the relations of all other nodes in the class. Then Pattison (1982) showed that such a partition was associated with an isotone homomorphism of the network semigroup.

DEFINITION. The partition *CR* on *X* satisfies the *central representatives condition* for the network **R** on *X* if, for each class C of *CR*, there exists some node $x^* \in C$, such that, for any $x \in C$,

1 $(x, y) \in R_i$ implies $(x^*, y) \in R_i$; and
2 $(y, x) \in R_i$ implies $(y, x^*) \in R_i$; $x, y \in X$; $i = 1, 2, \ldots, p$.

The node x^* is termed the *central representative* of class C.

The condition requires that there be an element in each class of C (a central representative) that possesses all of the connections of other class members. The elements to whom the other class members are linked are thus a subset of those to whom the central representative is connected. (Note that if any class contains two central representatives, then they must be structurally equivalent.)

THEOREM 5.7. *Let* P *be a partition on* X *satisfying the central representatives condition for the network* **R** *on* X, *and let* **T** *be the derived network of* **R** *induced by* P. *Then* S(**T**) *is an isotone homomorphic image of* S(**R**).

Proof: The proof is contained in Appendix B.

H. Kim and Roush's condition G_{im}. Finally, Kim and Roush (1984) have described the most general condition known on partitions of the node set of a network that guarantees the semigroup of the associated derived network to be an isotone homomorphic image of the original network semigroup. The condition is a generalisation both of the Central Representatives condition and Kim and Roush's condition G_i.

DEFINITION. The partition P on X satisfies *Kim and Roush's condition* G_{im} for the network **R** on X if, for each class C of P,

1 there exists a *central subset* S of C such that (a) $(x, y) \in R_j$ for some $x \in C$ implies $(x^*, y) \in R_j$ for some $x^* \in S$, and (b) $(y, x) \in R_j$ for some $x \in C$ implies $(y, x^*) \in R_j$ for some $x^* \in S$; $x, y \in X$, $j = 1, 2, \ldots, p$; and
2 if U denotes the union of the central subsets S, then the central subsets S for the classes of P satisfy Kim and Roush's condition G_i for relations on U.

The condition is equivalent to the central representatives condition when each subset S is constrained to consist of a single element. When each class C comprises a single node, it is equivalent to the condition G_i. Then Kim and Roush (1984) showed the following.

THEOREM 5.8. *Let* P *be a partition on the node set* X *of a network* **R** *that satisfies Kim and Roush's condition* G_{im}, *and let T be the derived network induced by* P. *Then the partially ordered semigroup* S(T) *is an isotone homomorphic image of* S(**R**).

Relations of generality among the conditions that we have described on partitions of the node set of a network are summarised in Figure 5.6, where a condition lies below and is connected to another condition if the first is a generalisation of the second.

Finally, we review several conditions on subsets of the node set of a network that also lead to semigroup homomorphisms.

I. Corollary to the condition G_{im}. A corollary of Kim and Roush's result for the condition G_{im} is that the subset U of X, consisting of the union of the central subsets for some partition P satisfying the condition G_{im}, also leads to a network **U** of **R** whose semigroup is an isotone homomorphic image of S(**R**).

DEFINITION. Let P be a partition on X satisfying the condition G_{im} for a network **R** on X. Then if U is the union of a set of central subsets for P, U is termed a *collection of central subsets* corresponding to P.

THEOREM 5.9. *The semigroup of the network induced by the subset* U *is an isotone homomorphic image of the semigroup of* **R**.

Proof: The proof is contained in Appendix B.

J. Receiver subset and sender subset conditions. A subnetwork may also be shown to possess a correspondence with an isotone homomorphic image of $S(\mathbf{R})$ in the following case.

DEFINITION. The subset Y of X is a *receiver* subset for the network $\mathbf{R}$ on X if, for each $z \in X \backslash Y$ and for each $i = 1, 2, \ldots, p$, $(w, z) \in R_i$ implies $w \in X \backslash Y$. [Alternatively, a receiver subset may be characterised as a subset of X for which YR_i is contained in Y, for all i, where $YR_i = \{x \in X : (y, x) \in R_i$ for some $y \in Y\}$.]

THEOREM 5.10. *Let* $\mathbf{T}$ *be the derived network induced by a receiver subset* Y *of* X. *Then* $S(\mathbf{T})$ *is an isotone homomorphic image of* $S(\mathbf{R})$.

Proof: The proof is given in Appendix B.

Dually, it may be established that if the subset Y satisfies the sender subset condition (defined next) then the induced subnetwork $\mathbf{T}$ of $\mathbf{R}$ is also such that $S(\mathbf{T})$ is an isotone homomorphic image of $S(\mathbf{R})$.

DEFINITION. The subset Y of X is a *sender* subset for the network $\mathbf{R}$ on X if, for each $x \in X \backslash Y$ and $i = 1, 2, \ldots, p$, $(x, w) \in R_i$ implies $w \in X \backslash Y$. [That is, a sender subset Y satisfies the condition that R_iY is contained in Y for all i, where $R_iY = \{x \in X : (x, y) \in R_i$ for some $y \in Y\}$.]

Generalisations of structural equivalance

The equivalence conditions we have described may all be viewed as different generalisations of structural equivalence. Each can be seen to yield descriptions of network nodes whose relations with other nodes possess some similar relational features. The nature of the similarity of relational features is determined by the condition and varies from one to another. For instance, two nodes belonging to the same class of an outdegree equivalence relation OE possess similar kinds of relations expressed *to* individuals in any other class. For an indegree equivalence IE, the similarity holds for relations received *from* individuals in other classes. For two individuals belonging to the same class of a partition satisfying the central representatives condition, a different property holds, namely, that their relations are a subset of those of some single member of their class.

The usefulness of any one of these algebraic conditions in determining the similarity of the relational features of network members depends on the context in which relational similarity is being assessed. One major distinction in the literature to date is that which has been drawn between

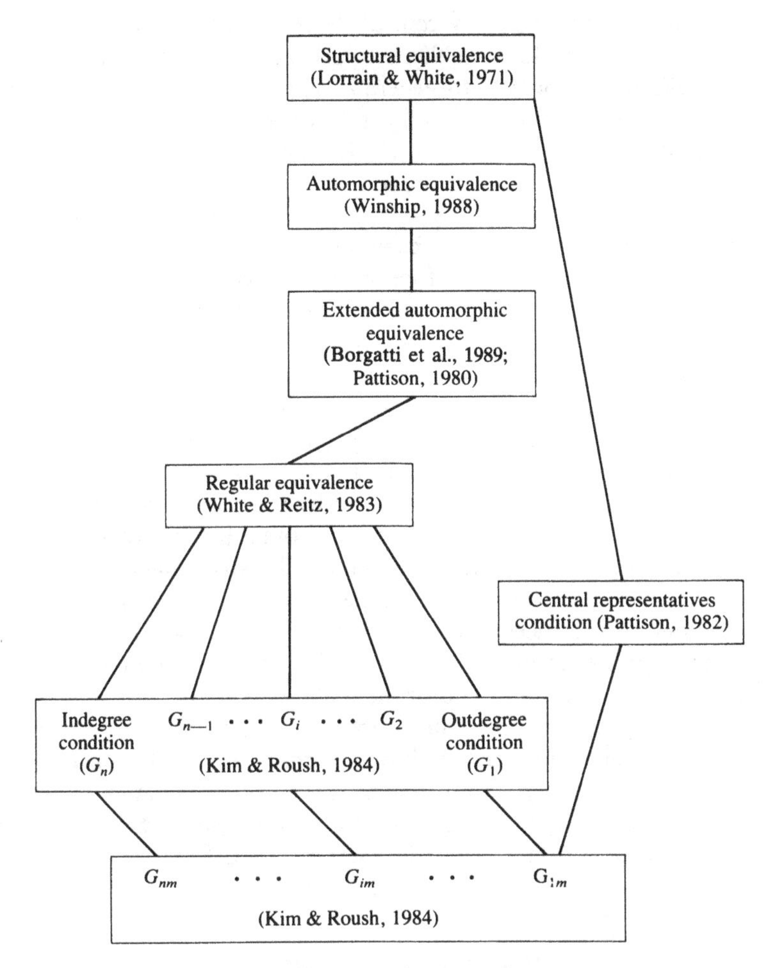

Figure 5.6. Relations among equivalence conditions

concrete and abstract equivalences (e.g., Winship, 1988) or between structural and "general" equivalences (Faust, 1988). The distinction is a basic one because it contrasts the notion of identity of social relations with every member of a population with that of parallel relations with similar types of members of a population. Indeed, many useful analyses of network positions have been made on the basis of this distinction (e.g., Doreian, 1988a; Faust, 1988). As Pattison (1988) has argued, however, other useful distinctions may be made in Figure 5.6, and only a limited exploitation of these has been accomplished. For instance, the indegree or outdegree condition might be used to analyse similarities in the receiving *or* sending of ties. Or the central representatives condition could be used to identify a set of central representatives whose interrelations model relational flows in the network in a more efficient way.

A different approach to generalising structural equivalence has been taken by Wasserman and colleagues (Anderson, Wasserman & Faust, 1992; Fienberg & Wasserman, 1981; Holland et al., 1983; Wasserman & Anderson, 1987). Their approach is stochastic and relies on the specification of some probability distribution over the class of networks on some set X of nodes. Their basic unit of analysis is dyadic, and refers to $\mathbf{D}_{ij}$, the random vector of network links of each type from node i to node j, and from node j to node i. They define two nodes i and j from X to be *stochastically equivalent* if the random dyadic variables $\mathbf{D}_{kl}$ are all statistically independent of each other and if $\mathbf{D}_{ik}$ and $\mathbf{D}_{jk}$ have the same probability distribution, for all nodes k distinct from i and j.

The definition thus relaxes the requirement that nodes i and j must have identical *observed* incoming and outgoing ties in order to be structurally equivalent. It is replaced by the requirement that they have identical *probabilities* of incoming and outgoing ties with each node: it is, in this sense, a stochastic version of the strict algebraic requirement of structural equivalence. Indeed, a "stochastic" version of the other algebraic equivalences could also be formulated. For instance, "stochastic automorphic equivalence" of two nodes i and j would require the existence of some permutation α of the node set X mapping i onto j such that $\mathbf{D}_{ik}$ and $\mathbf{D}_{j\alpha(k)}$ have the same probability distribution, for all nodes k.

The stochastic approach is clearly an important one for the development of strategies of analysis for network data subject to random variation, and it is to be hoped that the algebraic framework outlined here may serve as the basis for a useful class of stochastic models.

The correspondence definition

The conditions we have listed account for all known conditions under which a derived network has a partially ordered semigroup that is a homomorphic image of the semigroup of the original network. For the purposes of data analysis, though, it is useful to propose, in addition, some general conditions that should be satisfied if we wish to associate a derived network of **R** with a homomorphic image of $S(\mathbf{R})$. Our aim is a simple means of verifying whether an association between a derived network and a homomorphic image is plausible. Consider, for example, the network **X** of Table 5.1. The derived network **Y** associated with the mapping

$$\mu(1) = a = \mu(2); \qquad \mu(3) = b$$

Table 5.1. *The network* **X** *on* {1, 2, 3} *and the derived network* **Y** *on* {*a*, *b*}

Network	Element	Relation	
		P	*Q*
X	1	1 0 0	1 1 1
	2	1 1 0	1 1 0
	3	1 0 0	1 1 0
Y	a	1 0	1 1
	b	1 0	1 0

Table 5.2. *The partially ordered semigroup S*(**X**) *and the factors A and B of S*(**X**)

Label	Right mult. table				Partial order
			Generators		
	Element	Word	*P*	*Q*	
S(**X**)	1	*P*	1	3	1 0 0 0
	2	*Q*	4	3	1 1 0 0
	3	*PQ*	4	3	1 1 1 1
	4	*QP*	4	3	1 0 0 1
A	1	*P*	1	3	1 0 0
	2	*Q*	1	3	1 1 0
	3	*PQ*	1	3	1 1 1
B	1	*P*	1	2	1 0
	2	*Q*	2	2	1 1

is also displayed in Table 5.1. In Table 5.2, the semigroup *S*(**X**) is presented, and Figure 5.7 presents the lattice $L_\pi(S(\mathbf{X}))$ of π-relations of *S*(**X**). It can be seen from Figure 5.7 that $L_\pi(S(\mathbf{X}))$ has two atoms, π_1 and π_2, and that each of these atoms has a unique maximal complement, π_3 and π_1, respectively. Hence $\{\pi_1, \pi_3\}$ defines a unique factorisation of *S*(**X**). The isotone homomorphic images corresponding to π_1 and π_3 are presented in Table 5.2 and are labelled *A* and *B*, respectively. (Thus, $A = S(\mathbf{X})/\pi_1$ and $B = S(\mathbf{X})/\pi_3$.) The question we wish to be able to answer is this: Is there an association between the network **Y** and the image *A* of *S*(**X**)?

There are at least two means of investigating whether a derived network is associated with an isotone homomorphism of its semigroup. The first

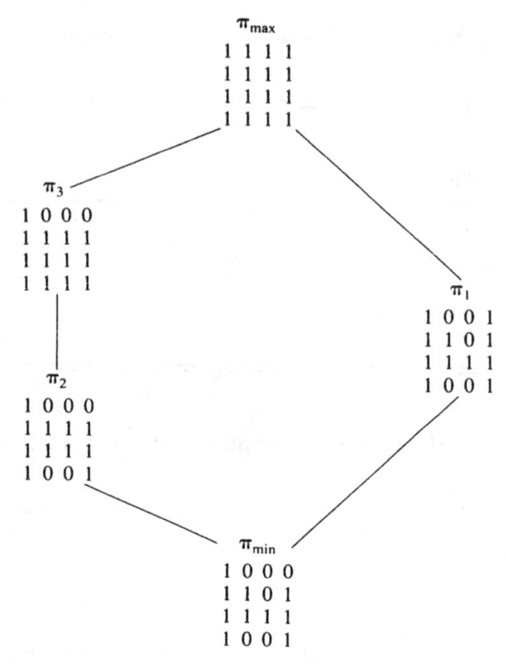

Figure 5.7. The lattice $L_\pi(S(\mathbf{X}))$ of π-relations of $S(\mathbf{X})$

Table 5.3. *Distinct relations in* $S(\mathbf{X})$

	Relation			
Element	*P*	*Q*	*PQ*	*QP*
1	100	111	111	110
2	110	110	111	110
3	100	110	111	110
a	10	11	11	10
b	10	10	11	10

is to determine whether the distinctions among relations made by the homomorphic image can also be seen when the relations are examined in the derived network. Consider, for instance, the distinct relations in the semigroup $S(\mathbf{X})$ for the network $\mathbf{X}$ of Table 5.1. The relations are listed in Table 5.3. Now, in the image A of $S(\mathbf{X})$, the relations P and PQ are distinguished. Are these relations also distinguished when we examine the derived relations induced by the mapping μ? The induced relations are presented in the second panel of Table 5.3. As Table 5.3 indicates, the distinctions among relations made in A are also made when attention is restricted to the relations induced by μ, so that there is some evidence of association of the derived set and the image. It can also be seen from Table 5.3 that P and QP are *not* distinguished when attention is restricted to relations induced by μ; similarly, P and QP are not distinguished in the image A (Table 5.2).

More generally, we can define the following partial order on a network semigroup in association with a derived network and its corresponding derived set. (In the following definitions, it is assumed that isotone homomorphisms are under consideration, but the definitions are easily modified to refer to homomorphisms more generally. Pattison, 1982, presented the analogous definitions for general homomorphisms.)

DEFINITION. Let $\mathbf{R}$ be a network defined on the node set X, and let Y be a set derived from X by the partial function μ. Also, let $\mu(s)$ denote the image under the partial function μ of the compound relation s (i.e., $(k, l) \in s$ implies $(\mu(k), \mu(l)) \in \mu(s)$), where $s, t \in S(\mathbf{R})$. Define a partial ordering $\leq_\mu$ on $S(\mathbf{R})$ by:

$$s \leq_\mu t \text{ if } (i, j) \in \mu(s) \text{ implies } (i, j) \in \mu(t),$$

for all $i, j \in Y$. The partial ordering $\leq_\mu$ is termed the *partial order on* $S(\mathbf{R})$ *induced by* μ.

We may observe that $s \leq t$ in S implies $s \leq_\mu t$, for any partial function μ, so that the partial order $\leq_\mu$ always contains the partial order of the semigroup. In fact, we can describe relations between the partial orderings on $S(\mathbf{R})$ using the nesting relation defined in chapter 2 for role algebras.

DEFINITION. Let $\leq_1$ and $\leq_2$ be two partial orders on a set P. Define $\leq_1$ to be *nested* in $\leq_2$ if $p \leq_2 q$ implies $p \leq_1 q$, for all $p, q \in P$.

According to the definition, the partial order $\leq_\mu$ is nested in the partial order of $S(\mathbf{R})$ for any partial function μ on the node set X of the network $\mathbf{R}$.

We can also identify any isotone homomorphism ϕ of the semigroup $S(\mathbf{R})$with the π-relation π_ϕ on $S(\mathbf{R})$ introduced in chapter 3. It is convenient to define a partial order $\leq_\phi$ on $S(\mathbf{R})$ that is the converse of π_ϕ.

DEFINITION. Let ϕ be an isotone homomorphism on the partially ordered semigroup S. Define the partial order $\leq_\phi$ by

$$t \leq_\phi s \text{ iff } \phi(t) \leq \phi(s),$$

for $s, t \in S$.

Clearly, $s \leq_\phi t$ if and only if $(t, s) \in \pi_\phi$. The partial order $\leq_\phi$ is also nested in the partial order for the semigroup S, for all isotone homomorphisms ϕ of S.

Table 5.4. *The partial orderings $\leq_\mu$ and $\leq_\phi$ associated with the mapping μ on the node set of* **X** *and the isotone homomorphism ϕ of S*(**X**)

$\leq_\mu$	$\leq_\phi$
1 0 0 1	1 0 0 1
1 1 0 1	1 1 0 1
1 1 1 1	1 1 1 1
1 0 0 1	1 0 0 1

Now the partial order corresponding to the derived set Y can be considered to preserve distinctions among relations made by ϕ whenever the partial order $\leq_\phi$ is nested in the partial order $\leq_\mu$, that is, whenever,

$$t \leq_\mu s \text{ implies } t \leq_\phi s$$

for all $s, t \in S$. For example, the partial ordering $\leq_\mu$ associated with the function μ given by

$$\mu(1) = a = \mu(2); \qquad \mu(3) = b$$

is shown on the left of Table 5.4. The right-hand side of the table contains the partial ordering $\leq_\phi$ associated with the isotone homomorphism from $S(\mathbf{X})$ onto the factor A. (The ordering is the converse of the π-relation π_1 which corresponds to factor A, as in Figure 5.7.) It can be seen in this case that the partial orders $\leq_\mu$ and $\leq_\phi$ are identical, and we can assert that the derived network associated with the mapping μ preserves the distinctions among relations encoded in the factor A.

A second means of assessing whether a derived set is associated with a homomorphic image of the semigroup of a network is to ask whether the corresponding derived network generates an algebra containing the image. That is, is the partially ordered semigroup corresponding to the homomorphism ϕ an isotone homomorphic image of the semigroup of the derived network?

In the approach to be outlined, we apply both of these considerations simultaneously. The necessity of the first condition, that the derived set should preserve distinctions among compounds made by the image, can be seen by considering the network $\mathbf{R} = \{A\}$ presented in Table 5.5. The semigroup S of $\mathbf{R}$ is shown in Table 5.6 together with its factors. The derived network associated with the subset $\{1, 2\}$ of nodes – $\mu(1) = 1$; $\mu(2) = 2$; $\mu(3)$; $\mu(4)$; $\mu(5)$ undefined – generates the second factor of the semigroup, yet its corresponding partial order on S does not preserve distinctions made by the factor. In particular, $A^2 = A$ in $\leq_\mu$, whereas $A^2 < A$ in the first factor of the semigroup. The derived set $\{1, 2\}$ is an integral

Table 5.5. *A network* **R** = {*A*} *on five elements*

The relation *A*
0 1 0 0 0
0 0 1 0 0
1 0 0 0 0
0 0 0 0 1
0 0 0 0 0

Table 5.6. *The partially ordered semigroup* $S(\mathbf{R})$ *of the network* **R** = {*A*} *and factors of* $S(\mathbf{R})$

	Right mult. table			
			Generator	
Label	Element	Word	*A*	Partial order
$S(\mathbf{R})$	1	*A*	2	1 0 0 1
	2	*AA*	3	0 1 0 0
	3	*AAA*	4	0 0 1 0
	4	*AAAA*	2	0 0 0 1
X	1	*A*	2	1 1
	2	*AA*	2	0 1
Y	1	*A*	2	1 0 0
	2	*AA*	3	0 1 0
	3	*AAA*	1	0 0 1

part of the cyclic structure displayed by the nodes {1, 2, 3} and represented by the second factor of the semigroup $S(\mathbf{R})$, and it cannot be separated from that structure in deriving associates of the factor corresponding to the single tie between nodes 4 and 5.

The second condition introduced is necessary to ensure that the identified derived set can generate the image in question; if it does not have that capacity, it is difficult to argue that the image characterises relational structure present in the derived network.

If the two conditions are taken together, associations of the following kind between derived sets Y of X and an isotone homomorphism ϕ of $S(\mathbf{R})$ can be proposed.

CORRESPONDENCE DEFINITION. Define an isotone homomorphism ϕ of $S(\mathbf{R})$, and its image $\phi(S(\mathbf{R}))$, to be *associated* with a derived set Y of X if

1 there is a partial function $\mu: X \to Y$;
2 $t \leq_\mu s$ implies $t \leq_\phi s$, for all $s, t \in S(\mathbf{R})$; and
3 there is an isotone homomorphism from the partially ordered semigroup $S(\mathbf{T})$ of the derived network $\mathbf{T}$ induced by μ onto $\phi(S(\mathbf{R}))$.

The definition asserts that a homomorphic image is associated with a derived set if the derived set generates an algebra containing the image and if all of the equations and orderings made on the derived set are also made in the image. It also requires that any distinctions among relations made in the image are also made on the derived set.

Now, in general, any particular image of a semigroup S will be associated with a number of different derived sets under the conditions of the Correspondence Definition. For instance, the derived set $Y = X$ is always associated with any image of S. It is useful, therefore, to seek minimal derived set associations of an image, where derived sets may be ordered according to the following definitions:

DEFINITION. Let $\mu_1: X \to Y_1$ and $\mu_2: X \to Y_2$ be partial functions. Then define a *partial ordering on derived sets* according to

$Y_2 \leq Y_1$ if there exists a partial function $\mu_3: Y_1 \to Y_2$ with $\mu_2(x) = \mu_3(\mu_1(x))$, for all $x \in X$.

DEFINITION. The *minimal derived set associations* of an image $\phi(S)$ of S are those sets Y for which

1 ϕ is associated with Y and with every Y' for which $Y \leq Y' \leq X$; and
2 if ϕ is associated with Y' and $Y' \leq Y$, then $Y = Y'$.

For instance, the derived network associated with the mapping $\mu(1) = a = \mu(2)$ and $\mu(3) = b$ is an example of a minimal derived set association for the factor A of $S(\mathbf{X})$ in Table 5.2. All further derivations of the derived network associated with μ lead to networks that fail to satisfy the requirements of the Correspondence Definition. The factor B of $S(\mathbf{X})$, on the other hand, has minimal associations with the subset $\{1, 2\}$ as well as with the derived sets $\{(13)\ (2)\}$ and $\{(1)\ (23)\}$.

As a second illustration of the definitions, consider the network $\mathbf{N} = \{L, A\}$ whose partially ordered semigroup $S(\mathbf{N})$ was analysed in chapter 4. The distinct compound relations in $S(\mathbf{N})$ are presented in Table 5.7. It was established in chapter 4 that $S(\mathbf{N})$ has three factors, corresponding to the π-relations π_4, π_5 and π_6 (Table 4.13). The isotone homomorphism corresponding to the first factor $S(\mathbf{N})/\pi_4$ of $S(\mathbf{N})$ has the partial order $\leq_\phi$ displayed in Table 5.8; the partial orderings corresponding to the derived sets $\{1, 2\}$ and $\{(134)\ (2)\}$ are shown in Table 5.9. It can be seen that all

Table 5.7. *Distinct relations generated by the network* **N**

L	*A*	*LA*	*AL*	*AA*	*LAL*
1100 1100 0010 0011	1011 1010 1001 1000	1011 1011 1001 1000	1111 1110 1111 1100	1011 1011 1011 1011	1111 1111 1111 1111

Table 5.8. *The partial order* $\leq_{\phi}$ *corresponding to the factor* $S(\mathbf{N})/\pi_4$ *of* $S(\mathbf{N})$

	Element					
Element	1	2	3	4	5	6
1	1	1	1	1	1	1
2	0	1	1	0	1	0
3	0	1	1	0	1	0
4	1	1	1	1	1	1
5	0	1	1	0	1	0
6	1	1	1	1	1	1

Table 5.9. *The partial orders corresponding to the derived sets* {1, 2} *and* {(134), (2)} *for the network* **N**

{1, 2}	{(134), 2}
111111 011010 011010 111111 011010 111111	111111 011010 011010 111111 011010 111111

three partial orders are identical; moreover, each of the corresponding derived networks generates a semigroup isomorphic to $S(\mathbf{N})/\pi_4$. Both derived sets are therefore associated with the factor according to the Correspondence Definition. Some other associations of the factor $S(\mathbf{N})/\pi_4$ with derived sets of X are listed in Table 5.10. Derived networks corresponding to minimal associations are shown in Table 5.11. Clearly, the factor describes the interrelations between block 2 and block 1 or between block 2 and block 1 in combination with other blocks. Those

Table 5.10. *Derived sets associated with the factor $S(\mathbf{N})/\pi_4$ of the semigroup $S(\mathbf{N})$*

Derived sets[a]	
{1, 2, 4}	{(134) (2)}*
{1, 2, 3}	{1, 2}*
{(13), 2, 4}	{(13) (2)}*
{(14), 2, 3}	{(14) (2)}*
{1, 2, (34)}	

[a] Those that are minimal are marked by an asterisk.

Table 5.11. *Derived networks corresponding to minimal derived set associations for the factor $S(\mathbf{N})/\pi_4$ of $S(\mathbf{N})$*

	Derived set							
Relation	134	2	1	2	13	2	14	2
L	1	1	1	1	1	1	1	1
	1	1	1	1	1	1	1	1
A	1	0	1	0	1	0	1	0
	1	0	1	0	1	0	1	0

interrelations are described by the equations and orderings that characterise the factor, namely,

$$LL = L = LA, \qquad AA = A = AL \qquad \text{and} \qquad A \leq L.$$

Examination of the derived networks corresponding to minimal derived set associations of the factor suggests that the network features corresponding to these algebraic relations are (a) the universality of Liking relations L in the region of the network defined by blocks 1 and 2 and (b) the distinction between block 2 and other blocks in the receiving of A ties: block 2 is the only block to receive none whereas block 1, for instance, receives A ties from all other blocks. This link between the factor of the partially ordered semigroup $S(\mathbf{N})$ and features of the network from which $S(\mathbf{N})$ was constructed is a quite precise statement of the network features giving rise to a particular algebraic feature.

The minimal derived set associates of the other factors of $S(\mathbf{N})$ are presented in Table 5.12, together with corresponding derived networks. The relational features giving rise to the factor $S(\mathbf{N})/\pi_5$ clearly depend on network ties involving block 3. In particular, it can be seen from Table 5.12 that block 3 expresses liking only among its own members

Table 5.12. *Minimal derived set associations and corresponding derived networks for other factors of S*(**N**)

Factor	Minimal derived set associations	Derived network	
		L	A
$S(\mathbf{N})/\pi_5$	{(14), 3}	1 1 0 1	1 1 1 0
	{(124), 3}	1 1 0 1	1 1 1 0
$S(\mathbf{N})/\pi_6$	{(13), 4}	1 0 1 1	1 1 1 0
	{(123), 4}	1 0 1 1	1 1 1 0

and there are no antagonistic relations amongst the members of block 3. For the factor $S(\mathbf{N})/\pi_6$, the ties of block 4 to other blocks are clearly paramount. As Table 5.12 indicates, block 4 is not liked by any other block although its members express liking among themselves, and it receives and expresses antagonistic ties with other blocks.

Searching for minimal derived set associations

The definition of minimal derived set associations may be used to define a search strategy for the minimal associations of any semigroup image. Suppose that we are seeking the minimal derived set associations of the isotone homomorphic image $\phi(S(\mathbf{R}))$ of the semigroup $S(\mathbf{R})$. Because X itself is always associated with $\phi(S(\mathbf{R})$, the search may begin at X. Consider all derived sets Y that are *covered* by X (i.e., for which $Y < X$ and for which $Y \leq Z \leq X$ implies $Z = Y$ or $Z = X$). Each such set Y may be checked for association with $\phi(S(\mathbf{R}))$ according to the Correspondence Definition. Define a path from X to any of the covered derived sets Y that is associated with $\phi(S(\mathbf{R}))$. Repeat the process for the endpoints of each path until no extensions to any of the existing paths can be found. The endpoints of the paths are then the minimal derived set associations for $\phi(S(\mathbf{R}))$. For instance, the outcome of the process for the factor $S(\mathbf{N})/\pi_4$ of $S(\mathbf{N})$ is illustrated in Figure 5.8.

Analysing entire networks

The factorisation technique for finite partially ordered semigroups identifies a collection of maximally independent homomorphic images of the

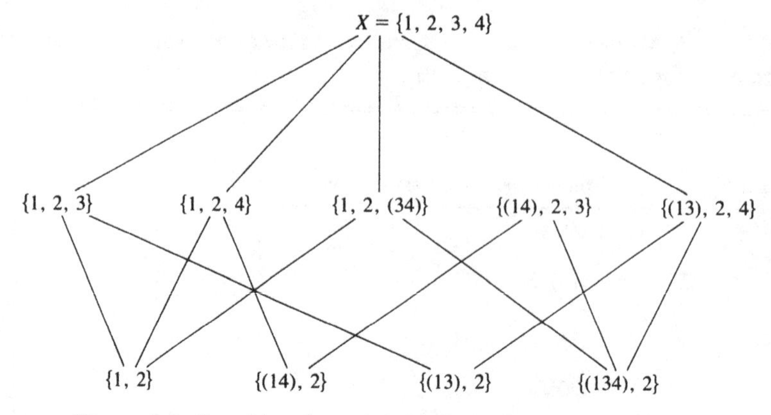

Figure 5.8. Searching for minimal derived set associations

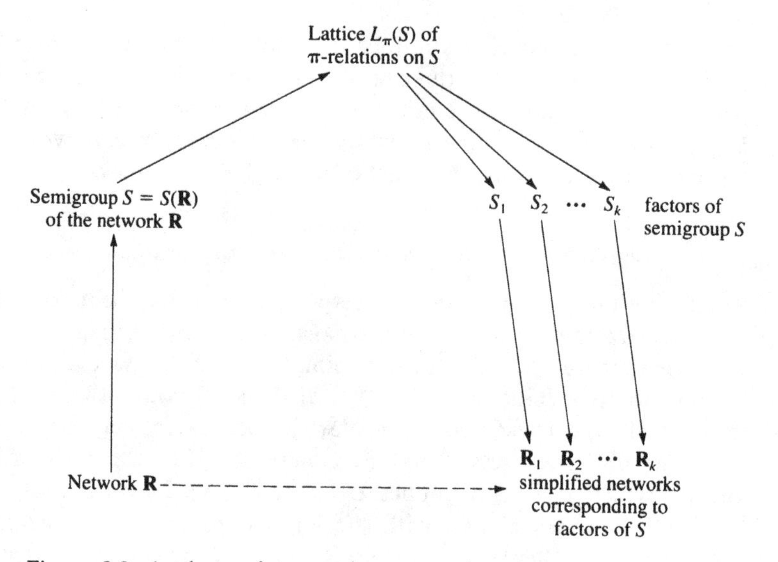

Figure 5.9. Analysis of a complete network

semigroup sufficient to describe its structure. The problem of describing features of the semigroup may, by use of the technique, be exchanged for that of describing features of its identified homomorphic images, or factors. In conjunction with the application of the Correspondence Definition, therefore, it may be used to provide an analysis of a network or blockmodel of the kind illustrated in Figure 5.9. The dashed arrow marks an interpretative path by which distinctive relational features of a network may be described. Though the path to that description is dependent upon the partially ordered semigroup representation (denoted by the solid arrows of Figure 5.9), its results may be reported in purely relational terms. The analysis suggested by Figure 5.9 thus produces a comprehensive but nonredundant account of the structural content of the system in terms of the original relational data. By contrast with many analytic strategies, the proposed analysis attempts to determine ways in which the data may be aggregated so as to emphasize its significant structural features. Like blockmodelling, therefore, it endorses "the philosophy that aggregation is to be inferred at the end of the analysis, not imposed at the beginning" (Arabie et al., 1978, p. 22).

In combination with the factorisation technique, the Correspondence Definition enables one to search for the "concrete" structural analogues of the abstract algebraic factors, or components, of the structure and so to identify the irreducible forms of relational interlock at the relational level. More importantly, perhaps, it yields a method by which to examine the ways that the various primary relational features are combined to provide an overall representation of relational structure.

An example: Relational structure in a self-analytic group

As part of a study demonstrating how one can integrate the concrete representation of social structure afforded by the blockmodel approach and the dimensional description of interpersonal relations developed by Bales and his collaborators (Bales & Cohen, 1979), Breiger and Ennis (1979) described a blockmodel for relations of Liking, Disliking and perceived Similarity among members of a self-analytical group. The blockmodel is reported in Table 5.13. Breiger and Ennis characterised the four blocks of the blockmodel in terms of their ratings on Bales and Cohen's SYMLOG dimensions and hence in terms of their typology of interpersonal behaviour. The dimensions are described as (a) Upward–Downward (*U–D*), reflecting a person's tendency towards dominant or submissive behaviour; (b) Positive–Negative (*P–N*), referring to the friendly or hostile orientation of a group member; and (c) Forward–Backward (*F–B*), indicating an orientation on a continuum from task-oriented instrumental behaviour to emotional, expressive behaviour. Block 1 was described as "type" *UP*, a group of members who were both ascendent and sociable and who were identified by Breiger and Ennis as the "positive kernel" of the group. Block 2 was of type *UN* and consisted of individuals who were dominating and hostile. Block 3 was identified as type *P*; it comprised a collection of positive but modest individuals who valued egalitarianism. Block 4 was of type *DN*, the

Table 5.13. *The Breiger–Ennis blockmodel for a self-analytical group*

Relation		
Liking	Disliking	Similarity
1 1 0 0	0 1 0 1	1 1 0 0
1 1 0 0	0 1 1 0	1 1 0 0
1 0 1 0	0 1 0 0	1 0 1 0
1 1 0 0	0 1 1 0	1 0 0 1

From Breiger & Ennis, 1979.

opposite of block 1, and was characterised by resentment of others and the rejection of social success as a salient value.

In their analysis of the relationship among the types of the four blocks and their social interrelationships, Breiger and Ennis (1979) demonstrated the usefulness of combining Bales and Cohen's (1979) dimensions with empirical data on social relations between group members. In particular, they showed that there was considerable interlocking between the description of types generated by Bales's generalised approach and the social relationships between the types summarised in the blockmodel. In the following discussion, the analysis of the partially ordered semigroup of the blockmodel is used to augment their account of the constraints between the type of a block and its embedding in a pattern of social relationships.

The partially ordered semigroup $BE1$ generated by the blockmodel of Table 5.13 is presented in Table 5.14. The π-relation lattice $L_\pi(BE1)$ of $BE1$ has three atoms, π_1, π_2 and π_3. Both π_2 and π_3 have unique maximal complements, π_6 and π_7, respectively and π_1 has the two possible maximal complements π_4 and π_5. The homomorphic images A, B, $C2$ and $C1$ corresponding to π_6, π_7, π_4 and π_5 are all displayed in Table 5.15. The full reduction diagram for $BE1$ is shown in Figure 5.10, and images of $BE1$ appearing in the diagram are reported in Table 5.16. The minimal derived set associations for some of the images in Figure 5.10 are shown in Table 5.17, and some derived networks corresponding to factors of $BE1$ appear in Table 5.18.

Table 5.17 indicates that the first factor of $BE1$ is associated with relations among blocks 1, 2 and 3, and the other factors are associated essentially with blocks 1, 2 and 4. From this it may be argued that the two dominant blocks 1 and 2 play a central role in defining the global relational structure in the blockmodel and that they interrelate with blocks 3 and 4 in essentially different ways. For instance, for blocks 1, 2 and 3,

Table 5.14. *The semigroup BE1 of the Breiger–Ennis blockmodel*

	Right mult. table			
	Generator			
Element	1	2	3	Partial order
L = 1	4	5	4	1 0 0 0 0 0 0 0 0 0 0 0 0 0
D = 2	6	7	8	0 1 0 0 0 0 0 0 0 0 0 0 0 0
S = 3	4	5	9	0 0 1 0 0 0 0 0 0 0 0 0 0 0
LL = 4	4	10	4	1 0 0 1 0 0 0 0 0 0 0 0 0 0
LD = 5	11	7	12	0 1 0 0 1 0 0 0 0 0 0 0 0 0
DL = 6	6	10	6	0 0 0 0 0 1 0 0 0 0 0 0 0 0
DD = 7	13	7	13	0 0 0 0 0 0 1 0 0 0 0 0 0 0
DS = 8	6	10	8	0 1 0 0 0 1 0 1 0 0 0 0 0 0
SS = 9	4	10	9	1 0 1 1 0 0 0 0 1 0 0 0 0 0
LLD = 10	13	7	14	0 1 0 0 1 0 1 0 0 1 0 0 0 0
LDL = 11	11	10	11	0 0 0 0 0 1 0 0 0 0 1 0 0 0
LDS = 12	11	10	12	0 1 0 0 1 1 0 1 0 0 1 1 0 0
DDL = 13	13	10	13	1 0 0 1 0 1 1 0 0 0 1 0 1 0
LLDS = 14	13	10	14	1 1 1 1 1 1 1 1 1 1 1 1 1 1

Table 5.15. *Factors of BE1*

	Right mult. table				
		Generator			
Factor	Element	1	2	3	Partial order
A	1, 3 = 1	4	5	4	1 0 0 0 0
	2, 6, 8 = 2	2	7	2	0 1 0 0 0
	4, 9 = 4	4	7	4	1 0 1 0 0
	5, 11, 12 = 5	5	7	5	0 1 0 1 0
	7, 10, 13, 14 = 7	7	7	7	1 1 1 1 1
B	1, 4, 6, 11, 13 = 1	1	5	1	1 0 0 0 1 0 0
	2 = 2	1	7	8	0 1 0 0 1 0 0
	3, 9 = 3	1	5	3	1 0 1 0 1 0 0
	5, 10 = 5	1	7	12	0 1 0 1 1 0 0
	7 = 7	1	7	1	0 0 0 0 1 0 0
	8 = 8	1	5	8	1 1 0 0 1 1 0
	12, 14 = 12	1	5	12	1 1 1 1 1 1 1
C2	1, 2, 4–14 = 1	1		1	1 1
	3 = 3	1		1	0 1
C1	1, 4, 6, 8, 9, 11–14 = 1	1	2	1	1 1 1
	2, 5, 7, 10 = 2	1	2	1	0 1 0
	3 = 3	1	2	1	0 1 1

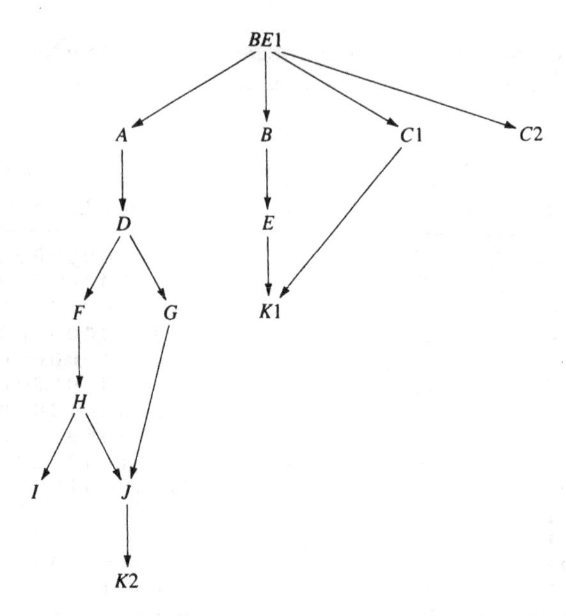

Figure 5.10. Reduction diagram for the Breiger–Ennis semigroup *BE*1

the Liking and Similarity relations are identical but intransitive, with blocks 1 and 2 forming a nucleus of individuals liked and perceived as similar. Members of block 3 like and perceive as similar other members of their own block, but they also like and see themselves as similar to the members of block 1. Thus, block 3 is somewhat peripheral to the nucleus comprising blocks 1 and 2. For blocks 1, 2 and 4, on the other hand, the relations Liking and Similarity are distinct; and Liking is a transitive relation with blocks 1 and 2 forming the "upper" level of a two-level hierarchy and block 4 the "lower" level. Though block 4 expresses Liking to blocks 1 and 2, its members do not express internal Liking ties, even though they recognize their similarities. Further, none of the positive relations expressed by block 4 to other blocks are reciprocated.

Further down the reduction diagram lie the isotone homomorphic images *K*1 and *K*2. The image *K*2 lies in the branch containing factor *A* and corresponds to relations among blocks 2 and 3. It records the property that blocks 2 and 3 have only internal Liking and Similarity ties, but that paths of Disliking ties link the members of each block to each other and to the members of the other block. Thus, in the region of blocks 2 and 3, Liking and Similarity ties are strong and are expressed only internally, whereas Disliking ties are expressed across the block boundaries as well

Table 5.16. *Other images of BE1 appearing in Figure 5.10*

Semigroup	Elements	Right mult. table: Generator 1	2	3	Partial order
D	1, 3 = 1	4	5	4	1 0 0 0 0
	2, 6, 8 = 2	2	7	2	0 1 0 0 0
	4, 9 = 4	4	7	4	1 0 1 0 0
	5, 11, 12 = 5	5	7	5	0 1 0 1 0
	7, 10, 13, 14 = 7	7	7	7	1 1 1 1 1
E	1, 3, 4, 6, 8, 9, 11–14 = 1	1	5		1 1 1 1
	2 = 2	1	7		0 1 0 1
	5, 10 = 5	1	7		0 1 1 1
	7 = 7	1	7		0 0 0 1
F	1, 3 = 1	4	5		1 0 0 0 0
	2, 6, 8 = 2	2	7		1 1 1 0 0
	4, 9 = 4	4	7		1 0 1 0 0
	5, 11, 12 = 5	5	7		1 1 1 1 0
	7, 10, 13, 14 = 7	7	7		1 1 1 1 1
G	1, 3, 4, 9 = 1	1	5		1 0 0
	2, 6, 8 = 2	2	5		0 1 0
	5, 7, 10–14 = 5	5	5		1 1 1
H	1, 3 = 1	4	5		1 0 0 0
	2, 6, 8 = 2	2	5		1 1 1 0
	4, 9 = 4	4	5		1 0 1 0
	5, 7, 10–14 = 5	5	5		1 1 1 1
I	1, 3 = 1	2	2		1 0
	2, 4–14 = 2	2	2		1 1
J	1, 3, 4, 9 = 1	1	5		1 0 0
	2, 6, 8 = 2	2	5		1 1 0
	5, 7, 10–14 = 5	5	5		1 1 1
*K*1	1, 3, 4, 6, 8, 9, 11–14 = 1	1	2		1 1
	2, 5, 7, 10 = 2	1	2		0 1
*K*2	1, 3, 4, 9 = 1	1	2		1 0
	2, 5–8, 10–14 = 2	2	2		1 1

Table 5.17. *Minimal derived set associations for some images of BE1 shown in Figure 5.10*

Image	Minimal derived set associations
A	{1, 2, 3} {1, (24), 3}
B	{1, 2, 4} {1, (23), 4}
*C*1	{1, 2, 4} {1, (23), 4} {1, 2, (34)}
*C*2	{1, 2, 4} {1, (23), 4} {1, 2, (34)} {(13), 2, 4}
*K*1	{1, (234)} {1, 2} {1, (23)} {1, (24)} {1, (34)}
*K*2	{2, 3} {(124), 3} {(12) (34)} {(24), 3}

Table 5.18. *Derived networks for associations with factors of BE1*

Factors	Derived set	Derived network		
A	{1, 2, 3}	1 1 0 1 1 0 1 0 1	0 1 0 0 1 1 0 1 0	1 1 0 1 1 0 1 0 1
A	{1, (24), 3}	1 1 0 1 1 0 1 0 1	0 1 0 0 1 1 0 1 0	1 1 0 1 1 0 1 0 1
B, C1, C2	{1, 2, 4}	1 1 0 1 1 0 1 1 0	0 1 1 0 1 0 0 1 0	1 1 0 1 1 0 1 0 1
B, C1, C2	{1, (23), 4}	1 1 0 1 1 0 1 1 0	0 1 1 0 1 0 0 1 0	1 1 0 1 1 0 1 0 1
C1, C2	{1, 2, (34)}	1 1 0 1 1 0 1 1 1	0 1 1 0 1 1 0 1 1	1 1 0 1 1 0 1 0 1
C2	{(13), 2, 4}	1 1 0 1 1 0 1 1 0	0 1 1 1 1 0 1 1 0	1 1 0 1 1 0 1 0 1

as within block 2. The image *K*1 lies in the branch containing factor *B* and has a "Last Letter" (Boorman & White, 1976; Lorrain, 1975) structure, with equations

$$LL = L = DL, \qquad LD = D = DD.$$

Its relational referents indicate that it reflects the absence of received Disliking ties for block 1, in accordance with its positive status on the *P–N* dimension.

Thus, the description of relational structure in the blockmodel suggested by the analytic scheme of Figure 5.9 may be related in a useful way to the account of the blocks given by Breiger and Ennis (1979).

Local networks

The questions that we have addressed for complete networks and their partially ordered semigroups may also be posed for local networks and their local role algebras. Under what conditions can simplifications of a local network be associated with structures nested in its local role algebra? When and how can a component of the factorisation of a local role algebra be associated with some derived local network?

As for complete networks, we examine these questions in turn. We first consider what conditions on simplifications of a local network guarantee that the resulting network has a role algebra nested in that of the original network. Then we describe a local network analogue of the Correspondence Condition, so that general associations between derived local networks and nested role algebras may be sought.

Derived local networks

Derived local networks can be constructed in a similar way to derived complete networks. Suppose that **R** is a local network on a set X, in which element 1 of X is the identified *ego*. Then a derived local network is associated with any partial function μ on X for which there is an image of element 1. Consider, for example, the network of Figure 5.1a, and think of it as a local network with node 1 as the identified *ego*. Each of the networks of Figures 5.1b, c and d may be seen as derived local networks of that of Figure 5.1a because each is a derived network in which node 1 has an image under the associated partial function μ. Figures 5.1c and d may not, however, be seen as an image of the local network of node 4 because node 4 does not have an image under the corresponding partial function.

DEFINITION. Let Y be a derived set of X corresponding to the partial function μ, and let μ be such that the identified *ego* of the local network, corresponding to the node 1 of X, has a non-null image under μ. Then the *derived local network* **T** on Y with identified element $\mu(1)$ is the network derived from **R** on X by μ.

Let n_Y be the number of nodes in Y. Then we may observe that the derived local network has relation vectors $1 * R_i$ of length n_Y with

$$(1 * R_i)[y] = 1 \text{ iff } (1, x) \in R_i \text{ for some } x \text{ such that } \mu(x) = y;\ R_i \in R;\ y \in Y.$$

The conditions under which a derived local network has a local role algebra nested in that of the original algebra have not been well studied. It is perhaps not surprising, however, to find that a number of the conditions reviewed earlier and appearing in Figure 5.6 lead to nested role algebras when applied to local derived networks. In particular, the result holds for regular equivalences and all of the conditions of which it is a generalisation.

THEOREM 5.11. *Let* P *be a partition on a set* X *on which a local network* **R** *is defined, and let* μ *be the partial function on* X *associated with* P. *If* P *is a regular equivalence, then the local role algebra of* $\mu(1)$ *in the* *derived local network* **T** *is nested in the local role algebra of node 1 in* **R**.

Proof: The proof is given in Appendix B.

It is interesting to observe that the nesting of role algebras was used by Mandel to determine whether derived networks constituted acceptable simplifications of a network. His condition was termed the *global/local criterion* and was defined as follows.

GLOBAL/LOCAL CRITERION. Let μ be a surjection of the node set X of a network **R**, and let **T** be the corresponding derived network. The partial function μ satisfies the *global/local criterion* if the local role algebra of the node $\mu(x)$ is nested in the local role algebra of x, for each $x \in X$.

For instance, the condition holds for the partition (12) (3) (4) of nodes of $X = \{1, 2, 3, 4\}$ in Figure 5.1, because the local role algebra Q_a of node a in Figure 5.1b is nested in the local role algebra Q_1 of node 1 in Figure 5.1a.

A correspondence definition for local role algebras

For local role algebras, a parallel approach to defining associations between factors of the local role algebra and regions of the local network can be constructed. That is, we can formulate a two-part correspondence definition of which one part requires that the identified region of the local network makes all of the relational distinctions characterising the factor and the other part ensures that the local region can generate an algebra containing the factor.

DEFINITION. Let Q be a local role algebra on a set X and let Y be a derived set of X associated with the partial function μ on X. Define a partial order $\leq_\mu$ according to

$s \leq_\mu t$ iff $(1 * s)[y] = 1$ implies $(1 * t)[y'] = 1$ for some y' such that $\mu(y') = \mu(y)$; for all $y \in Y$; and for any $s, t \in Q$.

Also, let T be a role algebra nested in Q. Define a partial order $\leq_T$ by

$s \leq_T t$ iff $s \leq t$ in T.

Both $\leq_\mu$ and $\leq_T$ are necessarily nested in the partial order of the role algebra Q. For instance, the local role algebra of block 1 in the network **N** (Table 2.7) has nested role algebras presented in Table 3.19. The partial order associated with the nested role algebra $T = Q_1$ of Table 3.19

Table 5.19. *The partial orders $\leq_\mu$ and $\leq_T$ for the local role algebra of block 1 of the network* **N**

$\leq_\mu$	$\leq_T$
1 1 1	1 1 1
0 1 0	0 1 0
1 1 1	1 1 1

is presented in Table 5.19, together with the partial order $\leq_\mu$ corresponding to the partial function

$$\mu(1) = a, \qquad \mu(2) = b, \qquad \mu(3) = a, \qquad \mu(4) = a.$$

Now the derived set Y preserves distinctions among relations in T if $\leq_T$ is nested in $\leq_\mu$, and we can formalise a Correspondence Definition.

CORRESPONDENCE DEFINITION FOR LOCAL NETWORKS. Define a role algebra T nested in the local role algebra Q of a local network on a set X to be *associated* with a derived local network corresponding to the partial function μ and the derived set Y if

1 $s \leq_\mu t$ implies $s \leq_T t$, for all $s, t \in Q$; and
2 T is nested in the local role algebra generated by the derived local network on Y.

One consequence of the definition is that, for each role algebra nested in a given local role algebra for a particular *ego*, the entire set X of members of *ego*'s local network satisfies the requirements of the Correspondence Definition. Indeed, one can partially order possible derived sets Y in exactly the same way as that described for the analysis for entire networks. Hence, one can search for minimal derived set associations of the factors of a local role algebra with the only procedural variation from the search outlined for entire networks being the restriction to derived sets in which *ego* has some representation.

For example, factors of the role algebra of block 1 of the network **N** may be identified with minimal derived sets of X as shown in Table 5.20. Some corresponding derived local networks are reported in Table 5.21. Factor A corresponds to the local network defined by block 1 (possibly in combination with blocks 3 and 4) and block 2, whereas factor B is associated with the local network comprising blocks 1 and 4, or blocks 1 and 3, as well as with some related combinations.

The analysis of a local network implied by the procedure is illustrated in Figure 5.11. Observe that the proposed scheme of analysis presented in Figure 5.11 bears considerable formal resemblance to that of entire

Table 5.20. *Derived set associations for the factors of the local role algebra of block 1 of* **N**

Factor	Derived set associations
A	{1, 2} {(134), 2} {(13), 2} {(14), 2}
B	{1, 3} {1, 4} {1, (34)} {(14), 3} {(13), 4} {(12), 4} {(12), 3} {(12), (34)} {(123), 4}

Table 5.21. *Derived local networks corresponding to some minimal derived set associations for the factors of the local role algebra of block 1*

		Derived local network	
Factor	Derived set	*L*	*A*
A	{1, 2}	1 1	1 0
		1 1	1 0
	{(134), 2}	1 1	1 0
		1 1	1 0
B	{1, 3}	1 0	1 1
		0 1	1 0
	{1, 4}	1 0	1 1
		0 1	1 0
	{(12), (34)}	1 0	1 1
		0 1	1 1

networks portrayed in Figure 5.9. Differences between the two schemes stem largely from the application of the global analysis to the entire network and of the local analysis to a single relation plane derived from the local network. Thus, although entire collections of matrices, representing all possible paths among all pairs of nodes, are implicitly analysed in the former scheme, the induced analysis in the latter case is of the set of role-relations in a relation plane, corresponding to paths emanating from a single fixed individual.

Several features of the analysis deserve comment. Firstly, the relationship between Mandel's (1978) two definitions of local role, elaborated separately in Mandel (1983) and Winship and Mandel (1983), is made apparent. The Mandel (1983) definition, which divides the relation plane into relation vectors (or path types), is the one adopted here for role

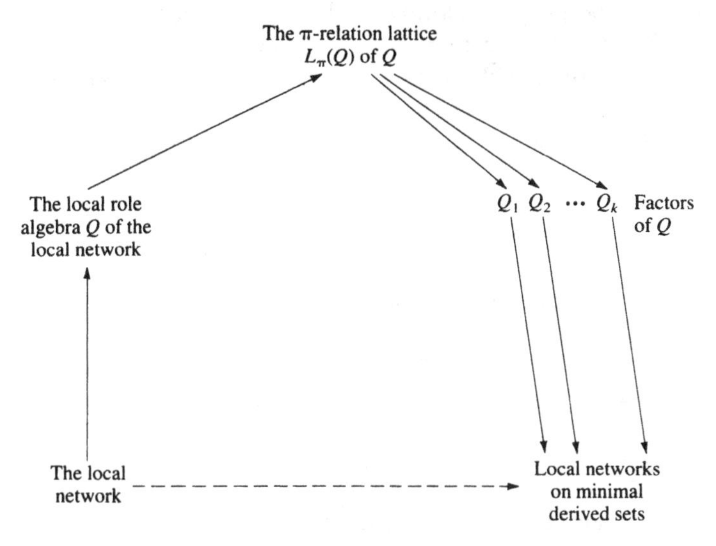

Figure 5.11. Analysis of a local network

algebra. The Winship and Mandel (1983) definition, on the other hand, characterises a relation plane in terms of its distinct role-relations; that is, it partitions the relation plane according to the relations between *ego* and other network members, recording for each member the collection of path types from *ego* to the member. The scheme proposed here relates to both of these characterisations, with an analysis of one inducing an analysis of the other.

Secondly, a comparison between Figures 5.9 and 5.11 demonstrates again the analogous analytic roles played: firstly by networks and local networks and secondly by partially ordered semigroups and role algebras. Mandel (1978) termed the role-set of an individual a "concrete" representation of local role and termed the local role algebra of an individual an "abstract" representation of role; his labelling is consistent with the idea of networks being concrete relational representations and algebras being abstract representations of role structures. Moreover, just as we have argued that both levels of representation are useful in the analysis of an entire network, so it can be argued that both local role formulations are useful. The essence of the claim is that an analysis at the more abstract level provides a means of selecting analyses at the concrete level, without which the analytic possibilities are too numerous. The search for independent components at the abstract level has the further advantage of making the concrete-level analysis an efficient one.

Thirdly, the analysis of local role algebras that we have developed makes it possible to describe local roles in terms of a number of elementary relational features. Different local roles may be compared by listing the features that they possess and thus by assessing their similarities and differences in terms of those features.

Some applications

The scheme of analysis for local networks portrayed in Figure 5.11 is illustrated in several ways. Firstly, a local role analysis is presented for each block in the Breiger–Ennis blockmodel. Then an analysis of the structure of two genuinely local networks is undertaken: the snowball network L introduced in chapter 2 and an example of a network identified by the General Social Survey network questions (Burt, 1984). Finally, the analysis is applied systematically to a class of "small" networks: namely, two-element two-relation networks.

Local roles in the Breiger–Ennis blockmodel

The local role analysis described in the preceding section is illustrated by its application to the four-blockmodel reported by Breiger and Ennis (1979) and analysed earlier from the global point of view. Here that analysis is augmented with one from the local perspective.

The local role algebras for each of the four blocks are presented in Table 5.22, and Table 5.23 identifies the factors of each local role algebra. Subsets of the block set X, corresponding to partial functions μ, for which the partial order of a factor is nested in $\leq_{\mu}$, are shown in Table 5.24.

Several conclusions may be drawn immediately from Table 5.24. Firstly, for each block, the local role algebra has a factor describing relations between itself, on the one hand, and blocks 1 and 4, on the other. That is, a component of the local role for each block is taken up with the pattern of relations to the two blocks of opposing type, and that component cannot be broken down further into a relational pattern for each block separately. Thus, the position of a block in relation to block 1 (of type *UP*) is inextricably linked with its position in relation to block 4 (of type *DN*), a finding that serves to reinforce further the conclusion by Breiger and Ennis (1979) of the strong relationship between the type of a block and its local social environment.

Secondly, all blocks except block 3 have a separate and independent component describing relations with block 3 (type *P*). Thus, for all blocks except block 3, relations with block 3 appear to be unconstrained by

Table 5.22. *Local role algebras for blocks in the Breiger–Ennis blockmodel*

Local role algebra	Elements	Right mult. table Generator 1	2	3	Partial order
A_1	$L, S = 1$	1	3	1	1 0 0 0 0 0 0
	$D = 2$	1	4	5	0 1 0 0 0 0 0
	$LD = 3$	6	4	7	0 1 1 1 0 0 0
	$DD = 4$	6	4	6	0 0 0 1 0 0 0
	$DS = 5$	1	3	5	1 1 0 0 1 0 0
	$LDL = 6$	6	3	6	1 0 0 1 0 1 0
	$LDD = 7$	6	3	7	1 1 1 1 1 1 1
A_2	$L, S = 1$	1	3	1	1 0 0 0 0
	$D = 2$	4	2	4	0 1 0 0 0
	$LD = 3$	4	2	5	0 1 1 0 0
	$DL = 4$	4	3	4	1 1 0 1 0
	$LDS = 5$	4	3	5	1 1 1 1 1
A_3	$L, S = 1$	3	4	3	1 0 0 0 0 0 0 0 0
	$D = 2$	5	6	5	0 1 0 0 0 0 0 0 0
	$LL = 3$	3	7	3	1 1 1 0 1 1 0 0 0
	$LD = 4$	5	6	8	0 1 0 1 0 0 0 0 0
	$DL = 5$	5	7	5	0 1 0 0 1 0 0 0 0
	$DD = 6$	3	6	3	0 1 0 0 0 1 0 0 0
	$LLD = 7$	3	6	9	0 1 0 1 0 1 1 0 0
	$LDS = 8$	5	7	8	0 1 0 1 1 0 0 1 0
	$LLDS = 9$	3	7	9	1 1 1 1 1 1 1 1 1
A_4	$L = 1$	1	4	1	1 0 0 0 0 0 0
	$D = 2$	5	2	5	0 1 0 0 0 0 0
	$S = 3$	1	4	6	0 0 1 0 0 0 0
	$LD = 4$	5	2	7	0 1 0 1 0 0 0
	$DL = 5$	5	4	5	1 1 0 0 1 0 0
	$SS = 6$	1	4	6	1 0 1 0 0 1 0
	$LDS = 7$	5	4	7	1 1 1 1 1 1 1

relations with other blocks. For block 3 itself, its internal relations are tied to its relations to block 2, suggesting a special relationship between blocks 2 and 3, which was also inferred by Breiger and Ennis (1979). This link between block 2 and block 3 relations is identified only in the local role algebra for block 3; in the local role algebras for blocks 1 and 2, block 2 is seen in the same undifferentiated terms, and block 4 has a separate component describing its relations to block 2.

Thirdly, some common components appear in the local role algebras for the four blocks. Blocks 2 and 3 have identical components describing their relations with blocks 1 and 4 (factor D). Thus, the blocks of type

Table 5.23. *Factors of the local role algebras of Breiger–Ennis blocks*

Local role algebra	Factor	Elements	Right mult. table: Generator 1	2	3	Partial order
A_1	A	1, 6 = 1	1	2	1	1 0 1 0
		2, 3 = 2	1	4	5	0 1 1 0
		4 = 4	1	4	1	0 0 1 0
		5, 7 = 5	1	2	5	1 1 1 1
	B	1, 2, 5 = 1	1	3	1	1 0
		3, 4, 6, 7 = 3	3	3	3	1 1
A_2	C	1 = 1	1	2	1	1 0
		2–5 = 2	2	2	2	1 1
	D	1, 4 = 1	1	3	1	1 1 0 0
		2 = 2	1	2	1	0 1 0 0
		3 = 3	1	2	5	0 1 1 0
		5 = 5	1	3	5	1 1 1 1
A_3	D	1, 3, 5 = 1	1	4	1	1 1 0 0
		2, 6 = 2	1	2	1	0 1 0 0
		4, 7 = 4	1	2	8	0 1 1 0
		8, 9 = 8	1	4	8	1 1 1 1
	E	1 = 1	3	2	3	1 0 0
		2, 4, 5, 8 = 2	2	3	2	0 1 0
		3, 6, 7, 9 = 3	3	3	3	1 1 1
A_4	C	1, 3, 6 = 1	1	2	1	1 0
		2, 4, 5, 7 = 2	2	2	2	1 1
	F	1, 5 = 1	1	4	1	1 1 0 0
		2 = 2	1	2	1	0 1 0 0
		3, 6, 7 = 3	1	4	3	1 1 1 1
		4 = 4	1	2	3	0 1 0 1
	G[a]	1, 2, 4–7 = 1	1	1	1	1 1
		3 = 3	1	1	1	0 1

[a] The factor G was not unique.

UN and of type *P* appear to relate to the blocks of opposing types *UP* and *DN* in the same way. Similarly, it can be seen that blocks 2 and 4 have the same component describing their relations to block 3 (factor *C*). From Table 5.24, the nature of these common components may be inferred. In relation to blocks 1 and 4, blocks 2 and 3 do not distinguish between Liking and Similarity, but for Disliking, they do not distinguish between blocks 1 and 4. (Thus, blocks 2 and 3 like and claim similarity to block 1 but not block 4, but they express dislike to neither block.) For the

Table 5.24. *Minimal subsets for which factors of the Breiger–Ennis local role algebras are nested in the subset partial order*

Local role algebra	Factor	Subset
A_1	A	{1, 4}
	B	{1, 3}
A_2	C	{2, 3}
	D	{2, 1, 4}
A_3	D	{3, 1, 4}
	E	{3, 2}
A_4	C	{4, 3}
	F	{4, 1}
	G	{4, 2}

relations from blocks 2 and 4 to block 3, Liking and Similarity ties are both absent and Disliking ties are present. In other words, blocks 2 and 4 adopt a similar stance in expressing resentment of the friendly and sociable role of block 3.

The other local role algebras in Table 5.23 may also be examined to add descriptive detail to the nature of other inter-block relations. Factor *A* deals with the relations of block 1 to itself and its opposite type and shows the disjunction between Liking and Similarity relations (expressed to itself and not block 4) and Disliking relations (expressed to block 4 and not itself). Factor *B* shows that block 1 ignores block 3 and is connected with it only through third parties. Factor *E* describes the relations of block 3 to itself and to block 2. In particular, Liking and Similarity ties are expressed to its own block members and not to those of block 2, whereas the reverse is true of Disliking ties. Block 2 can be reached by Liking ties through a third party, however, a feature that distinguishes this local role algebra component from factor *A*.

The relations of block 4 with block 2 and with block 1 are described by factors *G* and *F*, respectively. Block 4 does not claim Similarity with block 2 but expresses both positive and negative affect towards it. In relation to block 1, block 4 feels similar to itself and to block 1 members, and expresses Liking ties to only block 1 members and Disliking ties to neither.

From this description of local role algebras in the Breiger and Ennis (1979) blockmodel, several general points can be adduced. Firstly, the identification of similar components in different local role algebras can be made readily and is a useful feature of the descriptive procedure.

Table 5.25. *A local network from General Social Survey items*

	Relation *A* (association)							Relation *C* (close relation)					
	ego	*A*	*B*	*C*	*D*	*E*		*ego*	*A*	*B*	*C*	*D*	*E*
ego	0	1	1	1	1	1	*ego*	0	1	0	1	0	1
A	1	0	0	1	1	0	*A*	1	0	0	0	0	0
B	1	0	0	0	0	0	*B*	0	0	0	0	0	0
C	1	1	0	0	1	0	*C*	1	0	0	0	1	0
D	1	1	0	1	0	1	*D*	0	0	0	1	0	0
E	1	0	0	0	1	0	*E*	1	0	0	0	0	0

Adapted from Burt, 1984.

Secondly, the interdependence of role-relations is made explicit (e.g., the relations of each block with blocks 1 and 4 are linked with each other but are relatively independent of other role-relations). Thirdly, each local role algebra component is associated with a partial right multiplication table, a partial order and a collection of role-relations, all of which may be used to describe the component in considerable detail. This detail can be used to document further the relationship between social role and behavioural type initially demonstrated by Breiger and Ennis.

A General Social Survey network

Table 5.25 reports a local network similar to one used by Burt (1984) to illustrate some of the network data generated by the General Social Survey network items. The network reports associations and "especially close" relationships among a respondent (*ego*) and up to five of the first-named members of his or her local network. The local role algebra generated by the network of Table 5.25 is given in Table 5.26, and its reduction diagram is shown in Figure 5.12. Factors of the local role algebra and other role algebras mentioned in Figure 5.12 are listed in Tables 5.27 and 5.28, and minimal subsets having factors of the role algebra nested in their corresponding partial orders are given in Table 5.29.

It can be seen from Table 5.29 that *ego* and network members C and *D* feature in most of the minimal derived set associations, and that the interrelations of these three with members *E*, *B* and *A* are associated with the three respective factors. The third factor is also associated with relations among *ego*, *A*, *C* and *B*.

The factors *U* and *W* have the property that *ego* * *AA* is the maximal

Table 5.26. *The local role algebra of the General Social Survey network*

	Right mult. table		
	Generator		
Elements	1	2	Partial order
$A = 1$	3	4	1 1 0 0 0 0 0
$C = 2$	5	6	0 1 0 0 0 0 0
$AA = 3$	7	3	0 1 1 1 1 1 0
$AC = 4$	7	3	0 0 0 1 0 1 0
$CA = 5$	7	3	0 0 0 1 1 1 0
$CC = 6$	7	2	0 0 0 0 0 1 0
$AAA = 7$	7	3	1 1 1 1 1 1 1

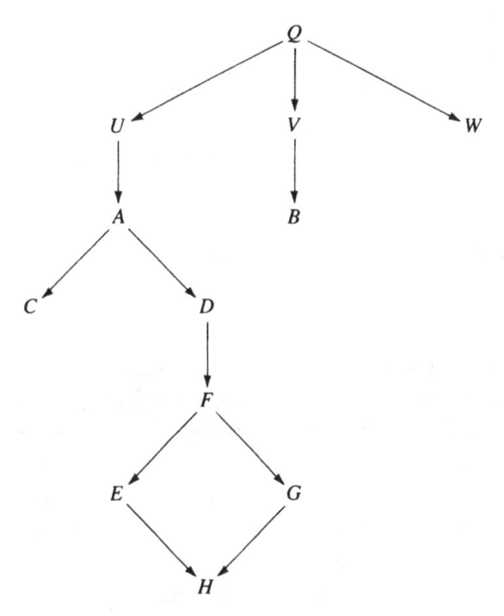

Figure 5.12. Reduction diagram for the local role algebra of the GSS network

relation vector. *Ego* is associated directly with each of D and C and indirectly with each of E and A through an intermediary; thus, *ego* has associates in common with any of D, C, E and A. Indeed, in factor W, each of *ego*'s associates is also an associate of one of *ego*'s close friends. In factor V, however, *ego* $*$ AAA is the maximal relation vector; *ego* and B have no other associates in common.

Table 5.27. *Factors for the role algebra of the GSS network*

		Right mult. table		
		Generator		
Role algebra	Elements	1	2	Partial order
U	1	3	4	1 1 0 0 0
	2	4	6	0 1 0 0 0
	37	3	3	1 1 1 1 1
	45	3	3	0 0 0 1 1
	6	3	2	0 0 0 0 1
V	1	3	3	1 1 0 0 0
	2	3	6	0 1 0 0 0
	345	7	3	0 1 1 1 0
	6	7	2	0 0 0 1 0
	7	7	3	1 1 1 1 1
W	1	3	4	1 1 0 0
	26	3	2	0 1 0 0
	357	3	3	1 1 1 1
	4	3	3	1 1 0 1

Factor W is also distinguished by the equation

$$ego * CC = ego * C$$

and by the distinction between the relations $ego * AC$ and $ego * CA$. So, for instance, for the subnetworks corresponding to factor W, namely, $\{ego, C, D, E\}$ and $\{ego, C, D, B\}$, the same members of the network can be reached for close friends of associates as for associates of close friends. Member A, however, is an associate of a close friend but not a close friend of an associate, hence the distinction of $ego * AC$ and $ego * CA$ in the factor corresponding to subnetworks containing member A.

Thus, we can distinguish these three overlapping regions of the network in terms of at least three properties:

1. the denseness of associates in the network: are *ego*'s associates also associates of some of *ego*'s associates (as in factors U and W)?
2. the closure of close ties: do close ties exist among *ego*'s close associates (as in factor W)?
3. the interlocking of associations and close ties: do the close ties of associates of *ego* define the same group of network members as associates of *ego*'s close friends (as in factors U and V)?

Table 5.28. *Other role algebras in the reduction diagram of Figure 5.12*

Role algebra	Right mult. table			Partial order
		Generator		
	Elements	1	2	
A	1	3	4	1 1 0 0 0
	2	4	6	0 1 0 0 0
	37	3	3	1 1 1 1 1
	45	3	3	0 1 0 1 1
	6	3	2	0 0 0 0 1
B	1	3	3	1 1 0 0 0
	2	3	6	0 1 0 0 0
	345	7	3	1 1 1 1 0
	6	7	2	0 0 0 1 0
	7	7	3	1 1 1 1 1
C	137	1	4	1 1 1 1
	2	4	6	0 1 0 0
	45	1	1	0 1 1 1
	6	1	2	0 0 0 1
D	1	3	3	1 1 0 0
	2	3	6	0 1 0 0
	3457	3	3	1 1 1 1
	6	3	2	0 0 0 1
E	13457	1	1	1 1 1
	2	1	6	0 1 0
	6	1	2	0 0 1
F	1	3	3	1 1 0 1
	2	3	6	0 1 0 0
	3457	3	3	1 1 1 1
	6	3	2	0 0 0 1
G	1	3	3	1 1 0
	26	3	2	0 1 0
	3457	3	3	1 1 1
H	13457	3	3	1 1
	26	3	2	0 1

Table 5.29. *Minimal subset associations for role algebras appearing in the reduction diagram of the GSS network*

Factor	Minimal subset associations
U	{*ego*, *C*, *D*, *E*}
V	{*ego*, *B*, *C*, *D*}
W	{*ego*, *A*, *C*, *B*} {*ego*, *A*, *C*, *D*}

Table 5.30. *The local role algebra generated by the snowball network* L

Elements	Right mult. table: Generator 1	Right mult. table: Generator 2	Partial order
C = 1	3	4	1 0 0 0 0 0 0 0
F = 2	5	6	1 1 0 0 0 0 0 0
CC = 3	3	6	1 0 1 0 0 0 0 0
C*F* = 4	7	6	1 1 1 1 1 0 1 1
*F*C = 5	8	6	1 0 1 0 1 0 0 0
FF = 6	7	6	1 1 1 1 1 1 1 1
C*F*C = 7	7	6	1 1 1 0 1 0 1 1
*F*CC = 8	5	6	1 1 1 0 0 0 0 1

Such features identify structural properties of a local network, and their presence or absence can be assessed in a large number of local networks. It remains to be seen what kinds of structural properties of local networks have some substantive significance, but the method of analysis that has been developed allows that question to be addressed with some care.

The snowball network L

The network L shown in Figure 2.1 in graph form and Table 2.2 in matrix form has the local role algebra reproduced in Table 5.30; the local role algebra has the reduction diagram shown in Figure 5.13. (Factors of the role algebra and other role algebras reported in Figure 5.13 are shown in detail in Table 5.31, and some derived sets whose partial orders contain the partial order of each factor are given in Table 5.32.)

Factor A describes the property that *ego* can reach all members of the network by a path of length 2, but not by a path of length 1. Factor *B* is associated with the network domain {*ego*, *b*, *g*, *h*}, in which *ego*'s associate (node *b*) is also the associate of one of *ego*'s close friends. The factor is distinguished by the equation *ego* * *AAC* = *ego* * *A*, reflecting the fact that node *b*, who is the only associate of *ego* in this region of the network, can also be reached as the close friend of one of *ego*'s associate's associates. Paths comprising two association relations define maximal connections in this (as well as all other) regions of the network. Factor C is identified with the network region including members *c*, *d*, *e* and *f*, where associates of close friends and associates of associates define maximal connections, distinct from all others. Close friends of associates, however, are also close friends; so that adding a "close friend"

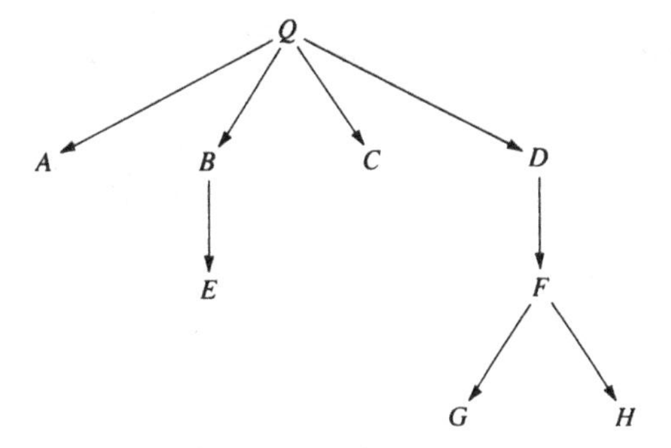

Figure 5.13. Reduction diagram for the local role algebra of the network **L**

Table 5.31. *Role algebras identified in Figure 5.13*

Role algebra	Elements	Right mult. table: Generator 1	Right mult. table: Generator 2	Partial order
A	1, 2 = 1	3	3	1 0
	3–8 = 3	3	3	1 1
B	1 = 1	2	2	1 0 0
	2–5, 7, 8 = 2	2	6	1 1 0
	6 = 6	2	6	1 1 1
C	1–3, 5, 7, 8 = 1	1	4	1 0
	4, 6 = 4	1	4	1 1
D	1, 3 = 1	1	4	1 0 0 0
	2, 8 = 2	5	4	1 1 0 0
	4, 6, 7 = 4	4	4	1 1 1 1
	5 = 5	2	4	1 0 0 1
E	1 = 1	2	2	1 0
	2–8 = 2	2	2	1 1
F	1, 3 = 1	1	4	1 0 0
	2, 5, 8 = 2	2	4	1 1 0
	4, 6, 7 = 4	4	4	1 1 1
G	1, 3 = 1	1	2	1 0
	2, 4–8 = 2	2	2	1 1
H	1–3, 5, 8 = 1	1	4	1 0
	4, 6, 7 = 4	4	4	1 1

Table 5.32. *Some derived set associations for factors of the network* L

Factor	Associated derived sets
A	$\{ego, (a, b, c, d, e, f, g, h)\}$
B	$\{ego, b, g, h\}$
C	$\{ego, (c, d), (e, f)\}$
D	$\{(ego, a, c, d, e, f, h), b, g\}$

link to a path in this region of the network does not extend the set of persons to whom *ego* is already linked by a "close friend" tie (because $ego * CC = ego * C$ and $ego * AC = ego * C$). The corresponding network region can be seen as having three components: (a) *ego*, the focal individual in the local network; (b) nodes *c* and *d*, who are close friends of *ego* as well as close friends with each other; and (c) nodes *e* and *f*, who are associates of *c* and *d*, but not directly tied to *ego*. This region of the network has the algebraic properties already described: close friend links are unable to connect *ego* to anyone other than his/her close friends, in contrast to the weaker links of association. Factor *D* is associated with relations of *ego* to members *b* and *g* and others. Associates of associates again define maximal connections in this region, as do associates of close friends; close friends of associates, however, do not.

Local role algebras for two-block two-generator models

Finally, we consider the application of the local role analysis to the collection of all two-element two-relation networks, also termed two-block two-generator (TBTG) models (Lorrain, 1973). The TBTG models systematically record all possible interrelationships between two nodes of two distinct types and so correspond to the simplest class of nontrivial interrelationships between two individuals or blocks. As such, one might expect the local role algebras for TBTG models to be largely irreducible, although cases in which they are not will present useful data for understanding how local role components can act independently.

The 57 TBTG models give rise to 114 local role algebras, 63 of which are distinct. Of these 63 distinct local role algebras, 59 are irreducible and a further 38 of those have no nontrivial nested role algebras. It is indeed the case, therefore, that for most of the elemental interaction patterns, local role algebras are indecomposable.

The four TBTG models giving rise to reducible local role algebras are presented in Table 5.33. In each case, the reduction of the local role

Table 5.33. *Reducible role algebras from two-element two-relation local networks*

	Right mult. table			
		Generator		
Local network[a]	Elements	1	2	Partial order
00 11	1	1	1	100
01 01	2	3	2	111
	21 = 3	3	3	101
10 01	1	1	2	100
11 01	2	3	2	010
	21 = 3	3	2	111
01 10	1	1	2	10
01 10	2	1	2	01
10 01	1	1	2	10
01 01	2	2	2	01

[a] *Ego* for the local network is the first network member.

algebra is associated with a decomposition of the role-set, with each role-relation being associated with exactly one factor. It is interesting to note that for all reducible local role algebras from TBTG models, the relation vectors for the two generators are complementary ones. Such a condition is not sufficient to give rise to a reducible role algebra, but it would be of interest to determine whether some analogue of it is necessary in the case of multiple blocks and/or multiple generators.

Summary

In this chapter, we have considered the question of how features of a complete or local network are related to components of the network algebra. We have reviewed some algebraic conditions under which useful connections are known to exist and we have developed a definition with which to investigate the association between components of the algebra of a network and its derived networks. The definition may be used in conjunction with the factorisation procedure of chapter 4 to obtain an efficient analysis of a complete or local network. The procedure was applied to several complete networks in the form of blockmodels, as well as to a number of local networks; and, in each case, it led to a detailed description of structure in the network.

6

Time-dependent social networks

Formal methods for dealing with change in social networks have been the subject of increasing interest among social scientists (e.g., Doreian, 1979, 1980, 1986; Hallinan, 1978; Holland & Leinhardt, 1977; Hunter, 1978; Iacobucci & Wasserman, 1988; Killworth & Bernard, 1976b; Runger & Wasserman, 1979; Wasserman, 1980; Wasserman & Iacobucci, 1988). Two distinct approaches to the development of these methods have emerged. In one, a social network is represented as a single relation matrix, and explicit "process" models for change in the constituent social relationships have been sought; for example, Wasserman's (1980) reciprocity model represents a relationship between any pair of network members as a stochastic function of characteristics of the relationship at an earlier time. In the other approach, a more complex structural representation of the social network is constructed, and the more modest goal in relation to change has been a language for the description of structural "evolution"; Doreian (1980), for example, presents an analysis of the changing structure in a group that is implied by Atkin's (1977) representation of structure using combinatorial topology.

The coupling of more complex structural representations for social network data and more descriptive approaches to accounts of their changing structure is not accidental. Not only are the analytic difficulties in developing complex structural models for change much more forbidding – because, in general, potential models are more numerous and undoubtedly less tractable – but also the structural models are usually constructed on the basis of some continuity in time. As a result, explicit structural models for change would need to be set within a macro rather than a micro time scale (Nadel, 1957). They would therefore need to reflect the impact of a number of influences external to the social system that cannot be assumed to be constant over long periods (Granovetter, 1979). Moreover, in practice, the primary objective of a structural model has been conceived as a representation of the "pathways" or "orbits" for social events (Nadel, 1957, p. 129), that is, as the "backcloth" for the flow of social "traffic" (Atkin, 1977; Doreian, 1980). The structure of the social system is not assumed to be invariant over long periods nor unchanged by the processes or traffic that it supports. Mechanisms for structural change are therefore best sought in the processes as well as in the prevailing structural patterns, so models for change require great complexity. As White, Boorman and Breiger (1976) argued, one must eventually be able to show how concrete social processes "shape and are shaped by" structure.

The aim of this chapter is to explore some descriptions of structural change that follow from the partially ordered semigroup and local role algebra representations for structure in social networks. In view of the difficulties in constructing models for structural change, which have just been outlined, a descriptive approach is adopted. It is one, however, that is intended to be sensitive to the long-term demands of a joint account of changing structure and process. Thus, in the first instance, the goal is simply a language for change, but it is a language in terms of which questions about the mechanisms of structural change may usefully be posed.

An example of such a question of change is the problem of "recruitment" of persons into roles and relationships (Nadel, 1957): how does a person with a particular set of characteristics and experiences come to adopt a particular pattern of relationships with those around? A related question is "the ancient problem of descent", that is, the problem of establishing the mechanisms for assimilating persons new to a social organization into "the social identities of their predecessors" (White, 1970, p. 329). One may hope that any useful description of change would be one in which these questions can be both framed and empirically addressed.

A language for change

The language for change that is associated naturally with the methods of analysis developed in the preceding chapters may be described in general terms as follows. Suppose that the representation of a structure at time 1 is the algebra S_1 and that the corresponding representation of structure at time 2 is S_2 (where S_1 and S_2 may be either partially ordered semigroups or local role algebras). Then, at each time, the algebraic structure may be decomposed into its components, and the relational referents of each component may be identified. The set of algebraic components that the two structures share yields a detailed description of those features of the structure that have remained invariant from time 1 to time 2, and the relational referents of those components locate the source of that invariance. Correspondingly, those components that are unique to one of the two structures provide a description of structural features that have been lost (in the case of S_1) or gained (in the case of S_2), and the relational source of the change is also made explicit. We may note a number of features of this proposed description of change.

Firstly, change is not assumed to occur uniformly throughout a network system; instead, it can be traced to a particular collection of network links. The fine-grained description of change that ensues is a necessary first step in the task of relating aspects of social processes and external events to structural change.

Secondly, change in a network link does not necessarily produce structural change. The results of earlier chapters document some of the conditions under which distinct systems at the relational level have identical algebraic structures. For example, increasing the size of a system by adding individuals who are structurally equivalent to individuals already in the system has no effect on the semigroup of the system. Similarly, adding individuals to a network so that their role-relation with respect to a given individual is the same as that of an existing individual, leaves that given individual's local role algebra unchanged. The identification of all such relational changes for which the algebraic representation is invariant is one of the mathematical tasks raised by the models and has been considered in chapter 3.

Thirdly, the language for change is defined by isotone homomorphic mappings in the case of semigroup algebras and by the nesting relation in the case of role algebras. In both cases, the relevant mappings between structures provide a way of comparing a more complex structural representation with a simpler one.

Such mappings have played a part in a number of speculations about structural change. Boyd (1969), for example, hypothesised in relation to group representations of marriage class systems that "if a group $G1$ evolves into a group $G2$, then $G1$ will be a homomorphic image of $G2$" (p. 139). For more general systems of social relations, he offers a relational mechanism that might prompt such structural "growth" (Boyd, 1980). White (1970) suggested a related but somewhat more general perspective on the development of structure when he argued

> One plausible hypothesis is that a structure of positions emerges as the skeleton deposited by, that is the residue in cultural terms from, repetitive enactment of orderly networks of relations among men. . . . Homomorphic images of relation mappings on a population may constitute the essential skeleton of social processes (pp. 329–30).

Some relational conditions for smooth change

Another mathematical task associated with this description of change is the enumeration of conditions under which change is smooth. Here *smooth* may be defined in a number of ways, but one possibility is that smooth change is reflected by algebraic extension, that is, by the structure at time 1 being a homomorphic image of the structure at time 2 (as Boyd, 1969, argued for marriage class systems). In chapter 5, we reviewed some of the relational conditions under which the semigroup of one network is a homomorphic image of the semigroup of another. For instance, the indegree and outdegree conditions, regular equivalence and the central representatives condition were interpreted as generalisations of the structural equivalence condition. If we think of one of the networks as preceding the other in time, then we can obtain from these conditions some useful schematic descriptions of smooth structural change.

Suppose, for example, that a network on a set X is expanded by adding new members to X and/or new ties between members. If the additions are made so that the new network is structurally equivalent to the old, then the relational structure of the two systems is the same; that is, no structural change has occurred. If, however, the changes occur so that the new system satisfies any of the indegree, outdegree or central representatives conditions and that the associated derived relational structure is isomorphic to the old one, then the structure of the new system is guaranteed to be an extension of that of the old. That is, the semigroup of the new system contains the semigroup of the old as an isotone homomorphic image. Thus, the conditions describe some of the ways in which elements and ties between elements can be added to a network without destroying its prevailing structural distinctions, although with the possible creation of new ones.

Moreover, the various conditions distinguish different ways in which this type of change may occur. In the indegree and outdegree cases, for instance, elements may be added so that their relationships are consistent with those of some existing element (in the sense described by the conditions). In the central representatives case, though, the new elements are added in ways that are strictly constrained by the ties of existing elements, that is, by their central representatives.

Similarly, conditions that guarantee the nesting of the role algebras of two role-sets can be interpreted as conditions for smooth structural change. The conditions include the local analogue of the regular equivalence condition, as well as any condition that leaves the original role-set as an intact subset of the new one.

All of these conditions provide descriptions of the ways in which structural change can be smooth, if the meaning of *smooth* is derived from the particular structural description that we have constructed. The question of whether change is, in fact, smooth in this sense is an empirical one, and in the next section we examine how the question may be addressed in empirical terms by illustrating how the analytic methods that have been developed may be applied to some time-dependent relational data.

An analysis of time-dependent blockmodels

Boorman and White (1976) described a partition of participants in Newcomb's (1961) fraternity study, into four blocks. The group (termed "Newcomb's Fraternity, Year 1" by Boorman & White) consisted of 17 male undergraduate students of the University of Michigan who shared a house for 16 weeks and, during that time, supplied information that included weekly interpersonal attraction ratings of each person for all of the others. The exact types of data obtained, as well as accounts of the original analyses performed upon them, may be found in Nordlie (1958) and Newcomb (1961); see also the summary description in Boorman and White (1976).

The partition of the group into four blocks described by Boorman and White (1976) was produced by application of the CONCOR algorithm (Breiger et al., 1975) to Newcomb's (1961) data. (Week 13 data were chosen by Boorman and White; as they argued, any of the later weeks could have been chosen for analysis with little change in results.) The resulting blockmodel images are labelled L (Liking) and A (Antagonism), respectively, and those for Weeks 1 to 15 are presented in Table 6.1. The blockmodels were constructed by Boorman and White (1976) by using the Week 13 blocking and a cutoff density of 0.20 throughout. (No data were collected for weeks 3, 4 and 10.)

The semigroup $S = S_{15} = S(\{L, A\})$, generated by the blockmodel for Week 15, is the semigroup $S(\mathrm{N})$ analysed in chapter 4 (Table 4.13). Its factorisation yielded three unique factors corresponding to the π-relations π_4, π_5 and π_6. The factors are shown in Table 4.13, and their minimal derived set associations are given in Tables 5.11 and 5.12.

S/π_4 is the Last Letter semigroup of order 2 (LL), whereas S/π_6 and S/π_5 have multiplication tables isomorphic to Boorman and White's (1976) target tables $T1$ and $T2$, respectively. Features of the relational structure in the Newcomb Fraternity, Year 1, Week 15 may be described by these factors and hence as follows.

Factor S/π_4. The Last Letter factor S/π_4 may be broadly interpreted as an indication that compound ties are determined by the connections of intermediaries (e.g., Breiger, 1979). More specifically, block 2 receives no antagonistic ties; hence compound ties with last letter A may be directed only to the aggregate block containing blocks 1, 3 and 4. Boorman and White (1976) argued that block 2 consists of the core

Table 6.1. *Blockmodels for Newcomb Fraternity, Year 1, Weeks 1 to 15*

	Blockmodel			Blockmodel	
Week	*L*	*A*	Week	*L*	*A*
1	1111	1101	9	1100	1011
	1111	1111		1100	1001
	1111	1001		0010	1001
	0011	1101		0011	1100
2	1111	1011	11	1100	1101
	1111	1001		1100	1011
	0111	1001		0010	1001
	1011	1001		1011	1100
5	1111	1001	12	1100	1111
	1110	1001		1100	1011
	0111	1001		0011	1001
	0011	1101		0011	1100
6	1111	1001	13	1100	1011
	0110	1001		1100	1011
	0011	1001		0010	1001
	1011	1100		0011	1100
7	1110	1011	14	1100	0011
	1100	1001		1100	1011
	0010	1001		0010	1001
	0011	1100		0011	1000
8	1110	1011	15	1100	1011
	1100	1011		1100	1010
	0110	1001		0010	1001
	0011	1100		0011	1000

Data based on Nordlie, 1958.

group of leaders of the dominant clique (whose members are persons in blocks 1 and 2). Their proposition, based upon a nonalgebraic analysis, is well supported here. Block 2 is the only block to whom no antagonism is directed; this fact clearly emerges from the complementary pair of basic structural units: the semigroup factor S/π_4 and the blockmodel images of L and A induced by the corresponding minimal derived sets.

Factor S/π_6. The $T1$ factor S/π_6 is characterised by the equations $LL = L$ and $LA = A$ and by the behaviour of AA as a left and right zero element. L is, in fact, idempotent and so is transitively closed, in the original semigroup S_{15}. Boorman and White (1976) discussed the implications of the equation $LA = A$ although their discussion was abstract in the sense that the equation was discussed independently of its associated network features. Table 5.12 shows the network features giving rise to this equation. Specifically, it suggests that the equation $LA = A$ results at least partially from the combination of facts that

1 block 3 sends no L ties to the aggregate block and no A ties to itself, and
2 block 3 sends L ties to itself and A ties to the aggregate block.

Together, (1) and (2) suggest the relatively high standing of block 3 in the group (Boorman and White's other core group of leaders). A comparison of the blockmodels corresponding to the partitions (134) (2) and (124) (3) indicates the major differences in the social operations of the two leader blocks: block 2 directs L ties to blocks other than itself and receives no A ties, whereas block 3 sends no L ties to blocks other than itself and receives a number of A ties. Block 2 has wider involvement in the positive affective aspects of the group life and is more generally popular.

Factor S/π_5. The $T2$ factor is characterised by the equations $LL = L$ and $AL = A$ and by the behaviour of AA as a zero element. The derived network associated with S/π_5 is, in fact, the converse of that associated with S/π_6 (see Table 5.12). The sole distinction between the two networks lies in the direction of the interblock L tie. Block 3 likes only itself, and the only block to like block 4 is itself. Black 4, therefore, is socially peripheral, consistent with Boorman and White's (1976) characterisation of its members as the "hangers-on to the subordinate clique" (p. 1432). The joint isotone homomorphic image of S/π_6 and S/π_5 is isomorphic to the semigroup comprising an identity element and a zero element. It is the semigroup that was argued by Breiger and Pattison (1978) to represent the interrelationship between strong and weak ties (Granovetter, 1973). In this image, L acts as the identity and A as the zero, pointing to the strength of L ties relative to A ties. In fact, this image has a minimal derived set association with the block partition (12) (34), which is the clique division to which Boorman and White (1976) refer: when considered from the perspective of the clique structure of the group, L indeed appears in its strongest form.

The development of relational structure

Boorman and White (1976) reported the occurrence of their eight target tables, $T1$ to $T8$, as images of the semigroups for Weeks 1 to 15 of the Newcomb, Year 1, data. Their analysis is extended here to examine how the "surface" structural, or relational, analogues of the factors of the

Table 6.2. *Incidence of factors of Week 15 semigroup as images of semigroups for earlier weeks*

	Semigroup image[a]			
Week	S/π_4	S/π_5	S/π_6	S/π_7
15	x	x	x	x
14	x	x	x	x
13	–	x	x	x
12	–	x	x	x
11	–	x	x	x
9	–	x	x	x
8	–	x	x	x
7	–	x	x	x
6	x	–	–	–
5	x	–	–	–
2	x	–	–	–
1	–	–	–	–

[a] x, image present; –, image absent.

Week 15 semigroup vary over the same time span. The occurrence of the factors S/π_4, S/π_5 and S/π_6 of the Week 15 semigroup as images of the semigroups of earlier weeks is summarised in Table 6.2, and corresponding minimal partitions and their derived networks are presented in Table 6.3. Data are also presented for the image S/π_7, where π_7 is the greatest lower bound (intersection) of π_5 and π_6. As Table 6.2 indicates, π_5 and π_6 always occur together and hence so does their intersection.

One immediate conclusion may be drawn from Tables 6.2 and 6.3. The stability of the factors over time is indeed a deep structural phenomenon; small surface realignments occur, but in a way entirely consistent with the underlying equations. Results of this kind will hopefully encourage tentative suggestions about the mechanisms of change. Once descriptions of structure and its development are of sufficient and relevant detail, consistent patterns associated with change might be expected to emerge.

Table 6.3 indicates that, with one exception, the factor S/π_5 is associated with the configuration corresponding to the partition (123) (4) in Week 15. It appears, therefore, that the position of block 4 in the group and its effect upon the total group structure is quite stable from Week 7 onward. By contrast, however, there is considerable variation among the relational associates of the factor S/π_6. Block 3, whom this factor largely concerns, could be the cause, a proposition that is not inconsistent with its position as the nondominant leader group. The negotiability of its

Table 6.3. *Minimal partitions associated with identified images of S_{15} and corresponding derived networks*

Image	Minimal associated partition	Corresponding derived network			Weeks
π_4	(134) (2)	(134)	1 1	1 0	2, 14, 15
		(2)	1 1	1 0	
π_4	(124) (3)	(124)	1 1	1 0	5, 6
		(3)	1 1	1 0	
π_5	(123) (4)	(123)	1 0	1 1	7, 8, 9, 11
		(4)	1 1	1 0	13, 14, 15
π_5	(12) (3) (4)	(12)	1 0 0	1 1 1	12
		(3)	0 1 1	1 0 1	
		(4)	0 1 1	1 0 0	
π_6	(124) (3)	(124)	1 1	1 1	7, 9, 13,
		(3)	0 1	1 0	14, 15
π_6	(14) (2) (3)	(14)	1 1 1	1 1 0	1 1
		(2)	1 1 0	1 0 1	
		(3)	0 0 1	1 0 0	
π_6	(13) (2) (4)	(13)	1 1 0	1 0 1	8
		(2)	1 1 0	1 0 1	
		(4)	1 0 1	1 1 0	
π_6	(12) (3) (4)	(12)	1 0 0	1 1 1	12
		(3)	0 1 1	1 0 1	
		(4)	0 1 1	1 0 0	
π_7	(12) (3) (4)	(12)	1 0 0	1 1 1	9, 13, 14
		(3)	0 1 0	1 0 1	15
		(4)	0 1 1	1 0 0	
π_7	(13) (2) (4)	(13)	1 1 0	1 0 1	7, 8, 9, 13
		(2)	1 1 0	1 0 1	
		(4)	1 0 1	1 1 0	
π_7	(12) (3) (4)	(12)	1 0 0	1 1 1	12
		(3)	0 1 1	1 0 1	
		(4)	0 1 1	1 0 0	
π_7	(12) (3) (4)	(12)	1 1 0	1 1 1	7
		(3)	0 1 0	1 0 1	
		(4)	0 1 1	1 0 0	

standing is also reflected by the partition connected to the factor S/π_4 for Weeks 5 and 6: for those weeks, but only those, block 3 received no antagonistic ties.

For the image S/π_7, which is reducible to S/π_6 and S/π_5, there exist variations not only in the partitions with which it is associated but also in the relational configurations corresponding to a given partition. For example, the partition (12) (3) (4) has three different corresponding networks between Weeks 7 and 15. Some of the deviations in the relations are the same as those responsible for fluctuations in *T*1 configurations; others are not. It can be seen, incidentally, that the consideration of factors has permitted the separation of at least one invariance (the position of block 4) from the much less clear picture associated with the semigroup S/π_7 and has also identified the detailed relational features connected with the latter.

An interesting feature of the three configurations associated with the partition (12) (3) (4) is that each has the same pattern. Variations occur only in the *L* relation and, moreover, only in *L* connections involving block 3. One may conjecture, therefore, that block 3 is the most active participant in the process of change during the later weeks, but that the types of change in which it is involved appear to be regulated by the invariant semigroup image S/π_7. Further, the Antagonism between blocks, from the perspective of the partition (12) (3) (4), is negotiated by Week 7, whereas Liking relationships undergo additional change. Antagonism is an invariant for this partition and Liking is more fluid, but the fluidity of the latter is once again bounded by the semigroup S/π_7 (with the exception of Week 11). All changes in Antagonism from Week 7 may thus be referred to the individual blocks 1 and 2; the only significant change is associated with the appearance of the factor S/π_4 at Week 14.

Thus, it has been possible to analyse in some detail the evolution of the factors appearing at Week 15. It has been demonstrated that relational fluctuations occur, though not uniformly, and that they may be constrained by underlying structural considerations. The significant changes have been able to be identified and those with structural import separated from those that appear to have a more homeostatic character. Contrast, for example, the *A* relations directed to block 2 for Weeks 13 and 14 with the *A* relations received by block 3 for the same weeks. Changes in the first case are associated with the appearance of the factor S/π_4, but in the second with only small and concrete alterations in the realization of the images S/π_5 and S/π_7. The analysis illustrates the detail with which relational changes may now be described and establishes the means for investigating the relationship between relational change and other individual- and group-related indicators of change (such as changes in attitudes).

A similar analysis has been conducted by Vickers (1981) for the social structure of a school classroom over three consecutive years. Vickers (1981) gathered relational data on Liking and Antagonism among members of an essentially intact classroom group in their seventh, eighth and ninth years of schooling. By a detailed analysis of the composition of blocks in

Table 6.4. *Local role algebras for blocks in the Newcomb blockmodel at Week 15*

	Right mult. table			
		Generator		
Block	Elements	1	2	Partial order
1	$L = 1$	1	2	1 0 0
	$A = 2$	3	2	0 1 0
	$AL = 3$	3	2	1 1 1
2	$L = 1$	1	3	1 0 0 0 0
	$A = 2$	4	3	0 1 0 0 0
	$LA = 3$	5	3	0 1 1 0 0
	$AL = 4$	4	3	1 1 0 1 0
	$LAA = 5$	5	3	1 1 1 1 1
3	$L = 1$	1	2	1 0 0 0
	$A = 2$	3	4	0 1 0 0
	$AL = 3$	3	4	1 1 1 1
	$AA = 4$	3	4	1 1 0 1
4	$L = 1$	1	3	1 0 0 0 0 0
	$A = 2$	4	5	0 1 0 0 0 0
	$LA = 3$	6	5	0 1 1 0 0 0
	$AL = 4$	4	5	0 1 0 1 0 0
	$AA = 5$	6	5	1 1 1 0 1 0
	$LAL = 6$	6	5	1 1 1 1 1 1

the blockmodel for each year, and by an analysis of the changing structure of the semigroup of the blockmodels over the three years, Vickers established the existence of a substantial number of structural features that were invariant over the three years. She also identified several instances of structural change, instances that were plausible in the light of additional interview data from members of the classroom.

A local role analysis of time-dependent blockmodels

A local role analysis of the blockmodel for Week 15 of the Newcomb fraternity data just analysed yields some similar conclusions to the analysis of the semigroup representation but serves to emphasize relations between individual blocks to a greater extent. It is presented in summary form to provide an additional comparative base for the local and entire network frameworks. The local role algebras for blocks of the Week 15 blockmodel are presented in Table 6.4, and their factors are shown in Table 6.5.

Table 6.5. *Factors of the local role algebras for the Week 15 blockmodel*

		Right mult. table			
			Generator		
Block	Factor	Elements	1	2	Partial order
1	*A*	1, 3 = 1	1	2	1 1
		2 = 2	1	2	0 1
	B	1 = 1	1	2	1 0
		2, 3 = 2	2	2	1 1
2	*A*	1, 4, 5 = 1	1	2	1 1
		2, 3 = 2	1	2	0 1
	B	1 = 1	1	2	1 0
		2–5 = 2	2	2	1 1
	C	1, 2, 4 = 1	1	3	1 0
		3, 5 = 2	3	3	1 1
3	*D*	1 = 1	1	2	1 0 0
		2 = 2	3	3	0 1 0
		3, 4 = 3	3	3	1 1 1
	E	1 = 1	1	2	1 0 0
		2, 4 = 2	3	2	1 1 0
		3 = 3	3	2	1 1 1
4	*D*[a]	1 = 1	1	2	1 0 0
		2, 3 = 2	4	4	0 1 0
		4, 5, 6 = 4	4	4	1 1 1
	E	1 = 1	1	2	1 0 0
		2, 3, 5 = 2	4	2	1 1 0
		4, 6 = 4	4	4	1 1 1
	F	1, 3, 5, 6 = 1	1	1	1 1
		2, 4 = 2	2	1	0 1

[a] The factor *D* was not unique.

Table 6.6 presents the incidence of factors of the Week 15 local role algebras as algebras nested in those at earlier times.

The correspondences of the factors with network regions is not presented in detail, but the analysis reinforces the conclusions drawn from the preceding analysis. Firstly, as before, minor relational variations are observed in the presence of greater structural stability, giving some weight to the algebraic representation as a robust description of interblock relations.

Table 6.6. *Incidence of Week 15 role algebra factors in earlier weeks*[a]

	Block 1 factors		Block 2 factors			Block 3 factors		Block 4 factors		
Week	*A*	*B*	*A*	*B*	C	*D*	*E*	*D*	E	*F*
1	–	–	–	–	–	–	–	–	–	–
2	x	–	x	–	–	–	–	–	–	–
5	x	–	x	–	–	–	–	–	–	–
6	x	–	x	–	–	–	–	–	–	–
7	–	x	–	x	–	x	–	x	x	–
8	–	x	–	x	–	x	–	–	x	–
9	–	x	–	x	–	x	–	–	x	–
11	–	x	–	x	–	x	–	–	x	–
12	–	x	–	x	–	–	–	–	x	–
13	–	x	–	x	–	x	–	–	x	–
14	–	x	x	x	–	x	x	–	x	–
15	x	x	x	x	x	x	x	x	x	x

[a] x, factor present; –, factor absent.

Secondly, the relations between each block and the block of "hangers-on to the subordinate clique" (White et al., 1976), block 4, remain relatively stable from Week 7 onward, indicating that positions with respect to that block were in place at an early time.

Thirdly, it can be demonstrated that block 1 and, to a lesser extent, block 2 have similar positions in relation to blocks 3 and 4 between Weeks 7 and 15: the relations between block 1 and block 3 are associated with the same factor as those between blocks 1 and 4 for five of the eight observations during that time. Thus, individuals in blocks 1 and 2 tend not to distinguish blocks 3 and 4 in terms of their relations. Block 3, on the other hand, tends to relate to block 4 in the same way that it does to block 1; from its position, blocks 1 and 4 stand in similar roles. It may be conjectured from this pattern of relations that the superordinate clique and its hangers-on see the other main faction in a similar and antagonistic light whereas the subordinate clique views the hangers-on to both cliques as having a similar unfavourable group position. This picture is not unlike that for the dominant and hostile block described for the Breiger–Ennis blockmodel, a suggestion that may be verified by comparing the local role algebra description of the two blocks.

Fourthly, some changes occur in the last few weeks in the relations between all blocks and blocks 1 and 2. Neither block 1 nor block 2 distinguishes Liking and Antagonism in relation to block 1, and relations with block 2 emerge in their final form only in the penultimate or final week. Blocks 3 and 4 tend to relate to block 1 in a manner that is linked with their relations to block 2, although, as noted earlier, blocks 1 and 4 are interchangeable from the perspective of block 3.

In all, a pattern of positions is found that is consistent with previous descriptions of the group. The local role analysis and the tracing of its features over the weeks preceding Week 15 give some additional support to the description and highlight the similarity of block relations from the perspectives of different block members. The analysis is clearly useful in identifying salient aspects of the Liking and Antagonism relations among different members of the group and their stability in time.

7

Algebras for valued networks

The partially ordered semigroups and local role algebras analysed in earlier chapters were constructed for complete and local networks of *binary* relations. In chapters 1 and 2, though, we also described complete and local networks that have *valued* links. In valued networks, a link from node i to node j on relation R has a value $v_R(i, j)$ indicating the value, or strength, of the relation of type R from i to j. For many date collection procedures, the values $v_R(i, j)$ may take only a finite set of possible values (and, of course, if the number of distinct possible values is just 2, the relations are binary). Both partially ordered semigroups and local role algebras may be defined for network data arising in this form, and in this chapter, we describe how to construct them. We also discuss the relationship between the algebraic structures derived from valued network data and the algebraic structures introduced in chapters 1 and 2 for binary data.

The semigroup of a valued network

For complete networks, valued relational information may be generated directly as a result of the methodology used in the network study, or it may arise as a result of some preliminary processing of the network data. The first type of data is generated when network ties are measured using a procedure that allows for ordinally scaled responses, with more than two possible response values. For instance, valued data may be obtained from questions eliciting information about network ties of the form "Indicate on the following four-point scale the degree to which you regard X as a friend", where the response scale values may be 0 = not at all, 1 = somewhat, 2 = quite strongly, 3 = strongly.

The second type of data may be obtained when a network is defined at an aggregated level, for example, at the level of "blocks" in a blockmodel. A valued network link between a pair of blocks may then be calculated as the density of the links from one block to the other. Thus, a valued network is defined by the collection of density matrices that form the intermediate stage between the original binary network data among persons and some final binary relational data among groups of persons (or blocks), as in a blockmodel. In most applications of blockmodelling, these density matrices are converted into binary relations by imposing a cutoff density value, above or equal to which densities are coded as 1 (tie present) and below which they are coded as 0 (tie absent). A cutoff value can also be used more generally to convert data in valued relational form to binary form. In either case, the binary relations produced may be used as generators of a semigroup (as in Boorman and White, 1976), providing a description of relational structure that is clearly contingent on the faithfulness of the binary relations from which it was constructed. The latter is dependent upon the existence and use of an appropriate cutoff value; in the case of blockmodel analysis, it is also contingent upon the validity of the blocks, in both their composition and number.

Although it may sometimes be appropriate to convert a valued network into a binary one using a single cutoff value, it is clearly also useful to consider whether such a simplification of the data is actually necessary for algebraic representation. Indeed, we show here that it is not. In particular, we show that we need only change the way in which we define paths in networks (i.e., the way we define the binary operation of relational composition) in order to construct the same kinds of algebraic structures as before.

Recall that the binary operation of composition in a network semigroup corresponds to the tracing of paths in the network. If R and S are two network relations, then we define a link RS from node i of the network to node j if and only if there is a path labelled RS from i to j through some node k. There may be a number of different nodes k through which such a path exists, and hence a number of different paths RS from i to j. In defining the composite relation RS, however, we simply record whether such a path exists, rather than the number of such paths. In the binary case, this procedure is consistent with recording only the presence or absence of primitive paths (i.e., paths of length one) in the network.

When the labelled links in network paths are valued, we find that labelled paths through intermediate nodes have a string of values attached. How, then, do we assign a value to the labelled and valued path RS from node i to node j? In fact, there are several ways in which the values can be assigned, and two that have been proposed in the literature (e.g., Boyd, 1989; Doreian, 1974; Peay, 1977b):

1 The value of the path RS through node k is equal to the *minimum* value of any link in the path – that is, min $(v_R(i, k), v_s(k, j))$ – and the value of the path RS over all k is equal to the largest value of path RS through any k. (Thus, $v_{RS}(i, j) =$

$\max_k\{\min(v_R(i,k), v_S(k,j))\}$.) This rule for determining the value of a labelled, valued path may be termed a *max-min* product rule.

2 The value of the path RS through node k is equal to the *product* of the values of links in the path; and over all k, the path has value equal to the sum of these values (so that $v_{RS}(i,j) = \Sigma_k v_R(i,k)v_s(k,j)$). This rule may be termed an *ordinary* product rule.

Other rules are also possible (e.g., Dubois & Prade, 1980; Peay, 1977b), but rules 1 and 2 are the ones most commonly proposed (e.g., Boyd, 1989; Doreian, 1974). The two rules differ in both their assumptions and their properties. More specifically, rule 1 assumes only ordinally scaled responses and is equivalent to relational composition when the values of the links are restricted to the set $\{0, 1\}$ (i.e., when the data are binary). Rule 2, on the other hand, assumes responses on an interval scale (so that equal differences in scale scores signify equal differences in relational values) and is equivalent to a path-counting measure when the data are binary. In addition, if we define a comparison relation among complex labelled valued paths U and V of the form

$$U \leq V \text{ iff } v_U(i,j) \leq v_V(i,j), \text{ for all } i, j \in X,$$

and an equivalence relation by

$$U = V \text{ iff } U \leq V \text{ and } V \leq U,$$

then rule 1 generates a finite number of distinct labelled valued paths in a network, whereas rule 2 may generate an infinite number. In the following development of algebraic structures for valued networks, rule 1 is selected because (a) it assumes only ordinal measurements of relation values, (b) it is equivalent to relational composition when the data are binary and (c) it can be used to generate finite algebraic structures. It should be noted, though, that for some applications, the ordinary product rule 2 may be preferred. For instance, if the values are frequency-based measures, such as densities, then the assumption of interval measurement may be reasonable and the ordinary product rule may, with some additional assumptions, be given a probabilistic interpretation.

More formally, we will assume that the complete network data comprise a collection of valued relations $\mathbf{V} = \{V_1, V_2, \ldots, V_p\}$ on a set $X = \{1, 2, \ldots, n\}$; we denote by $v_k(i,j)$ the value of the tie of type k directed from node i to node j $(i, j \in X)$ and assume that $v_k(i,j)$ is nonnegative. The max-min product V_kV_l of two valued relations V_k and V_l is then defined as

$$v_{kl}(i,j) = \max\{\min[v_k(i,1), v_l(1,j)], \ldots, \min[v_k(i,n), v_l(n,j)]\}.$$

Table 7.1. *A valued network* V = {*A*, *B*}

A		*B*	
0.83	0.17	0.33	0.67
0.33	1.00	0.17	0.00

Table 7.2. *Some max-min products for the valued relations A and B of the valued network* V

AB		*BA*		*BB*		*ABB*	
0.33	0.67	0.33	0.67	0.33	0.33	0.33	0.33
0.33	0.33	0.17	0.17	0.17	0.17	0.33	0.33

It can be argued that the max-min product of V_k and V_l represents the largest potential "flow" of paths labelled $V_k V_l$ in the valued network. In particular, $v_{kl}(i, j)$ may be interpreted as the value of the strongest path from node i to node j in any one path labelled by relation k and relation l (in that order), where the value of a path is defined as the value of its weakest constituent link. That is, if one had to choose the path whose weakest link was strongest, a path leading to a maximal value in the preceding definition would be selected and the value of the path equated to the "weakest" value. For example, the valued relations *A* and *B* of the valued network V of Table 7.1 are the density matrices for a blockmodel comprising two blocks and two relations. The compound relation *AB* generated using the max-min product rule is shown in Table 7.2, together with some other compound relations.

Given this composition rule, we may now proceed to define a partially ordered semigroup for the network in exactly the same way as before. That is, we define

1 the collection *FS*(V) of all labelled paths of finite length constructed from elements of V, with the values for each path *U* in *FS*(V) calculated using the max-min rule;
2 a comparison relation among labelled paths, namely:

$$U \le V \text{ iff } v_U(i, j) \le v_U(i, j), \text{ for all } i, j \in X$$

as well as an equivalence relation:

$$U = V \text{ iff } U \le V \text{ and } V \le U;$$

Table 7.3. *The partially ordered semigroups* $S(\mathbf{V})$ *and* $S(\mathbf{B})$ *generated by the valued network* $\mathbf{V}$ *and the blockmodel* $\mathbf{B}$

	Right mult. table				
			Generator		
	Element	Word	*A*	*B*	Partial order
$S(\mathbf{V})$	1	*A*	1	3	1 0 0 0 0 0
	2	*B*	4	5	0 1 0 0 0 0
	3	*AB*	3	6	0 1 1 1 1 1
	4	*BA*	4	5	0 1 0 1 1 0
	5	*BB*	5	5	0 0 0 0 1 0
	6	*ABB*	6	6	0 0 0 0 1 1
$S(\mathbf{B})$	1	*A*	1	2	1 0 1
	2	*B*	2	3	0 1 1
	3	*BB*	3	3	0 0 1

3 a binary operation

$$C_U C_V = C_{UV}$$

and a partial ordering

$$C_U \leq C_V \text{ iff } U \leq V$$

on the classes of the equivalence relation, where C_U denotes the class containing U.

Then it can readily be established that the max-min composition rule is associative (e.g., Kaufman, 1975) and that it has the property that

$$U \leq V \text{ implies } UW \leq VW \text{ and } WU \leq WV, \text{ for any } W \in FS(\mathbf{V}).$$

As a result,

THEOREM 7.1. *The classes of the equivalence relation just defined for valued labelled paths, together with the associated binary operation and partial ordering, form a partially ordered semigroup* $S(\mathbf{V})$.

In practice, the semigroup $S(\mathbf{V})$ may be constructed using an algorithm very similar to that presented in chapter 1 for the partially ordered semigroup of a complete binary network. The only modification to the binary network algorithm that is required is that Boolean products of relations need to be replaced by max-min products. For example, the valued network of Table 7.1 yields the distinct valued relations presented in Table 7.2. Hence the partially ordered semigroup $S(\mathbf{V})$ generated by $\mathbf{V}$ is that shown in Table 7.3.

Binary network semigroups from valued networks

The semigroup $S(\mathbf{V})$ represents the structure of labelled valued paths in the network $\mathbf{V}$. It does so without the loss of information entailed by converting the valued network data into binary form; consequently, it is likely to be complex. It is of some interest, therefore, to determine the relationship between the semigroup $S(\mathbf{V})$ and the semigroup that we would have obtained if the valued data had been converted to binary form using a cutoff of α. That is, we investigate the relationship between $S(\mathbf{V})$ and the partially ordered semigroup $S(\alpha)$ generated by the binary network $\mathbf{V}_\alpha$.

DEFINITION. Let $\mathbf{V} = \{V_1, V_2, \ldots, V_p\}$ be a valued network on a set X. For each $k = 1, 2, \ldots, p$, define the binary relation $(V_k)_\alpha$ obtained from V_k using the cutoff α:

$$(i, j) \in (V_k)_\alpha \text{ iff } V_k(i, j) \geq \alpha.$$

We write $(V_k)_\alpha (i, j) = 1$ if $(i, j) \in (V_k)_\alpha$, and $(V_k)_\alpha(i, j) = 0$, otherwise. The binary relation $(V_k)_\alpha$ obtained by applying the cutoff α to the valued relation V_k is termed the *component of* V_k *at level* α. The binary network $\mathbf{V}_\alpha$ is given by $\{(V_1)_\alpha, (V_2)_\alpha, \ldots, (V_p)_\alpha\}$.

We may express a valued relation V_k in terms of its components in the following way.

THEOREM 7.2. *Decomposition Theorem for Fuzzy Relations (e.g., Kaufman, 1975). A valued relation* V_k *may be decomposed into binary relations according to*

$$V_k = \cup\, \alpha((V_k)_\alpha)$$

where $\alpha(V_\alpha)$ *indicates that elements of* V_α *are multiplied by* α, *and where* $\cup$ *denotes the fuzzy union of valued relations, given by*

$$(A \cup B)(i, j) = max\{A(i, j), B(i, j)\}.$$

The union is over all distinct nonzero values α *taken by relations in* $\mathbf{V}$.

For example, the relations A and B of the valued network $\mathbf{V}$ of Table 7.1 have the components presented in Table 7.4.

Now if we apply the cutoff value α to *all* relations in $S(\mathbf{V})$, we induce a binary relation π_α on the elements of the semigroup $S(\mathbf{V})$. The relation is defined as follows:

DEFINITION. Define a relation π_α on $S(\mathbf{V})$ by

$$(U, V) \in \pi_\alpha \text{ iff } V_\alpha \leq U_\alpha.$$

Table 7.4. *Components of the valued relations A and B of the valued network* **V**

α	A	B
1.00	0 0 0 1	0 0 0 0
0. 83	1 0 0 1	0 0 0 0
0.67	1 0 0 1	0 1 0 0
0.33	1 0 1 1	1 1 0 0
0.17	1 1 1 1	1 1 1 0

Table 7.5 *Components of the valued relations in S*(**V**) *for the valued network* **V**

α	A	B	AB	BA	BB	ABB
1.00	0 0 0 1	0 0 0 0	0 0 0 0	0 0 0 0	0 0 0 0	0 0 0 0
0.83	1 0 0 1	0 0 0 0	0 0 0 0	0 0 0 0	0 0 0 0	0 0 0 0
0.67	1 0 0 1	0 1 0 0	0 1 0 0	0 1 0 0	0 0 0 0	0 0 0 0
0.33	1 0 1 1	1 1 0 0	1 1 1 1	1 1 0 0	1 1 0 0	1 1 1 1
0.17	1 1 1 1	1 1 1 0	1 1 1 1	1 1 1 1	1 1 1 1	1 1 1 1

The relation is termed the *filtering relation at level* α.

For instance, the components at each possible level of α of the elements of S(**V**) for the valued network of Table 7.1 are presented in Table 7.5. The corresponding filtering relations at each level are shown in Table 7.6.

Now it may readily be established that the relation π_α is actually a π-relation on S(**V**) (chapter 3) and is consequently associated with an isotone homomorphism of S(**V**). Indeed, the isotone homomorphic image to which it corresponds is simply $S(\alpha)$. That is,

Table 7.6. *Filtering relations for the semigroup* $S(\mathbf{V})$

α	π_{α}
1.00	1 1 1 1 1 1 0 1 1 1 1 1 0 1 1 1 1 1 0 1 1 1 1 1 0 1 1 1 1 1 0 1 1 1 1 1
0.83	1 1 1 1 1 1 0 1 1 1 1 1 0 1 1 1 1 1 0 1 1 1 1 1 0 1 1 1 1 1 0 0 0 0 0 1
0.67	1 0 0 0 0 1 0 1 1 1 1 1 0 1 1 1 1 1 0 1 1 1 1 1 0 0 0 0 1 1 0 0 0 0 1 1
0.33	1 0 0 0 0 0 0 1 0 1 1 0 1 1 1 1 1 1 0 1 0 1 1 0 0 1 0 1 1 0 1 1 1 1 1 1
0.17	1 1 1 1 1 1 0 1 0 0 0 0 1 1 1 1 1 1 1 1 1 1 1 1 1 1 1 1 1 1 1 1 1 1 1 1

THEOREM 7.3. *The collection* $\{S(\alpha), \alpha > 0\}$ *of semigroups generated by the binary networks* $\{(V_1)_\alpha, (V_2)_\alpha, \ldots, (V_p)_\alpha\}$ *derived from* $\{V_1, V_2, \ldots, V_p\}$ *using cutoff values* $\alpha > 0$ *comprise a set of isotone homomorphic images of the semigroup* $S(\mathbf{V})$ *of the valued network generated by the max-min product rule. The collection of binary networks generating these homomorphic images of* $S(\mathbf{V})$ *are derived from the Decomposition Theorem for Fuzzy Relations (Kaufman, 1975).*

For instance, the filtering relation $\pi_{.67}$ presented in Table 7.6 leads to an isotone homomorphism from $S(\mathbf{V})$ onto $S(\mathbf{B})$, the partially ordered semigroup of the binary network **B** (obtained from **V** by using a cutoff value of $\alpha = 0.67$). The semigroup $S(\mathbf{B})$ is presented in Table 7.3.

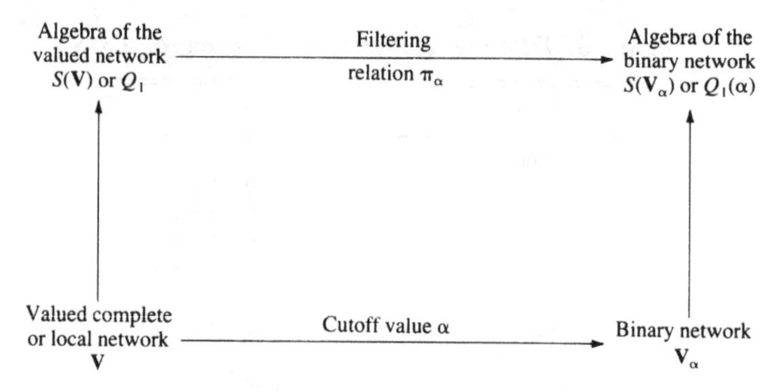

Figure 7.1. The decomposition theorem for valued network semigroups

Indeed, it is clear that the partial order π_{min} for the semigroup $S(\mathbf{V})$ of a valued network $\mathbf{V}$ may be expressed as the intersection of the relations π_α, for $\alpha > 0$. This may be established by observing that each relation π_α is nested in π_{min} and if $(U, V) \in \pi_\alpha$ for all values of α, then $(U, V) \in \pi_{min}$. It follows, therefore, that the following result holds.

THEOREM 7.4. *$S(V)$ is a subdirect product of the semigroups $S(V_\alpha)$, $\alpha > 0$.*

The representation of $S(\mathbf{V})$ in terms of the subdirect components $S(\mathbf{V}_\alpha)$ may not be as efficient a representation as a factorisation of the full partially ordered semigroup $S(\mathbf{V})$ of the valued network $\mathbf{V}$. Nonetheless, the result establishes that all of the structural distinctions among labelled and valued paths in the valued network $\mathbf{V}$ are represented in the semigroups $S(\mathbf{V}_\alpha)$ of the binary components of $\mathbf{V}$.

Theorems 7.3 and 7.4 are illustrated in Figure 7.1. Each semigroup $S(\alpha)$, corresponding to a cutoff, or filtering, value α, is an isotone homomorphic image of the full semigroup S; furthermore, the class of semigroups $\{S(\alpha); 0 \leq \alpha \leq 1\}$ thus induced is the class of images of S corresponding to the collection of relational components of the valued network identified by the Decomposition Theorem for Fuzzy Relations. In the case of valued networks that are the density matrices for relations among blocks, these relational components are the collection of all possible α-blockmodels for the set of blocks. Consequently, the use of a filtering density to convert valued matrices into binary ones is consistent with the generation of a semigroup from the valued matrices using max-min composition. The later procedure for semigroup generation maintains a great deal more relational information, namely, ordinal relations among ties of a given type; but it is useful to know that a clearly defined and comprehensive set of abstractions of its complex structure, in the form of homomorphic images $S(\alpha)$, may be obtained without its prior construction.

Local role algebras in valued local networks

For valued local networks, the same valued composition operations may be defined, including the max-min and ordinary product rules. As for complete networks, the max-min rule is adopted here because it generalises the binary composition rule in a manner that assumes only ordinally scaled values, and it yields a finite algebra.

To construct the local role algebra of a valued local network, we need only replace the *binary* composition or path construction operation by the *max-min* composition operation for valued networks. That is, the path of a particular type from the *ego* of a valued local network to some other network member i is assigned a value in the following way. The value of a particular labelled path from *ego* to node i is assigned the *minimum* value of its constituent links, and the *value* of a labelled path of a certain type from *ego* to a member of the network is defined as the *maximum* value of all such paths with that label from *ego* to node i. Consider, for instance, the valued network presented in Table 7.7. The identified *ego* of the network (node 1) has an F relation of value 4 to node 2, which in turn has an N relation of strength 4 to node 4. Hence the path FN from *ego* to node 4 through node 2 has a minimum value of 4. There is also a path labelled FN from node 1 to node 4 through node 3. The F link has value 2, and the N link has value 3; hence the value of the FN path from node 1 to node 4 through node 3 is 2. Thus, the maximum value of any path labelled FN from *ego* to node 4 is 4, and this is the value entered into the relation vector $1 * FN$ in the position corresponding to node 4. Other values may be computed in the same way to obtain $1 * FN = (0\ 0\ 0\ 4)$.

Indeed, relation vectors may be computed in this way for any possible path label, and the collection of all possible paths emanating from *ego* may be partially ordered according to

$$A \leq B$$

if and only if

$$v(1 * A)[i] \leq v(1 * B)[i],$$

for each $i \in X$ (where $v(1 * A)[i]$ is the value of the ith entry of the relation vector $1 * A$, and where $v(1 * A)[i] = v_A(1, i)$. As a result, a local role algebra for the valued local network may be defined. That is,

Table 7.7 *A valued local network*

Relation	
F	N
0 4 2 0 3 3 3 0 3 3 3 0 0 0 0 0	0 0 0 4 0 0 0 4 0 0 0 3 3 3 0 0

THEOREM 7.5. *Let* V *be a valued local network on a set* $X = \{1, 2, \ldots, n\}$, *where element 1 is the identified* ego *of the network. Then, defining the value of a labelled path* UV *by*

$$v_{UV}(1, i) = max_{j \in X}\{min\ v_U(1, j), v_V(j, i)\},$$

and an ordering on labelled paths by

$$U \leq V \text{ iff } v_U(1, i) \leq v_V(1, i), \text{ for all } i \in X$$

leads to a role algebra Q_1, *termed the* local role algebra of the local valued network *V*.

The algorithm described in chapter 2 for constructing local role algebras may be modified for valued local networks: the Boolean products of the algorithm are simply replaced by max-min products. For example, the valued local network of Table 7.7 yields the distinct relation vectors presented in Table 7.8 and the corresponding local role algebra shown in Table 7.9.

There is also a simple relationship between the algebra constructed from valued local network data and the algebra obtained when a cutoff value of α is used to convert the valued data into binary form. The relationship is similar in form to that for complete networks and is summarised in Theorem 7.6.

THEOREM 7.6. *The local role algebra* $Q_1(\alpha)$ *constructed from the binary local network* $V_\alpha = \{(V_1)_\alpha, (V_2)_\alpha, \ldots, (V_p)_\alpha\}$ *is nested in the local role algebra* Q_1 *constructed from the valued local network* V. *The* π-*relation on* Q_1 *corresponding to the nesting is given by*

$$(U, V) \in \pi_\alpha \text{ iff } v_{V(\alpha)}(1, i) \leq v_{U(\alpha)}(1, i), \text{ for all } i \in X;$$

where $v_{U(\alpha)}(1, i) = 1$ *if* $v_U(1, i) \geq \alpha$, *and 0, otherwise.*

It may similarly be shown that

Table 7.8. *Distinct relation vectors from the valued local network of Table 7.7*

Relation vectors	
1 * *F*	0 4 2 0
1 * *N*	0 0 0 4
1 * *FF*	3 3 3 2
1 * *NF*	0 0 0 0
1 * *NN*	3 3 0 0
1 * *FFF*	3 3 3 3
1 * *FFN*	2 2 0 3
1 * *NNF*	3 3 3 0
1 * *NNN*	0 0 0 3
1 * *FFFN*	3 3 0 3
1 * *FFNF*	2 2 2 0
1 * *FFNN*	3 3 0 2
1 * *FFNFF*	2 2 2 2
1 * *FFNFN*	0 0 0 2
1 * *FFNFFN*	2 2 0 2
1 * *FFNFNN*	2 2 0 0

Table 7.9. *The local role algebra of node 1 in the valued local netwok of Table 7.7*

	Right mult. table			
		Generator		
Element	Class	*F*	N	Partial order
1	*F*	3	2	1 0 0 1 0 0 0 0 0 0 0 0 0 0 0 0
2	*N*	4	5	0 1 0 1 0 0 0 0 1 0 0 0 0 1 0 0
3	*FF*	6	7	0 0 1 1 1 0 0 1 0 0 1 1 1 1 1 1
4	*NF*	4	4	0 0 0 1 0 0 0 0 0 0 0 0 0 0 0 0
5	*NN*	8	9	0 0 0 1 1 0 0 0 0 0 0 0 0 0 0 1
6	*FFF*	6	10	0 0 1 1 1 1 1 1 1 1 1 1 1 1 1 1
7	*FFN*	11	12	0 0 0 1 0 0 1 0 1 0 0 0 0 1 1 1
8	*NNF*	6	9	0 0 0 1 1 0 0 1 0 0 1 0 0 0 0 1
9	*NNN*	4	5	0 0 0 1 0 0 0 0 1 0 0 0 0 1 0 0
10	*FFFN*	8	10	0 0 0 1 1 0 1 0 1 1 0 1 0 1 1 1
11	*FFNF*	13	14	0 0 0 1 0 0 0 0 0 0 1 0 0 1 0 1
12	*FFNN*	8	7	0 0 0 1 1 0 0 0 0 0 0 1 0 1 1 1
13	*FFNFF*	13	15	0 0 0 1 0 0 0 0 0 0 1 0 1 1 1 1
14	*FFNFN*	4	16	0 0 0 1 0 0 0 0 0 0 0 0 0 1 0 0
15	*FFNFFN*	11	15	0 0 0 1 0 0 0 0 0 0 0 0 0 1 1 1
16	*FFNFNN*	11	14	0 0 0 1 0 0 0 0 0 0 0 0 0 0 0 1

THEOREM 7.7. *The local role algebra Q_l of a valued local network V is a subdirect product of the local role algebras $Q_{l(\alpha)}$ constructed from the binary components V_α of the local network V.*

Using valued network algebras

We have shown in this chapter how valued networks give rise to algebraic structures in a way that generalises the constructions for binary network data. In these valued network algebras, two labelled paths *U* and *V* are equated if, for every relevant pair of network nodes, the values of *U* and *V* coincide. For complete networks the condition applies to all pairs of nodes, whereas for local networks it applies to all pairs having *ego* as their first member. The condition for the equation of two labelled paths is therefore a strict one; as a result, relatively small valued networks can give rise to large and complex algebraic structures. To make practical use of valued network algebras, it is likely that we need to follow one of several possible courses of action in order to deal with this complexity.

If the algebra is not too large, then it can be analysed using the procedures described in chapters 4 and 5. That is, its factors can be obtained, and the relational referents of each factor can be sought in the binary components of the network relations. If factorisation is precluded by the size of the semigroup concerned, then other analytic strategies must be adopted. One possibility is provided by Theorems 7.3 and 7.6: the valued data may be converted to binary form, perhaps using several different cutoff values α, and an algebra can be constructed from the binary data. Theorems 7.3 and 7.6 guarantee that the algebras thereby obtained are approximations to the valued network algebras in a well-defined mathematical sense. Further, Theorems 7.4 and 7.7 ensure that all of the structure of the valued networks can be described if we use all possible cutoff values α in the process.

Another possible strategy, which is yet to be fully developed, is that of defining a partial algebra from valued network data. A partial multiplication table and a partial order table may be constructed for only those paths that do not exceed some predetermined length. This strategy was illustrated in chapter 2 for local networks, and it was shown that many of the algebraic constructions that have been useful for analysing network algebras may also be invoked for partial algebras.

8

Issues in network analysis

Two major themes have guided the development of the representations of structure in social networks that we have been considering. One is the search for an account of the relational context in which individual behaviour takes place; the other is the need to describe the structural framework on which a variety of social processes occur. In this chapter we review the progress towards these goals afforded by the representations and the analytic methods that have been developed for them.

Describing social context: Positions and roles

In describing the relational context for behaviour, we have attempted to characterise the patterns of relations that exist in a local network or in an entire group. How far do the representations enable us to describe the patterns of relations surrounding an individual, and what form do the descriptions take? The representations themselves are expressed in terms of orderings and equations among paths in networks. In the case of entire networks, the orderings and equations hold for any paths in the network with the same source and endpoint. For local networks, the orderings and equations pertain only to paths having the identified ego as their source. The representations make no reference to specific individuals in the network and so can be used to make comparisons from one network to another. In fact, a consequence of the representation is that two individuals or groups have the same relational context if the orderings and equations among their paths are the same. That is, if the existence of one type of path between two individuals always entails the existence of a path of another particular type, and if all such entailments are the same in the two networks, then we assert that the relational patterns in the two networks are the same. More generally, we can compare two networks by comparing their collections of orderings and equations: this procedure will be described in greater detail later.

Thus, the form of the description of relational context is in terms of path comparisons: it is precisely those orderings among paths that hold for all relevant paths in the network that are included in the description. This descriptive form is modified in an important way by the suggested means of analysing the representation. In particular, the analysis leads to a description of relations among paths in terms of a collection of simpler sets of path comparisons with more limited domains of application. Factorisation of the algebraic representation yields sets of path comparisons that do not hold universally in the network but apply instead only to certain regions of the network that are identified using the Correspondence Definition. Taken together, these various simplified sets are sufficient to define the path orderings that hold everywhere, but it is deemed useful to work with the collection of simplified sets for two major reasons. The first is a practical one: the components or factors are more simply described, and it is quite easy to decide whether two networks share any particular component. The second is more theoretical in flavour: it is a useful step to admit the possibility that relational structure is not everywhere constant. Heterogeneity in relational context is an important motivation for the descriptive exercise on which we are embarked, and it is therefore helpful to construct a representation that recognizes it explicitly. Of course, it is also useful to describe those relational forms that do hold universally in a network: such forms may have cultural significance, as Bonacich (1980) has argued. But for many of the situations in which descriptions of relational context are required, we need to recognise and then describe the variations that do occur. I would argue that the scheme of analysis that has been proposed admits this goal by obtaining a representation of relational context as the intersection of maximally heterogeneous relational forms.

Positions and roles

Intertwined with this discussion of the nature of relational context revealed by a local or entire network are notions that are relevant to the description of positions and roles. The representations that we have considered for both local and entire networks have often been seen as providing relational operationalisations of the notions of position and role (White et al., 1976; Winship & Mandel, 1983). A position is usually associated with a niche in the network, and as many authors have suggested, two positions are similar to the extent that they share interrelations of the same kind with the same or similar other positions (Burt, 1976; Faust, 1988; White et al., 1976; White & Reitz, 1983; Winship & Mandel, 1983). The social role associated with a position is usually deemed to include the pattern of relations obtaining in the vicinity of the position, that is, its associated relational forms. Following Mandel (1983), we have characterised this pattern by the collection of path comparisons relevant to the position.

In an entire network, the analysis that has been proposed provides an implicit description of positions and their associated roles. Each component of the partially ordered semigroup identifies a simple relational form, and by virtue of the Correspondence Definition, this form can be identified with a region of the original network. The simple relational form describes an interlocking of the roles associated with the positions in the region. The role of each individual position is decribed by the relational patterns linking the position to other identified positions. If a particular network element is identified with a position in two or more such regions, then its social position overall is described as the aggregate of positions participating in multiple forms of role interlock. Thus, implicitly, the analysis of an entire network identifies, for each network member, a potential multiplicity of positions and associated roles in the network.

For a local network of an individual, the role of the individual is identified explicitly with the network's local role algebra. Factorisation of the local role algebra yields a collection of simple components of that potentially complex role, and each role component is associated with a restricted local network domain. The relations of ego to other network members in this domain give a relational basis to the role component. One way of viewing this analysis of the role of the individual is as a decomposition into maximally heterogeneous relational forms, and so as the expression of the multiple positions and roles held by an individual in terms of their simple constituents.

Thus, the proposed analysis has the capacity to provide substantial structural detail in the description of positions and roles in networks. It allows the identification of "elemental" forms of role interlocking in entire networks and of role components in local networks. The realisation of these forms in terms of actual network relationships can also be described, and there is clearly a need for the analysis of much network data in these terms.

The structure and content of relations

The application of the proposed analyses to a variety of empirical networks of different kinds will yield a catalogue of the relational forms that occur for relations of different types, together with some indication of the conditions under which they occur. If relations of particular types interlock frequently and almost universally in particular relational forms, then there is evidence for a strong link between the content of relations and their structure. Such a link suggests the presence of a culturally defined relational form. Indeed, an early motivation underlying the construction of a representation of the "relationships between social relations" was the notion that relationships of specified types may, in certain circumstances, interrelate in similar ways (Lorrain, 1975; Lorrain & White, 1971; Nadel, 1957). The argument was made that, if invariances of relational structure were expected to exist, then it would be at the level of relational interlock that they would be likely to be found.

In fact, a number of invariances in relational form for particular types of network relations have been suggested in the literature, and they are summarised in Table 8.1. They take the form of models for relational structure in different network types. The occurrence and domain of application of any proposed model can be determined for a given empirical network using the procedures that have been developed here. The analysis yields not only a detailed account of whether, and in what regions of the network, a particular model applies, but also what alternative relational forms are found. This is in contrast to many of the previous attempts to investigate the fit of such models. In earlier work, model fit has mainly been evaluated either by calculating indices of the fit of model to data, such as balance and transitivity indices (e.g., Harary & Kommel, 1979; Peay, 1977a) or by comparing such an observed index with its expected distribution in a particular random graph population (Frank, 1980; Frank & Harary, 1980; Holland & Leinhardt, 1970, 1978). In either case, the outcome of the analysis was an indication of the level of fit of the model, with no information about where in the network the model failed to fit, nor what a useful alternative model might be. Of course, the analysis that has been proposed here really requires error-free data. Pattison, Caputi and Breiger (1988) have attempted to recast the problem of fitting models of the kind listed in Table 8.1 to network data in a way that admits error but also identifies regions of lack of fit. Boyd (1991) has also demonstrated how a simulated annealing algorithm (e.g., Press, Flannery, Teukolsky & Vetterling, 1986) can be applied in some cases to fit a model expressed in algebraic terms.

Models for relational interlock can be derived in some instances on theoretical grounds. Such an undertaking has been illustrated by Breiger and Pattison (1978) for pairs of relations whose content is conceptualised as strong and weak, using the relational arguments originally proposed by Granovetter (1973). An extension of their model is derived in the next section as an illustration of the approach, and similar pursuits for other relational types are also summarised.

There are at least two ways in which a knowledge of the empirical constraints between the content of relations and their structure might be applied. One is to use predictions about the structural form that the interrelating of particular types of relations is likely to take in the development of models for social processes. A second is to derive hypotheses about the content of relations from their structure. The meaning of a given relation may be seen as at least partially defined by the context in which it occurs. Thus, one may use the existence of a particular empirical form

Table 8.1. *Some models for networks*

Model	Right mult. table: Element	Generator			Partial order
Transitivity		1			
	$R = 1$	1			1
Classical		1	2		
balance	$P = 1$	1	2		1 0
	$N = 2$	2	1		0 1
Complete		1	2		
clustering	$P = 1$	1	2		1 0 0
model	$N = 2$	2	3		0 1 0
	$NN = 3$	3	3		1 1 1
Complete		1	2	3	
ranked	$M = 1$	1	2	3	1 0 0 0 0 0 . . . 0 0 1
cluster	$A = 2$	2	5	2	0 1 0 0 1 1 . . . 1 1 1
model[a]	$N = 3$	3	2	4	0 0 1 0 0 0 . . . 0 0 1
	$NN = 4$	4	2	4	1 0 1 1 0 0 . . . 0 0 1
	$AA = 5$	5	6	5	0 0 0 0 1 1 . . . 1 1 1
	$AAA = 6$	6	7	6	0 0 0 0 0 1 . . . 1 1 1
	⋮	⋮	⋮	⋮	⋮
	$A^{k-3} = k$	k	k	k	0 0 0 0 0 0 . . . 0 0 1
Complete		1	2	3	
ranked	$M = 1$	1	2	3	1 0 0 0 0 . . . 0 1
2-cluster	$A = 2$	2	4	2	0 1 0 1 1 . . . 1 1
model[b]	$N = 3$	3	2	1	0 0 1 0 0 . . . 0 1
	$AA = 4$	4	5	4	0 0 0 1 1 . . . 1 1
	⋮	⋮	⋮	⋮	⋮
	$A^{k-2} = k$	k	k	k	0 0 0 0 0 . . . 0 1
Strong–		1	2		
Weak tie	$S = 1$	1	2		1 0 0 . . . 0 0
model	$W = 2$	2	3		1 1 0 . . . 0 0
(version 1)	$WW = 3$	3	4		1 1 1 . . . 0 0
	⋮	⋮	⋮		⋮
	$W^{k-1} = k$	k	k		1 1 1 . . . 1 1
Strong–		1	2		
Weak tie	$S = 1$	1	2		1 0
model	$W = 2$	2	2		1 1
(version 2)					

[a] Assuming at least 3 clusters per level and k–3 levels.
[b] Assuming exactly 2 clusters per level and k–2 levels.

of relational interlock to generate hypotheses about the substantive nature of the relations themselves. Speculation of this kind was undertaken by Breiger and Pattison (1978) and demonstrated how "the cultural content of the social structure becomes a question for empirical research rather than a matter of definition" (White et al., 1976, p. 770).

Some models for relational structure

The models appearing in Table 8.1 are expressed in a form applicable to entire networks: they assert that the orderings and/or equations among relations by which they are characterised apply to all network members. They may equally well be seen as appropriate models for the analysis of local networks. Indeed, a useful analysis for an entire network in relation to a particular model is the investigation of model fit in the collection of local networks generated by taking each network member as a local network ego. For the sake of simplicity, though, the models are described here in relation to entire networks only.

We may observe that the relational expression of the model given in Table 8.1 is not necessarily equivalent to the form in which it was first introduced. For example, while the original formulation of the balance model implies the partially ordered semigroup representation of Table 8.1, the two forms are not equivalent. Rather, the form presented in Table 8.1 is a generalisation of the original model, in that there exist networks whose semigroups are identical to that presented in the first panel of the table but which do not conform to the original model. Table 8.1 should therefore be seen as a summary of proposals for particular relational interpretations of the models concerned.

We may also note that the collection of semigroup structures for which the establishment of a structure-content link has been attempted is rather limited. Part of the methodological and substantive challenge of the semigroup representation is the extension of this list, perhaps in the manner illustrated for networks of strong and weak relations.

Strong and weak ties

The way in which the form of relational interlock may be theoretically derived is illustrated by considering Granovetter's (1973) account of the configurations that one might expect relationships of strong and weak ties to form. Granovetter presented a persuasive argument for the inference of certain network properties of interpersonal ties from a knowledge of their "strength". His basic contention was that the stronger the tie between two individuals, the larger the proportion of individuals to whom they are both connected by either a strong or weak tie. A consequence of this claim is that strong ties should tend to be densely concentrated within groups whereas weak ties should be less densely concentrated but more likely to link members of different groups.

Taking S to represent a strong tie and W a tie that is at least weak, Breiger and Pattison (1978) argued that the Granovetter hypothesis implies the approximate truth of the equations

$$SW = W, \tag{8.1}$$

$$WS = W, \tag{8.2}$$

$$S^2 = S, \tag{8.3}$$

because individuals A and B linked by a strong tie S have most of their ties in common, and strong ties tend to exhibit inbreeding. It may also be argued that the following equations and orderings are consistent with Granovetter's hypothesis:

$$W^{k+1} = W^k, \text{ for some } k \geq 1, \tag{8.4}$$

because at some point, very long paths constructed from weak ties are likely to create no new ties among individuals; and

$$S \leq W \leq W^2 \leq W^3 \leq \cdots \leq W^k, \tag{8.5}$$

because (a) for small k, weak ties tend to create new connections among individuals, and (b) provided there is a sufficient distribution of W ties, it is reasonable to assume that any W tie may be replaced by a pair of W ties.

In their original formulation, Breiger and Pattison (1978) suggested the equation

$$W^2 = W$$

in place of equations 8.4 and 8.5; this equation emerges in an isotone homomorphic image of the Strong–Weak tie structure implied by equations 8.1–8.5. The Breiger and Pattison version of the Strong–Weak tie model may therefore be seen as an approximation of the more general model in Table 8.1. Both models are consistent with the banning of Granovetter's forbidden triad, that is, a triad of elements in which two elements are connected to a third by a strong tie but have no strong or weak relation to one another. Hypothesising that this triad does not occur is equivalent to asserting that the ordering

$$S^2 \leq W$$

holds. The ordering is consistent with both formulations of strong and weak tie structure.

Equations 8.1–8.5 describe a partially ordered semigroup with an identity element S and a zero element equal to some power of W (W^k). The relations written in the order

$$S, W, W^2, \ldots W^k$$

are increasingly "weak"; taken in the reverse order, they are increasingly "strong". The identification of the weak tie W with the semigroup zero W^k induces the homomorphism from the semigroup of Table 8.1 onto the Breiger and Pattison model.

It may be observed that this model predicts that compounds constructed from weak ties are increasingly weak as path length increases. In many empirical network semigroups, certain compound ties act as semigroup zeros, and others nearly so; compound ties are almost always weak in relation to primary ties because few observed network relations are strong in Granovetter's strict sense.

The balance model

The original application of the balance model to social networks was to relations describing positive and negative affect among individuals (Cartwright & Harary, 1956). This application was later considered in relation to networks of positive and null relations by Davis, Holland and Leinhardt in a series of comparisons of the balance model with more general models (e.g., Davis, 1979). The model predicts that a set of network members can be divided into two groups in such a way that positive relations exist only between members within a group and negative relations exist only between members from different groups. If all individuals within a group are linked by positive relations and if all individuals from different groups are linked by negative relations, then the network structure for positive and negative relations presented in Table 8.1 ensues. If some of the predicted within- and between-group relations are absent, then a structure containing that of Table 8.1 as a homomorphic image is generated, provided that a sufficiently rich set of connections are present. For instance, if the division of the network into two groups is a partition satisfying one of Kim and Roush's (1984) conditions and if the derived network has positive ties linking each group to itself and negative ties linking each group to the other, then the balance model of Table 8.1 emerges as a homomorphic image.

The complete clustering model

Analyses of empirical networks of positive and negative relations led Davis (1967) to generalise the balance model to the clustering model. The clustering model predicts that a network can be partitioned into two *or more* groups so that positive ties occur only within groups and negative ties occur only between groups. In the complete version of the clustering model, all possible positive and negative ties conforming to this condition are present (e.g., Johnsen, 1986). The complete clustering model for a network with three or more groups has the network structure shown in Table 8.1. As for the balance model, Kim and Roush's (1984) conditions may be used to derive conditions under which the general clustering model gives rise to a network structure that can be mapped homomorphically onto that for the complete clustering model.

The transitivity model

Holland and Leinhardt (1970, 1971, 1972) and Johnsen (1985) have also developed a series of models that generalise the balance and clusterability models further. Most of these models are expressed in terms of a single network relation and its partitioning into mutual, asymmetric and null components. All of the models can be described by listing those triads in the triad census of the graph that are permitted to occur (e.g., Johnsen, 1985). They are usually aligned with the balance and clustering models by the assertion that mutual or asymmetric ties are "positive" and that null ties are "negative". In fact, different decompositions of a single relation are clearly possible, so that different statements of predicted network structure are also possible. For instance, the transitivity model may be expressed in the form

$$(R^*)^2 = R^*,$$

where R^* is the union of the identity relation, the mutual ties in a relation R and the asymmetric ties. Alternatively, taking the mutual and asymmetric components of R separately, we obtain

$$(M^*)^2 = M^* \qquad \text{and} \qquad M^*A = AM^* = A^2 = A,$$

where M^* is the union of the mutual component M of R and the identity relation, and where A is the asymmetric component of R. Note that the multiplication table for the latter form of the model is the same as that for the Breiger and Pattison (1978) version of the Strong–Weak tie model; its partial order is different, though, as the relations M and A have a null intersection.

Other triad-based models

The derivation of the network structure of other triad-based models, such as the ranked-cluster model (Davis & Leinhardt, 1972), the hierarchical cliques model (Johnsen, 1985) and the "39+" model (Johnsen, 1985) is yet to be generalised, but some special cases are presented in Table 8.1. They are the complete ranked-cluster model with at least three clusters per level, and the ranked 2-cluster model with exactly two clusters per level.

The First and Last Letter laws

Lorrain (1975) discussed the First and Last Letter semigroups for network structure; they have occurred quite commonly in data analyses (e.g., Boorman & White, 1976; Breiger, 1979; Lorrain, 1975). In the First Letter table, compound relations have paths identical to those of their first relational constituent; and in the Last Letter table, the compound is equal to the last relation in the path. In the First and Last Letter table, any compound relation is determined by both the first and the last relation in the chain. Several interpretations of these patterns have been attempted, but none has taken into account the partial orderings among the relations, and so they can probably be elaborated by the additional information. For example, Boorman and White (1976) suggested that the First Letter table (*FL*) is associated with interlock between an objective (though positive) sort of tie and a positive affect tie ("thus one's ties to any kind of contact of one's business associate take on the colour of a business association, whereas one views in affective terms any kind of contact through a friend", p. 1414).

Lorrain established differences in receiver characteristics for structures manifesting the Last Letter (*LL*) structure and differences in emitter characteristics for configurations having the *FL* structure. In the context of business relations and community affairs relations, the *LL* table was interpreted by Breiger (1979) as indicating the ability of network members to assimilate their contact's quality of tie to a third party in their own ties with third parties.

Permutation models for kinship structures

White (1963) described permutation group models for kinship structures and assessed their fit in a number of Australian aboriginal groups. The network relations he used were marriage and descent rules: in his model, societies were partitioned into clans, and the marriage and descent rules specified the clan membership of the wife of men in each clan and the clan membership of the children of men in given clans. The network structure generated by these relations describes the structure among kinship relations in the society; particular versions of the structures can be used as models for kinship relations in a society.

Some other models

Other cases that have been discussed in the literature include structures whose multiplication tables are some kind of inverse semigroup. Some of these structures are generated by out-trees and their converses (Pattison, 1980), and others may be described as possessing a centre–periphery structure.

Describing common structure

The appropriate means of describing the relational structure common to two networks has been the subject of considerable debate. On the one hand, Boorman and White (1976) have proposed the joint homomorphic image of the semigroups of two networks as the representative of shared structure. Breiger and Pattison (1978) adopted this definition in an analysis of the structure shared by two community elites. In chapter 3, a joint isotone homomorphic image was also defined for the partially ordered semigroups of two networks, and it was shown how this construction is the analogue of the *JNTHOM* for partially ordered structures.

Bonacich and McConaghy (1979; Bonacich, 1980; McConaghy, 1981), however, have claimed that the common structure semigroup records the shared structure in two networks, and like the *JNTHOM*, the common structure semigroup can be defined for both abstract and partially ordered semigroups. In the partially ordered case, the lattice $L(\mathbf{R})$ of semigroups with generators $\mathbf{R} = \{R_1, R_2, \ldots, R_p\}$, partially ordered by the relation

$$S_1 \leq S_2$$

if and only if there is an isotone homomorphism from S_2 onto S_1, the *JNTIHOM* of two semigroups is their greatest lower bound, and the common isotone structure semigroup (*CISS*) is their least upper bound. An analogous lattice $A(\mathbf{R})$ may be constructed for abstract homomorphisms, and Boorman and White and also Bonacich and McConaghy introduced their definitions in the lattice $A(\mathbf{R})$. For consistency with earlier discussions, though, we will refer to the lattice $L(\mathbf{R})$ based on isotone homomorphisms.

Each of these alternatives is a useful construction, but the interpretations accompanying them are very different. In particular, the common structure semigroup records equations and orderings that hold universally for every element in each network. That is, it specifies orderings and equations among paths that are true for relations linking any pair of elements in either network. The common structure semigroup is the semigroup generated by the network constructed as the disjoint union of the two separate networks (the formal definition of network union was given in chapter 3). If universally or culturally shared forms of relational structure are a focus, then the common structure semigroup is clearly a useful construction. Because the equations and orderings that define it apply everywhere in both networks, they take the form of relational "laws" for the networks concerned. The common structure semigroup ignores, however, those relational forms that occur in only some parts of the combined network: information about heterogeneity in relational form is lost.

The joint isotone homomorphic image has complementary properties. Its equations and orderings do not necessarily hold for all elements in both networks, but they do enable us to identify the simple relational forms that occur in both networks. Consider, once more, the analysis of a network that has been developed. It describes a network as a collection of simple relational forms, each associated with a restricted network domain of application. The *JNTIHOM* of two networks has components that are simple relational forms of this kind, and the Correspondence Definition may be used to identify the restricted domain in which those components apply *in each network*. The result is a statement of the simple forms that occur in both networks and a matching of the relational domains for which they hold. Where heterogeneity in relational structure is of interest, this is clearly a useful result. The *JNTIHOM* therefore enables us to identify regions in the two networks that have some forms of interrelating in common, as well as the nature of those common forms.

We now illustrate this application of the *JNTIHOM:* first, in relation to the blockmodels for two self-analytic groups, one of which was analysed in detail in chapter 5, and then in relation to two community elite blockmodels, which have been analysed from both perspectives in the literature (Breiger & Pattison, 1978; McConaghy, 1981; Pattison, 1981).

Common relational forms in two self-analytic groups

Ennis (1982) reported the analysis of data from a self-analytic group of a type similar to that described by Breiger and Ennis (1979). In each of the cases reported by Breiger and Ennis (1979) and Ennis (1982), a group of individuals reported ties of Liking, Disliking and perceived Similarity among them; also, they were all assessed in terms of Bales and Cohen's (1979) three dimensions of interpersonal behaviour. The blockmodel described by Breiger and Ennis was analysed in some detail in chapter 5; here, we outline briefly the blockmodel reported by Ennis and use the

Table 8.2. *The Ennis blockmodel*

Relation		
Liking	Disliking	Similarity
1 0 0 1 1 0 1 0 1	0 1 1 0 0 1 0 1 0	1 0 0 1 1 0 1 0 1

From Ennis, 1982

Table 8.3. *The semigroup BE2 of the Ennis blockmodel*

	Right mult. table		
	Generator[a]		
Element	1	2	Partial order
$L = 1$	1	3	1 0 0 0 0 0 0
$D = 2$	4	5	0 1 0 0 0 0 0
$LD = 3$	6	3	0 1 1 0 1 0 0
$DL = 4$	4	3	0 1 0 1 0 0 0
$DD = 5$	7	2	0 0 0 0 1 0 0
$LDL = 6$	6	3	1 1 1 1 1 1 1
$DDL = 7$	7	3	1 0 0 0 1 0 1

[a] The third generator S is equal to L.

joint isotone homomorphism of the semigroups of the two blockmodels to illustrate the analysis of their common relational forms.

The blockmodel reported by Ennis (1982) has three blocks and is shown in Table 8.2. Its partially ordered semigroup $BE2$ is given in Table 8.3. The joint isotone homomorphic image of $BE1$ and $BE2$ is the semigroup K presented in Table 8.4; K has two factors $K1$ and $K2$ also reported in Table 8.4. The minimal derived set associations for $K1$ and $K2$ in each of the two blockmodels are listed in Table 8.5. The full reduction diagram for $BE2$ appears in Figure 8.1; images of $BE2$ in Figure 8.1 are reported in detail in Table 8.6.

The relations of Liking and perceived Similarity are equated in the joint isotone image K and are contrasted with the Disliking relation (as they are in $BE2$). The factor $K1$ is the Last Letter semigroup (Table 8.1) and records the tendency for paths ending in Liking (or Disliking) relations to resemble simple Liking (or Disliking) relations. In the Breiger–Ennis

Table 8.4. *The joint isotone homomorphic image K of BE1 and BE2 and its factors K1 and K2*

		Right mult. table		
		Generator[a]		
Semigroup	Element	1	2	Partial order
K	*L* = 1	1	2	1 0 0
	D = 2	3	2	0 1 0
	DL = 3	3	2	1 1 1
*K*1	*L* = 1	1	2	1 1
	D = 2	1	2	0 1
*K*2	*L* = 1	1	2	1 0
	D = 2	2	2	1 1

[a] The third generator *S* is equal to *L*.

Table 8.5. *Derived set associations of K1 and K2 in the Breiger–Ennis and Ennis blockmodels*

Semigroup	Blockmodel	Minimal derived set associations
*K*1	Breiger–Ennis	{1, (234)} {1, 2} {1, (23)} {1, (24)} {1, (34)}
	Ennis	{1, (23)}
*K*2	Breiger–Ennis	{2, 3} {(124), 3} {(12), (34)} {(24), 3}
	Ennis	{(12), 3} {(13), 2} {1, (23)}

blockmodel, this feature may be traced to that region of the blockmodel describing relations of block 1 with other blocks and especially with the more hostile block 2. More specifically, Liking relations are expressed universally within and between blocks 1 and 2, but Disliking relations are expressed only towards block 2. In the Ennis blockmodel, *K*1 may be associated with the derived set {1, (23)}, that is, with the relations between the "leader" block 1 and the other two blocks. Block 1 is characterised in the Ennis blockmodel also by the absence of received Disliking ties; it is a common feature of the two blockmodels that the two positive and somewhat dominant blocks receive no Disliking relations from other group members.

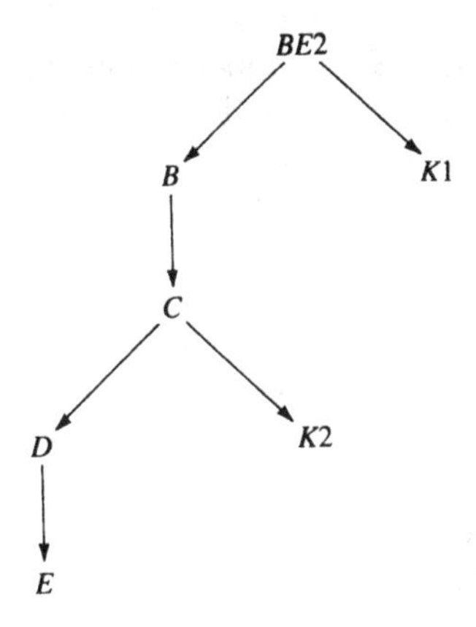

Figure 8.1. Reduction diagram for the Ennis semigroup

Table 8.6. *Images of BE2 appearing in Figure 8.1*

	Right mult. table			
		Generator		
Semigroup	Element	1	2	Partial order
K1	1, 4, 6, 7 = 1	1	2	1 1
	2, 3, 5 = 2	1	2	0 1
B	1	1	3	1 0 0 0 0 0
	2	4	5	0 1 0 0 0 0
	3, 6 = 3	3	3	1 1 1 1 1 1
	4	4	3	0 1 0 1 0 0
	5	7	2	0 0 0 0 1 0
	7	7	3	1 0 0 0 1 1
C	1	1	3	1 0 0 0 0
	2	4	5	0 1 0 0 0
	3, 6 = 3	3	3	1 1 1 1 1
	4, 7 = 4	4	3	1 1 0 1 1
	5	4	2	0 0 0 0 1
D	1, 3, 4, 6, 7 = 1	1	1	1 1 1
	2	1	5	0 1 0
	5	1	2	0 0 1
E	1, 3, 4, 6, 7 = 1	1	1	1 1
	2, 5 = 2	1	2	0 1
K2	1	1	2	1 0
	2, 3, 4, 5, 6, 7 = 2	2	2	1 1

The factor *K*2 records the strength of Liking and Similarity ties in relation to Disliking ties. In the Breiger–Ennis blockmodel, the factor may be traced principally to relations between the dominant and hostile block 2 and the more positive and egalitarian block 3. Liking relations among these two blocks occur only within blocks, whereas block 2 expresses Disliking for block 3. The feature occurs more commonly in the Ennis blockmodel; indeed, it occurs in the relations between any block and the other pair of blocks. The strength of the Liking relation comes from the occurrence of Liking ties that are expressed only internally or towards block 1; Disliking relations, one the other hand, are expressed by other blocks to both blocks 2 and 3. In both blockmodels, this factor reflects the tendency of the more dominant blocks to express Disliking rather than Liking for the less dominant ones.

We may observe that this analysis of relational forms common to the two blockmodels is somewhat consonant with the description of the blocks in terms of Bales and Cohen's (1979) dimensions of interpersonal behaviour. In particular, it seems to describe forms of interrelating that are characteristic of the positive leader blocks as similar in the two blockmodels. More data are required before we can achieve a clearer understanding of the constraints between the typical modes of interacting in a group that an individual adopts and the actual social relations that the individual comes to hold; the kinds of results illustrated by the preceding analysis, however, demonstrate the usefulness of the proposed analysis in assessing these data. It is particularly important for an enterprise of this kind that we assess the common relational forms in a way that does not assume homogeneity of relational form. Features of the two blockmodels that are distinctive can be ascertained by examining the reduction diagrams for the semigroups of the blockmodels, that is, Figure 5.10 for *BE*1 and Figure 8.1 for *BE*2.

Common relational forms in two community elites

Laumann and associates (Laumann, Marsden & Galaskiewicz, 1977; Laumann & Pappi, 1973, 1976; Laumann, Verbrugge & Pappi, 1974) have studied the community influence systems in two small regional cities of comparable industrial composition, one in West Germany and one in the United States of America. As part of their investigations, they undertook a network analysis of the identified influential persons in each town, asking each elite member to indicate the three others in the town with whom the member had the closest business or professional relations, the most frequent discussion of community affairs and the most frequent social meetings. [A detailed account of the survey

Table 8.7. *The Altneustadt blockmodel*

Block	Business/Professional ties	Community Affairs ties	Social ties
*A*1	1 0 0 0	1 0 0 0	1 0 0 0
*A*2	0 1 0 0	1 1 0 0	0 1 0 0
*A*3	0 0 1 1	0 0 1 0	0 0 1 0
*A*4	0 0 0 1	1 0 0 0	1 0 0 1

Table 8.8. *The Towertown blockmodel*

Block	Business/Professional ties	Community Affairs ties	Social ties
*T*1	1 1 0 0	1 1 0 0	1 0 0 0
*T*2	1 1 0 0	1 1 0 0	1 1 0 0
*T*3	1 0 0 1	0 0 1 1	0 0 1 1
*T*4	0 1 0 1	0 1 0 1	0 1 0 0

procedures may be found in Laumann and Pappi (1976) and Laumann et al. (1977).]

The three resulting networks in each community elite have been subject to a number of blockmodel and related analyses (Breiger, 1979; Breiger & Pattison, 1978; Burt, 1977; Laumann & Pappi, 1973, 1976; McConaghy, 1981; Pattison, 1981). The analysis reported here is of the CONCOR-generated blockmodels for the Altneustadt and Towertown elites adopted by Breiger and Pattison (1978) and McConaghy (1981); the blockmodels are reported in Tables 8.7 and 8.8. Altneustadt and Towertown blocks are identified as *A*1, *A*2, *A*3, *A*4 and *T*1, *T*2, *T*3, *T*4, respectively.

The semigroups generated by the Altneustadt and the Towertown blockmodels are presented in Tables 8.9 and 8.10 and are labelled *A* and *T*, respectively. The joint isotone homomorphism *L* of the two semigroups is reported in Table 8.11 and is associated with the π-relations π_α on *A* and π_τ on *T* reported in Table 8.12. The minimal derived sets associated with the joint image *L* are listed in Table 8.13, together with corresponding derived networks. The joint image *L* is subdirectly irreducible and may be interpreted, using the arguments previously advanced in relation to the interlocking of strong and weak ties, as indicating the strength of Social ties in relation to Business/Professional or Community Affairs ties. The domain of the Altneustadt blockmodel corresponding to this relational form is in the boundary between blocks *A*1 and *A*4, on the one hand, and blocks *A*2 and *A*3,

Table 8.9. *The Altneustadt semigroup A*

Element	Right mult. table: Generator 1	2	3	Partial order
$B = 1$	1	4	5	1 0 0 0 0 0 0 0
$C = 2$	6	2	2	0 1 0 0 0 0 0 0
$S = 3$	7	2	3	0 0 1 0 0 0 0 0
$BC = 4$	8	4	4	0 1 0 1 0 0 0 0
$BS = 5$	5	4	5	1 0 1 0 1 0 1 0
$CB = 6$	6	4	8	0 1 0 0 0 1 0 0
$SB = 7$	7	4	5	1 0 1 0 0 0 1 0
$BCB = 8$	8	4	8	0 1 0 1 0 1 0 1

Table 8.10. *The Towertown semigroup T*

Element	Right mult. table: Generator 1	2	3	Partial order
$B = 1$	4	4	5	1 0 0 0 0 0 0 0 0 0 0 0
$C = 2$	4	6	7	0 1 1 0 0 0 0 0 0 0 0 0
$S = 3$	8	7	9	0 0 1 0 0 0 0 0 0 0 0 0
$BB = 4$	4	4	5	1 0 0 1 1 0 0 1 0 0 0 0
$BS = 5$	5	5	5	0 0 0 0 1 0 0 0 0 0 0 0
$CC = 6$	4	10	11	0 1 1 0 0 1 1 0 1 0 0 0
$CS = 7$	8	11	11	0 0 1 0 0 0 1 0 1 0 0 0
$SB = 8$	8	8	5	0 0 0 0 1 0 0 1 0 0 0 0
$SS = 9$	8	11	12	0 0 1 0 0 0 0 0 1 0 0 0
$CCC = 10$	4	10	11	1 1 1 1 1 1 1 1 1 1 1 1
$CCS = 11$	8	11	11	0 0 1 0 1 0 1 1 1 0 1 1
$SSS = 12$	8	11	12	0 0 1 0 0 0 0 0 1 0 0 1

Table 8.11. *The joint isotone homomorphic image L of the semigroups A and T*

Element	Right mult. table: Generator[a] 1	3	Partial order
$B = 1$	1	1	1 1
$S = 3$	1	3	0 1

[a] The generator C is equal to the generator B.

Table 8.12. *The π-relations π_α on A and π_τ on T corresponding to the joint isotone homomorphic image L*

π_α	π_τ
1 1 1 1 1 1 1 1	1 1 1 1 1 1 1 1 1 1 1 1
1 1 1 1 1 1 1 1	1 1 1 1 1 1 1 1 1 1 1 1
0 0 1 0 0 0 0 0	0 0 1 0 0 0 0 0 1 0 0 1
1 1 1 1 1 1 1 1	1 1 1 1 1 1 1 1 1 1 1 1
1 1 1 1 1 1 1 1	1 1 1 1 1 1 1 1 1 1 1 1
1 1 1 1 1 1 1 1	1 1 1 1 1 1 1 1 1 1 1 1
1 1 1 1 1 1 1 1	1 1 1 1 1 1 1 1 1 1 1 1
1 1 1 1 1 1 1 1	1 1 1 1 1 1 1 1 1 1 1 1
	0 0 1 0 0 0 0 0 1 0 0 1
	1 1 1 1 1 1 1 1 1 1 1 1
	1 1 1 1 1 1 1 1 1 1 1 1
	0 0 1 0 0 0 0 0 1 0 0 1

Table 8.13. *Minimal derived set associations with L in the Altneustadt and Towertown blockmodels and corresponding derived networks*

Blockmodel	Derived set	Derived network		
Altneustadt	{(14), (23)}	1 0	1 0	1 0
		1 1	1 1	0 1
Towertown	{1, (234)}	1 1	1 1	1 0
		1 1	1 1	1 1
	{1, 2}	1 1	1 1	1 0
		1 1	1 1	1 1
	{1, (23)}	1 1	1 1	1 0
		1 1	1 1	1 1
	{1, (24)}	1 1	1 1	1 0
		1 1	1 1	1 1

on the other. Blocks *A*1 and *A*4 express no ties of any kind to blocks *A*2 and *A*3; blocks *A*2 and *A*3 report Business/Professional and Community Affairs ties, but not Social ties, as links to blocks *A*1 and *A*4. In the Towertown blockmodel, the relational form corresponding to *L* may be traced to the links between block *T*1 and block *T*2, the latter sometimes in combination with other blocks. Block *T*1 expresses Social ties only among its own members whereas Business/Professional and Community Affairs ties are seen as linking members of block *T*1 to those of block *T*2.

Thus, the common relational forms in the two networks may be described in terms of a single component recording the strength of Social ties in comparison to other assessed network links. The network domains possessing this relational form have been identified in each elite blockmodel. All other relational forms characterising relational structure in the blockmodels are distinct; as for the analyses of the self-analytic groups, these distinctive forms can be described using the method of analysing a single network system that has been advanced.

Social structure

How can the representations that have been described be used as models for the structure of social relations? The representations encode relations between paths in networks, and paths define the possible routes by which social processes flow. A number of models for social processes have been discussed in the literature – for instance for such phenomena as the transfer of job information (Boorman, 1975; Granovetter, 1973), alcohol consumption (Skog, 1986), the spread of infection (Rapoport, 1983; Klovdahl, 1985) and the diffusion of technical innovations (Coleman, Katz & Menzel, 1957; Fennell & Warnecke, 1988). These models derive the course and outcome of the diffusion processes that they represent, for example, the proportion of people knowing about a fact *X*, adopting a method *Y*, or being in a state *Z*. The derivation of these results requires assumptions about the probabilities of transmissions and about network structure. As noted earlier, the assumptions made about network structure are critical to the results obtained, but most reported models have used very simple ones. For instance, Rapoport (1979) discussed generalisations of the assumption of uniform connectivity in a single relation to admit biases to reciprocity and transitivity in diffusion models, and Boorman (1975) assumed uniform connectivity for strong and weak relations in his model of job information transmission. These assumed forms for the structure of network relations have received at best moderate empirical support, and the uniform connectivity model seems quite implausible in the light of most of the literature on network structure. Thus, the challenge in constructing social process models is to use empirically plausible assumptions about network structure in the place of assumptions such as that of uniform connectivity. For example, can the calculations made by Boorman (1975) on job information transmission in networks of strong and weak ties be repeated, using one of the Strong–Weak tie models of Table 8.1 instead of uniform connectivity? Can diffusion processes in a single large network be assessed if we assume that the network conforms to Johnsen's "39+" model or to a related model expressed in the form of a partially ordered semigroup?

These questions can only be answered with (a) a much better knowledge of the applicability of models for network structure such as those listed in Table 8.1 and (b) some substantial mathematical developments. They are important questions to consider, however, because the ability to model faithfully a wide range of diffusion processes is important in understanding many social phenomena. The focus of the representations that we have been considering on labelled paths in networks should assist the project because the processes in such models may be postulated to traverse certain types of network paths.

Analysing large networks

Almost all of the analyses presented here have been for small networks or for large networks summarised in the form of a blockmodel. There have been two reasons for this: for ease of presentation and computational simplicity, and because the representation and its analysis lead to a full and detailed description of the network. The analysis is neither a means of extracting only some of the relational forms present in the network nor a procedure for summarising its most "salient" forms: instead, it yields a description of *all* maximally heterogeneous relational forms.

Of course, useful algorithms for performing an initial summary of network structure have been developed for several applications. For instance, there are algorithms for grouping members of an entire network who share similar network relations in the sense of being structurally equivalent (Breiger et al., 1975; Burt, 1976), automorphically equivalent (Borgatti, Boyd, & Everett, 1989; Pattison, 1988; Winship, 1988), regularly equivalent (White & Reitz, 1983, 1989), local role equivalent (Mandel, 1983; Winship & Mandel, 1983) or indegree or outdegree equivalent (Pattison, 1988). The results of each of these clustering algorithms may be used to define a blockmodel of relations among the clusters of network members or abstracted network positions (Pattison, 1988), and this blockmodel may be submitted to the detailed analysis that has been developed. For local networks, variety of procedures, including those described by Mandel (1983), Winship and Mandel (1983) and Breiger and Pattison (1986), may be adapted to aggregate members of a local network and hence to construct a smaller abstracted local network, as illustrated by Pattison (1988).

Another useful approach for a large entire network may be to perform exact analyses for each of the local networks defined by its members. The partially ordered semigroup of the entire network is a two-sided role algebra, and we can think of that role algebra as the intersection of the local role algebras of the individual network members. Whereas the upper limit on semigroup size for a network of n members is 2^k, where $k = 2^n$, the upper limit on distinct relations in any individual role algebra is 2^n. Thus, a considerable reduction in the size of the algebras with which we need to deal can be achieved by this strategy. The analysis will produce a list of features of local networks – some common to more than one local network and some unique to a particular network – that when used in conjunction with the Correspondence Definition will identify the nature and location of many of the relational forms contributing to structure in the entire network.

In the case of very large networks, it may also be useful to consider a model-fitting approach rather than a descriptive one. The steps taken by Pattison et al. (1988) may be particularly helpful in this regard. Their procedure may be used not only to assess the level of fit of network data to a model but also to identify regions in the network where fit is poor. An approach developed by Krackhardt (1988) may also be used to fit network models expressed in linear terms. In either case, a more descriptive local approach could be applied where the fit of the model is poor. Detailed analysis of a partial algebra such as the partial role algebra defined in chapter 2 is another possibility, but the merits and hazards of this strategy require further exploration.

Nonetheless, full analysis of larger networks may sometimes be of interest, and it will be necessary to improve the computational efficiency of the algorithms that have been presented for the analysis of entire and partial networks. There are no major computational obstacles to extending the algorithms to somewhat larger networks and somewhat larger semigroups and role algebras. One major limitation of the current implementation of the factorisation algorithm is an upper limit on semigroup or role algebra size of about 40. This upper limit could be increased by more efficient use of storage. A second area where additional computational improvements can be made is in the development of an efficient lattice-searching algorithm for determining minimal derived set associations for a component of a partially ordered semigroup or a role algebra.

Though the representations and analyses that have been proposed for entire and local networks have made some progress towards the representational goals that were identified in chapter 1, there is still a great need for development. Not only do we need to develop more efficient algorithms to conduct the analyses that have been described and to apply them to a great deal more network data, but it seems particularly important to address the problem of constructing network process models sympathetic to the outcomes of such analyses.

References

Abell, P. (1969). Some problems in the theory of structural balance: Towards a theory of structural strain. In M. Lane (Ed.), *Structuralism* (pp. 389–409). London: Jonathon Cape.

Alba, R. D. (1973). A graph-theoretic definition of a sociometric clique. *Journal of Mathematical Sociology, 3*, 113–126.

Anderson, C. J., Wasserman, S., & Faust, K. (1992). Building stochastic blockmodels. *Social Networks, 14*, 137–161.

Anderson, J. G., & Jay, S. J. (1985). Computers and clinical judgement: The role of physician networks. *Social Science of Medicine, 10*, 969–979.

Arabie, P. (1977). Clustering representations of group overlap. *Journal of Mathematical Sociology, 5*, 113–128.

Arabie, P. (1984). Validation of sociometric structure by data on individuals' attributes. *Social Networks, 6*, 373–403.

Arabie, P., & Boorman, S. A. (1982). Blockmodels: Developments and prospects. In H. C. Hudson & Associates, *Classifying Social Data: New Applications of Analytic Methods for Social Science Research* (pp. 177–198). San Francisco: Jossey-Bass.

Arabie, P., Boorman, S. A., & Levitt, P. R. (1978). Constructing blockmodels: How and why. *Journal of Mathematical Psychology, 17*, 21–63.

Arabie, P., Hubert, L. J., & Schleutermann, S. (1990). Blockmodels from the Bond Energy approach. *Social Networks, 12*, 99–126.

Atkin, R. H. (1977). *Combinatorial connectivities in social systems*. Basel: Birkhauser.

Baker, W. E. (1984). The social structure of a national securities market. *American Journal of Sociology, 89*, 775–811.

Bales, R. F., & Cohen, S. P. (1979). *SYMLOG: A system for the multiple level observation of groups*. New York: The Free Press.

Barnes, J. A. (1954). Class and committees in a Norwegian island parish. *Human Relations, 7*, 39–58.

Barnes, J. A. (1969a). Networks and political process. In J. C. Mitchell (Ed.), *Social networks in urban situations: Analyses of personal relationships in Central African towns* (pp. 51–76). Manchester: Manchester University Press.

Barnes, J. A. (1969b). Graph theory and social networks: A technical comment on connectedness and connectivity. *Sociology, 3*, 215–232.

Batchelder, W. H. (1989). Inferring meaningful global network properties from individual actor's measurement scales. In L. C. Freeman, D. R. White &

A. K. Romney (Eds.), *Research methods in social network analysis* (pp. 89–134). Fairfax, VA: George Mason University Press.

Batchelder, W. H., & Lefebvre, V. (1982). A mathematical analysis of a natural class of partitions of a graph. *Journal of Mathematical Psychology, 26*, 124–148.

Bavelas, A. (1948). A mathematical model for group structure. *Applied Anthropology, 7*, 16–30.

Bernard, H. R., Johnsen, E. C., Killworth, P. D., McCarty, C., Shelley, G. A., & Robinson, S. (1990). Comparing four different methods for measuring personal social networks. *Social Networks, 12*, 179–215.

Bernard, H. R., & Killworth, P. D. (1973). On the social structure of an ocean-going research vessel and other important things. *Social Science Research, 2*, 145–184.

Bernard, H. R., & Killworth, P. D. (1977). Informant accuracy in social network data. II. *Human Communication Research, 4*, 3–18.

Bernard, H. R., & Killworth, P. D. (1978). On the structure and effective sociometric relations in a closed group over time. *Connections, 1*, 44.

Bernard, H. R., Killworth, P. D., Kronenfeld, D., & Sailer, L. (1984). The problem of informant accuracy: The validity of retrospective data. *Annual Review of Anthropology, 13*, 495–517.

Bernard, P. (1973). Stratification sociometrique et réseaux sociaux. *Sociologie et Sociétes, 5*, 127–150.

Bienenstock, E. J., Bonacich, P., & Oliver, M. (1990). The effect of network density and homogeneity on attitude polarisation. *Social Networks, 12*, 153–172.

Birkhoff, G. (1948). *Lattice theory* (rev. ed.). Providence, RI: American Mathematical Society.

Birkhoff, G. (1967). *Lattice theory* (3rd ed.). Providence, RI: American Mathematical Society.

Blau, P. M. (1977). *Inequality and heterogeneity: A primitive theory of social structure.* New York: The Free Press.

Boissevain, J. (1974). *Friends of friends: Networks, manipulators and coalitions.* Oxford: Blackwell.

Boissevain, J., & Mitchell, J. C. (1973). *Network analysis: Studies in human interaction.* The Hague: Mouton.

Bolland, J. M. (1988). Sorting out centrality: An analysis of the performance of four centrality models in real and simulated networks. *Social Networks, 10*, 233–253.

Bonacich, P. (1980). The "common structure semigroup", an alternative to the Boorman and White "joint reduction". *American Journal of Sociology, 86*, 159–166.

Bonacich, P. (1983). Representations for homomorphisms. *Social Networks, 5*, 173–192.

Bonacich, P. (1989). What is a homomorphism? In L. C. Freeman, D. R. White & A. K. Romney (Eds.), *Research methods in social network analysis* (pp. 255–293). Fairfax, VA: George Mason University Press.

Bonacich, P., & McConaghy, M. (1979). The algebra of blockmodeling. In K. F. Schuessler (Ed.), *Sociological Methodology 1980* (pp. 489–532). San Francisco: Jossey-Bass.

Boorman, S. A. (1975). A combinatorial optimization model for transmission of job information through contact networks. *Bell Journal of Economics*, 6, 216–249.

Boorman, S. A. (1977). Informational optima in a formal hierarchy: Calculations using the semigroup. *Journal of Mathematical Sociology*, 5, 129–147.

Boorman, S. A., & White, H. C. (1976). Social structures from multiple networks: II. Role structures. *American Journal of Sociology*, 81, 1384–1446.

Borgatti, S. P., Boyd, J. P., & Everett, M. G. (1989). Iterated roles – mathematics and application. *Social Networks*, 11, 159–172.

Borgatti, S. P., & Everett, M. G. (1989). The class of all regular equivalences: Algebraic structure and computation. *Social Networks*, 11, 65–88.

Borgatti, S. P., Everett, M. G., & Freeman, L. (1991). *UCINET IV*. Columbia, SC: Analytic Technologies.

Bott, E. (1957). *Family and social networks*. London: Tavistock.

Boyd, J. P. (1969). The algebra of group kinship. *Journal of Mathematical Psychology*, 6, 139–167.

Boyd, J. P. (1980). The universal semigroup of relations. *Social Networks*, 2, 91–117.

Boyd, J. P. (1983). Structural similarity, semigroups and idempotents. *Social Networks*, 5, 157–172.

Boyd, J. P. (1989). Social semigroups and Green relations. In L. C. Freeman, D. R. White & A. K. Romney (Eds.), *Research methods in social network analysis* (pp. 215–254). Fairfax, VA: George Mason University Press.

Boyd, J. P. (1991). *Social semigroups: A unified theory of scaling and blockmodelling as applied to social networks*. Fairfax, VA: George Mason University Press.

Boyd, J. P., Haehl, J. H., & Sailer, L. D. (1972). Kinship systems and inverse semigroups. *Journal of Mathematical Sociology*, 2, 37–61.

Boyle, R. P. (1969). Algebraic systems for normal and hierarchical sociograms. *Sociometry*, 32, 99–119.

Bradley, R. T., & Roberts, N. C. (1989). Network structure from relational data: Measurement and inference in four operational models. *Social Networks*, 11, 89–134.

Breiger, R. L. (1974). The duality of persons in groups. *Social Forces*, 53, 181–190.

Breiger, R. L. (1976). Career attributes and network structure: A blockmodel study of a biomedical research specialty. *American Sociological Review*, 41, 117–135.

Breiger, R. L. (1979). Toward an operational theory of community elite structures. *Quality and Quantity*, 13, 21–57.

Breiger, R. L. (1981). Structures of economic interdependence among nations. In P. M. Blau & R. K. Merton (Eds.), *Continuities in structural inquiry* (pp. 353–380). Beverly Hills, CA: Sage.

Breiger, R. L., Boorman, S. A., & Arabie, P. (1975). An algorithm for clustering relational data with applications to social network analysis and comparisons with multidimensional scaling. *Journal of Mathematical Psychology*, 12, 328–382.

Breiger, R. L., & Ennis, J. (1979). Personae and social roles: The network structure of personality types in small groups. *Social Psychology Quarterly*, *42*, 262–270.

Breiger, R. L., & Pattison, P. E. (1978). The joint role structure in two communities' elites. *Sociological Methods and Research*, *7*, 213–226.

Breiger, R. L., & Pattison, P. E. (1986). Cumulated social roles: The duality of persons and their algebras. *Social Networks*, *8*, 215–256.

Brown, G. W., & Harris, T. (1978). *Social origins of depression: A study of psychiatric disorder in women.* London: Tavistock.

Burt, R. S. (1976). Positions in networks. *Social Forces*, *55*, 93–122.

Burt, R. S. (1977). Positions in multiple network systems. Part two: stratification and prestige among elite decision-makers in the community of Altneustadt. *Social Forces*, *56*, 106–131.

Burt, R. S. (1983). Distinguishing relational contents. In R. S. Burt, M. J. Minor & Associates, *Applied network analysis: A methodological introduction* (pp. 35–74). Beverly Hills, CA: Sage.

Burt, R. S. (1984). Network items and the General Social Survey. *Social Networks*, *6*, 293–339.

Burt, R. S. (1986a). A cautionary note. *Social Networks*, *8*, 205–212.

Burt, R. S. (1986b). A note on sociometric order in the General Social Survey network data. *Social Networks*, *8*, 149–174.

Burt, R. S. (1987a). A note on the General Social Survey's ersatz density item. *Social Networks*, *9*, 75–85.

Burt, R. S. (1987b). A note on strangers, friends and happiness. *Social Networks*, *9*, 311–331.

Burt, R. S., & Guilarte, M. G. (1986). A note on scaling the GSS network items response categories. *Social Networks*, *8*, 387–396.

Campbell, K. E., Marsden, P. V., & Hurlbert, J. S. (1986). Social resources and socioeconomic status. *Social networks*, *8*, 97–117.

Cannon, J. J. (1971). Computing the ideal structure of finite semigroups. *Numerische Mathematik*, *18*, 254–266.

Cantor, N., & Kihlstrom, J. F. (Eds.). (1981). *Personality, cognition and social interaction.* Hillsdale, NJ: Lawrence Erlbaum & Associates.

Capobianco, M. F., & Molluzzo, J. C. (1980). The strength of a graph and its application to organizational structure. *Social Networks*, *2*, 275–283.

Carroll, J. D., & Chang, J. J. (1970). Analysis of individual differences in multidimensional scaling via an N-way generalization of "Eckart-Young" decomposition. *Psychometrika*, *35*, 283–319.

Cartwright, D., & Harary, F. (1956). Structural balance: A generalisation of Heider's theory. *Psychological Review*, *63*, 277–293.

Clifford, A. H., & Preston, G. B. (1961). *The algebraic theory of semigroups, vol. 1.* Providence, RI: American Mathematical Society.

Coates, D. L. (1987). Gender differences in the structure and support characteristics of black adolescents' social networks. *Sex Roles*, *17*, 667–687.

Cohen, S., & Syme, S. L. (Eds.). (1985). *Social support and health.* New York: Academic.

Coleman, J. S., Katz, E., & Menzel, H. (1957). *Medical innovation.* New York: Bobbs-Merrill.

Curcione, N. R. (1975). Social relations among inmate addicts. *Journal of Research in Crime and Delinquency, 12*, 61–74.

Davis, A., Gardner, B., & Gardner, M. R. (1941). *Deep South*. Chicago: Chicago University Press.

Davis, J. A. (1967). Clustering and structural balance in graphs. *Human Relations, 20*, 181–187.

Davis, J. A. (1970). Clustering and hierarchy in interpersonal relations: Testing two graph-theoretical models on 742 sociomatrices. *American Sociological Review, 35*, 843–851.

Davis, J. A. (1979). The Davis/Holland/Leinhardt studies: An overview. In P. W. Holland & S. Leinhardt (Eds.), *Perspectives on social network research* (pp. 51–62). New York: Academic.

Davis, J. A., & Leinhardt, S. (1972). The structure of positive interpersonal relations in small groups. In Berger, J. (Ed.), *Sociological theories in progress, II* (pp. 218–251). Boston: Houghton Mifflin.

Donninger, C. (1986). The distribution of centrality in social networks. *Social Networks, 8*, 191–203.

Doreian, P. (1974). On the connectivity of social networks. *Journal of Mathematical Sociology, 3*, 245–258.

Doreian, P. (1979). Structural control models for group processes. In P. W. Holland & S. Leinhardt (Eds.), *Perspectives on social network research* (pp. 201–221). New York: Academic.

Doreian, P. (1980). On the evolution of group and network structure. *Social Networks, 2*, 235–252.

Doreian, P. (1986). On the evolution of group and network structure II: Structures within structures. *Social Networks, 8*, 33–64.

Doreian, P. (1988a). Using multiple network analytic tools for a single social network. *Social Networks, 10*, 287–312.

Doreian, P. (1988b). Equivalences in a social network. *Journal of Mathematical Sociology, 13*, 243–282.

Dubois, D., & Prade, H. (1980). *Fuzzy sets and systems: Theory and applications*. New York: Academic.

Ennis, J. G. (1982). Blockmodels and spatial representations of group structures: Some comparisons. In H. C. Hudson & Associates, *Classifying social data: New applications of analytic methods for social science research* (pp. 199–214). San Francisco: Jossey-Bass.

Erickson, B. H. & Nosanchuk, T. A. (1983). Applied network sampling. *Social Networks, 3*, 367–382.

Everett, M. G., & Borgatti, S. (1988). Calculating role similarities: An algorithm that helps determine the orbits of a graph. *Social Networks, 10*, 77–91.

Fararo, T. J. (1981). Biased networks and social structure theorems. *Social Networks, 3*, 137–159.

Fararo, T. J. (1983). Biased networks and the strength of weak ties. *Social Networks, 5*, 1–11.

Fararo, T. J., & Doreian, P. (1984). Tripartite structural analysis: Generalising the Breiger–Wilson formalism. *Social Networks, 6*, 177–192.

Fararo, T. J., & Skvoretz, J. (1984). Biased networks and social structure theorems: Part II. *Social Networks*, *6*, 223–258.

Faust, K. (1988). Comparison of methods for positional analysis: Structural and general equivalences. *Social Networks*, *10*, 313–341.

Faust, K., & Romney, A. K. (1985). Does STRUCTURE find structure? A critique of Burt's use of distance as a measure of structural equivalence. *Social Networks*, *7*, 77–103.

Faust, K., & Wasserman, S. (1992). Blockmodels: Interpretation and evaluation. *Social Networks*, *14*, 5–61.

Faust, K., & Wasserman, S. (1993). Centrality and prestige: A review and synthesis. *Journal of Quantitative Anthropology* (to appear).

Feiring, C., & Coates, D. (1987). Social networks and gender differences in the life space of opportunity: Introduction. *Sex Roles*, *17*, 611–620.

Fennell, M. L., & Warnecke, R. B. (1988). *The diffusion of medical innovations: An applied network analysis*. New York: Plenum Press.

Fienberg, S. E., Meyer, M. M., & Wasserman, S. (1985). Statistical analysis of multiple sociometric relations. *Journal of the American Statistical Association*, *80*, 51–67.

Fienberg, S. E., & Wasserman, S. (1981). Categorical data analysis of single sociometric relations. In S. Leinhardt (Ed.), *Sociological methodology 1981* (pp. 156–192). San Francisco: Jossey-Bass.

Fischer, C. (1982). *To dwell among friends: Personal networks in town and city*. Chicago: University of Chicago Press.

Fischer, C., Jackson, R. M., Stueve, C. A., Gerson, K., & McAllister Jones, L. (1977). *Networks and places: Social relations in the urban setting*. New York: The Free Press.

Fiske, S. T., & Taylor, S. E. (1990). *Social cognition* (2nd ed.). New York: McGraw-Hill.

Flament, C. (1963). *Applications of graph theory to group structure*. Englewood Cliffs, NJ: Prentice-Hall.

Ford, L. R., & Fulkerson, D. (1962). *Flows in networks*. Princeton: Princeton University Press.

Forsythe, G. E. (1955). SWAC computes 126 distinct semigroups of order 4. *Proceedings of the American Mathematical Society*, *6*, 443–447.

Foster, C. C., & Horvath, W. J. (1971). A study of a large sociogram. III. Reciprocal choice probabilities as a measure of social distance. *Behavioral Science*, *16*, 429–435.

Foster, C. C., Rapoport, A., & Orwant, C. J. (1963). A study of a large sociogram. II. Elimination of free parameters. *Behavioral Science*, *8*, 56–65.

Frank, O. (1971). *Statistical inference in graphs*. Stockholm: Forsvarets Forkininganstalt.

Frank, O. (1979). Estimation of population totals by use of snowball samples. In P. W. Holland & S. Leinhardt (Eds.), *Perspectives on social network research* (pp. 319–347). New York: Academic.

Frank, O. (1980). Transitivity in stochastic graphs and digraphs. *Journal of Mathematical Sociology*, *7*, 199–213.

Frank, O., & Harary, F. (1980). Balance in stochastic signed graphs. *Social Networks, 2,* 155–163.

Frank, O., & Strauss, D. (1986). Markov graphs. *Journal of the American Statistical Association, 81,* 832–842.

Freeman, L. C. (1979). Centrality in social networks: Conceptual clarification. *Social Networks, 1,* 215–239.

Freeman, L. C. (1983). Spheres, cubes and boxes: Graph dimensionality and network structure. *Social Networks, 5,* 139–156.

Freeman, L. C. (1989). Social networks and the structure experiment. In L. C. Freeman, D. R. White, & A. K. Romney (Eds.), *Research methods in social network analysis* (pp. 11–40). Fairfax, VA: George Mason University Press.

Freeman, L. C., & Romney, A. K. (1987). Words, deeds and social structure. *Human Organization, 46,* 330–334.

Freeman, L. C., Romney, A. K., & Freeman, S. C. (1986). Cognitive structure and informant accuracy. *American Anthropologist, 89,* 310–325.

Friedell, M. F. (1967). Organizations as semilattices. *American Sociological Review, 32,* 46–54.

Friedkin, N. (1980). A test of structural features of Granovetter's strength of weak ties theory. *Social Networks, 2,* 411–420.

Friedkin, N. (1981). The development of structure in random networks: An analysis of the effects of increasing network density on five measures of structure. *Social Networks, 3,* 41–52.

Fuchs, L. L. (1963). *Partially ordered algebraic systems.* Oxford: Pergamon.

Gaines, B. R. (1977). System identification, approximation and complexity. *International Journal of General Systems, 3,* 145–174.

Goodman, L. (1961). Snowball sampling. *Annals of Mathematical Statistics, 32,* 148–170.

Gould, R. V. (1987). Measures of betweenness in non-symmetric networks. *Social Networks, 9,* 277–282.

Granovetter, M. (1973). The strength of weak ties. *American Journal of Sociology, 78,* 1360–1380.

Granovetter, M. (1974). *Getting a job: A study of contacts and careers.* Cambridge, MA: Harvard University Press.

Granovetter, M. (1976). Network sampling: Some first steps? *American Journal of Sociology, 81,* 1287–1303.

Granovetter, M. (1979). The theory-gap in social network analysis. In P. W. Holland & S. Leinhardt (Eds.), *Perspectives on social network research* (pp. 501–518). New York: Academic.

Granovetter, M. (1982). The strength of weak ties: A network theory revisited. In P. V. Marsden & N. Lin (Eds.), *Social structure and network analysis* (pp. 105–130). Beverly Hills, CA: Sage.

Granovetter, M. (1985). Economic action and social structure: The problem of embeddedness. *American Journal of Sociology, 91,* 481–510.

Grossman, I., & Magnus, W. (1964). *Groups and their graphs.* New York: Random House.

Guttman, L. (1979). A definition of dimensionality and distance for graphs. In

J. C. Lingoes, E. E. Roskam & I. Borg (Eds.), *Geometric representations of relational data* (2nd ed., pp. 713–723). Ann Arbor, MI: Mathesis.

Hall, A., & Wellman, B. (1985). Social networks and social support. In S. Cohen & S. L. Syme (Eds.), *Social support and health* (pp. 23–41). Orlando, FL: Academic.

Hallinan, M. (1974). *The structure of positive sentiment.* New York: Elsevier.

Hallinan, M. (1978). The process of friendship formation. *Social Networks, 1,* 193–210.

Hallinan, M., & Felmlee, D. (1975). An analysis of intransitivity in sociometric data. *Sociometry, 38,* 195–212.

Hammer, M. (1984). Explorations into the meaning of social network interview data. *Social Networks, 6,* 341–371.

Harary, F. (1959a). Graph-theoretic methods in the management sciences. *Management Science, 5,* 387–403.

Harary, F. (1959b). Status and contrastatus. *Sociometry, 22,* 23–43.

Harary, F., & Kommel, H. J. (1979). Matrix measures for transitivity and balance. *Journal of Mathematical Sociology, 6,* 199–210.

Harary, F., Norman, R., & Cartwright, D. (1965). *Structural models: An introduction to the theory of directed graphs.* New York: Wiley.

Harre, R., & Secord, P. F. (1972). *The explanation of social behaviour.* Oxford: Blackwell.

Heil, G. (1983). *Algorithms for network homomorphism: Blockmodeling as a structural analytic method for social structure.* Structural Analysis Programme, Occasional Paper No. 2. Department of Sociology, University of Toronto.

Heil, G., & White, H. C. (1976). An algorithm for finding simultaneous homomorphic correspondences between graphs and their image graphs. *Behavioral Science, 21,* 26–35.

Henderson, S., Byrne, D. G., & Duncan-Jones, P. (1981). *Neurosis and the social environment.* New York: Academic.

Herman, N. J. (1984). Conflict in the Church: A social network analysis of an Anglican congregation. *Journal for the Scientific Study of Religion, 23,* 60–74.

Higgins, C. A., McLean, R. J., & Conrath, D. W. (1985). The accuracy and biases of diary communication data. *Social Networks, 7,* 173–187.

Holland, P. W., Laskey, K. B., & Leinhardt, S. (1983). Stochastic blockmodels: Some first steps. *Social Networks, 5,* 109–137.

Holland, P. W., & Leinhardt, S. (1970). A method for detecting structure in sociometric data. *American Journal of Sociology, 76,* 492–513.

Holland, P. W., & Leinhardt, S. (1971). Transitivity in structural models of small groups. *Comparative Group Studies,* 2, 107–124.

Holland, P. W., & Leinhardt, S. (1972). Some evidence on the transitivity of positive interpersonal sentiment. *American Journal of Sociology, 77,* 1205–1209.

Holland, P. W., & Leinhardt, S. (1973). The structural implications of measurement error in sociometry. *Journal of Mathematical Sociology, 3,* 85–111.

Holland, P. W., & Leinhardt, S. (1975). Local structure in social networks. In

D. Heise (Ed.), *Sociological methodology 1976* (pp. 1–45). San Francisco: Jossey-Bass.
Holland, P. W., & Leinhardt, S. (1977). A dynamic model for social networks. *Journal of Mathematical Sociology, 5*, 5–20.
Holland, P. W., & Leinhardt, S. (1978). An omnibus test for social structure using triads. *Sociological Methods & Research*, 7, 227–255.
Holland, P. W., & Leinhardt, S. (1981). An exponential family of probability distributions for directed graphs. *Journal of the American Statistical Association*, 76, 33–50.
Homans, G. C. (1951). *The human group*. London: Routledge & Kegan Paul.
Huang, G., & Tausig, M. (1990). Network range in personal networks. *Social Networks, 12*, 261–268.
Hubert, L. J. (1987). *Assignment methods in combinatorial data analysis*. New York: Marcel Dekker.
Hunter, J. E. (1978). Dynamic sociometry. *Journal of Mathematical Sociology*, 6, 87–138.
Iacobucci, D., & Wasserman, S. (1988). A general framework for the statistical analysis of sequential dyadic interaction data. *Psychological Bulletin, 103*, 379–390.
Iacobucci, D., & Wasserman, S. (1990). Social networks with two sets of actors. *Psychometrika, 55*, 707–720.
Johnsen, E. C. (1985). Network macrostructure models for the Davis–Leinhardt set of empirical sociomatrices. *Social Networks*, 7, 203–224.
Johnsen, E. C. (1986). Structure and process agreement models for friendship formation. *Social Networks*, 8, 257–306.
Kadushin, C. (1982). Social density and mental health. In P. V. Marsden & N. Lin (Eds.), *Social structure and network analysis* (pp. 147–158). Beverly Hills, CA: Sage.
Kapferer, B. (1969). Norms and the manipulation of relationships in a work context. In J. C. Mitchell (Ed.), *Social networks in urban situations: Analyses of personal relationships in Central African towns* (pp. 181–244). Manchester: Manchester University Press.
Kapferer, B. (1972). *Strategy and transaction in an African factory*. Manchester: Manchester University Press.
Katz, L. (1953). A new status index derived from sociometric analysis. *Psychometrika, 18*, 39–43.
Kaufman, A. (1975). *Introduction to the theory of fuzzy subsets. I*. New York: Academic.
Kent, D. (1978). *The rise of the Medici: Faction in Florence, 1426–1434*. Oxford: Oxford University Press.
Kerckhoff, A., Back, K., & Miller, N. (1965). Sociometric patterns in hysterical contagion. *Sociometry*, 28, 2–15.
Kessler, R. C., & McLeod, J. D. (1985). Social support and mental health in community samples. In S. Cohen & S. L. Syme (Eds.), *Social support and health* (pp. 219–240). Orlando, FL: Academic.
Kessler, R. C., Price, R. H., & Wortaman, C. B. (1985). Social factors in psychopathology. *Annual Review of Psychology*, 36, 531–572.

Killworth, P. D., & Bernard, H. R. (1976a). Informant accuracy in social network data. *Human Organization, 35*, 269–286.

Killworth, P. D., & Bernard, H. R. (1976b). A model of human group dynamics. *Social Science Research, 5*, 173–224.

Killworth, P. D., & Bernard, H. R. (1979). Informant accuracy in social network data, III. *Social Networks*, 2, 10–46.

Kim, K. H. (1982). *Boolean matrix theory and applications*. New York: Marcel Dekker.

Kim, K. H., & Roush, F. W. (1978). Two-generator semigroups of binary relations. *Journal of Mathematical Psychology, 17*, 236–246.

Kim, K. H., & Roush, F. W. (1983). *Applied abstract algebra*. Chichester: Ellis Horwood.

Kim, K. H., & Roush, F. W. (1984). Group relationships and homomorphisms of Boolean matrix semigroups. *Journal of Mathematical Psychology, 28*, 448–452.

Klovdahl, A. S. (1985). Social networks and the spread of infectious diseases: The AIDS example. *Social Science and Medicine*, 21, 1203–1216.

Knoke, D., & Kuklinski, J. H. (1982). *Network analysis*. Newbury Park: Sage.

Kotler, T., & Pattison, P. E. (1977). *Representing interpersonal relations in families, I. A structural analysis* (Melbourne Psychology Reports, No. 29). University of Melbourne, Department of Psychology.

Krackhardt, D. (1987). Cognitive social structures. *Social Networks, 9*, 109–134.

Krackhardt, D. (1988). Predicting with networks: Nonparametric multiple regression analyses of dyadic data. *Social Networks, 10*, 359–381.

Krackhardt, D., & Porter, L. W. (1987). The snowball effect: Turnover embedded in communication networks. *Journal of Applied Psychology, 71*, 50–55.

Krantz, D. H., Luce, R. D., Suppes, P., & Tversky, A. (1971). *Foundations of measurement, Vol. 1. Additive and polynomial representations*. New York: Academic.

Krohn, K., & Rhodes, J. (1965). Algebraic theory of machines. I. Prime decomposition theorem for finite semigroups and machines. *Transactions of the American Mathematical Society, 116*, 450–464.

Krohn, K., Rhodes, J., & Tilson, B. R. (1968). The prime decomposition theorem of the algebraic theory of machines. In M. A. Arbib (Ed.), *Algebraic theory of machines, languages and semigroups* (pp. 81–125). New York: Academic.

Kurosh, A. G. (1963). *Lectures on general algebra* (K. A. Hirsch, Trans.). New York: Chelsea.

Landau, H. G. (1951). On dominance relations and the structure of animal societies. I. *Bulletin of Mathematical Biophysics, 13*, 1–19.

Langeheine, R., & Andresen, N. (1982). A sociometric status index without subgroup membership bias. *Social Networks, 4*, 233–242.

Laumann, E. (1973). *Bonds of pluralism*. New York: Wiley.

Laumann, E. O., Marsden, P. V., & Galaskiewicz, J. (1977). Community-elite influence structures: Extension of a network approach. *American Journal of Sociology*, 83, 594–631.

Laumann, E., Marsden, P. V., & Prensky, D. (1983). The boundary specification problem in network analysis. In R. S. Burt, M. J. Minor & Associates, *Applied network analysis: A methodological introduction* (pp. 18–34). Beverly Hills, CA: Sage.

Laumann, E. O., Marsden, P. V., & Prensky, D. (1989). The boundary specification problem in network analysis. In L. C. Freeman, D. R. White & A. K. Romney (Eds.), *Research methods in social network analysis* (pp. 61–87). Fairfax, VA: George Mason University Press.

Laumann, E. O., & Pappi, F. U. (1973). New directions in the study of community elites. *American Sociological Review, 38*, 212–230.

Laumann, E. O., & Pappi, F. U. (1976). *Networks of collective action: A perspective on community influence systems.* New York: Academic.

Laumann, E. O., Verbrugge, L., & Pappi, F. U. (1974). A causal modelling approach to the study of a community elite's influence structure. *American Sociological Review*, 39, 162–174.

Lee, N. H. (1969). *The search for an abortionist.* Chicago: Chicago University Press.

Leinhardt, S. (1972). Developmental change in the sentiment structure of children's groups. *American Sociological Review, 37*, 202–212.

Leung, S., Pattison, P. E., & Wales, R. J. (1992). *The semantics of friend.* Unpublished manuscript, University of Melbourne, Department of Psychology.

Lievrouw, L. A., Rogers, E. M., Lowe, C. U., & Nadel, E. (1987). Triangulation as a research strategy for identifying invisible colleges among biomedical scientists. *Social Networks*, 9, 217–248.

Light, J. M., & Mullins, N. C. (1979). A primer on blockmodeling procedure. In P. W. Holland & S. Leinhardt, (Eds.), *Perspectives on social network research* (pp. 85–118). New York: Academic.

Lin, N., Dayton, P., & Greenwald, P. (1978). Analysing the instrumental use of social relations in the context of social structure. *Sociological Methods and Research*, 7, 149–166.

Lin, N., Dean, A., & Ensel, W. (Eds.). (1986). *Social support, life events and depression.* New York: Academic.

Lin, N., & Dumin, M. (1986). Access to occupations through social ties. *Social Networks, 8*, 365–385.

Lin, N., Ensel, W. M., & Vaughn, J. C. (1981). Social resources and strength of ties: Structural factors in occupational status attainment. *American Sociological Review, 46*, 393–405.

Linzell, J. (1975). [Review of *Mathematical Structure in Human Affairs*]. *Quality and Quantity*, 9, 371–376.

Lorrain, F. P. (1972). *The network-organisation of social systems and cultural modes of classification.* Doctoral Dissertation. Department of Sociology, Harvard University.

Lorrain, F. P. (1973). *Handbook of two-block–two-generator models.* Unpublished manuscript, Michigan Society of Fellows.

Lorrain, F. P. (1975). *Réseaux sociaux et classifications sociales.* Paris: Hermann.

Lorrain, F. P., & White, H. C. (1971). Structural equivalence of individuals in social networks. *Journal of Mathematical Sociology, 1*, 49–80.

Luce, R. D. (1950). Connectivity and generalised cliques in sociometric group structure. *Psychometrika, 15*, 169–190.

Luce, R. D., & Perry, A. (1949). A method of matrix analysis of group structure. *Psychometrika, 14*, 94–116.

Magnusson, D., & Endler, N. (Eds.). (1977). *Personality at the crossroads: Current issues in interactional psychology*. Hillsdale, NJ: Lawrence Erlbaum & Associates.

Mandel, M. J. (1978). *Roles and networks: A local approach*. Unpublished B. A. Honours thesis, Department of Applied Mathematics, Harvard University.

Mandel, M. J. (1983). Local roles and social networks. *American Sociological Review, 48*, 376–386.

Marsden, P. V. (1983). Restricted access in networks and models of power. *American Journal of Sociology, 88*, 686–717.

Marsden, P. V. (1986). Core discussion networks of Americans. *American Sociological Review, 52*, 122–131.

Marsden, P. V. (1988). Homogeneity in confiding relations. *Social Networks, 10*, 57–76.

Marsden, P. V. (1990). Network data and measurement. *Annual Review of Sociology*, 16, 435–463.

Marsden, P. V., & Lin, N. (Eds.). (1982). *Social structure and network analysis*. Beverly Hills, CA: Sage.

Mayer, A. C. (1977). The significance of quasi-groups in the study of complex societies. In S. Leinhardt (Ed.), *Social networks: A developing paradigm* (pp. 293–318). New York: Academic.

McConaghy, M. J. (1981). The common role structure: Improved blockmodeling methods applied to two communities' elites. *Sociological Methods & Research*, 9, 267–285.

Meyer, J. F. (1972). Algebraic isomorphism invariants for graphs of automata. In R. C. Read (Ed.), *Graph theory and computing* (pp. 123–152). New York: Academic.

Mintz, B. (1984). *Bank hegemony: Corporate networks and intercorporate power*. Chicago: University of Chicago Press.

Mintz, B., & Schwartz, M. (1981). Interlocking directorates and interest group formation. *American Sociological Review*, 46, 851–869.

Mitchell, J. C. (Ed.). (1969). *Social networks in urban situations: Analysis of personal relationships in Central African towns*. Manchester: Manchester University Press.

Mizruchi, M., & Schwartz, M. (Eds.). (1987). *Intercorporate relations: The structural analysis of business*. New York: Academic.

Mokken, R. J. (1979). Cliques, clubs and clans. *Quality and Quantity, 13*, 161–173.

Moscovici, S. (1972). Society and theory in social psychology. In J. Israel & H. Tajfel (Eds.), *The context of social psychology: A critical assessment* (pp. 17–68). New York: Academic.

Mullins, N., Hargens, L., Hecht, P., & Kick, E. (1977). The group structure of cocitation clusters: A comparative study. *American Sociological Review*, 42, 552–562.

Murray, S. O., & Poolman, R. C. (1982). Strong ties and scientific literature. *Social Networks*, 4, 225–232.

Nadel, S. F. (1957). *The theory of social structure.* Melbourne: Melbourne University Press.
Naylor, A. W. (1981). On decomposition theory: Generalised dependence. *IEEE Transactions on Systems, Man and Cybernetics, SMC-11*, 699–713.
Naylor, A. W. (1983). On decomposition theory: Duality. *IEEE Transactions on Systems, Man and Cybernetics, SMC-13*, 215–221.
Newcomb, T. M. (1961). *The acquaintance process.* New York: Holt, Rinehart & Winston.
Nieminen, U. J. (1973). On status in an organization. *Behavioral Science, 18*, 417–419.
Nordlie, P. B. (1958). *A longitudinal study of interpersonal attraction in a natural group setting.* Ann Arbor, MI: University Microfilms. No. 58–7775.
Oeser, O., & Harary, F. (1964). A mathematical model for structural role theory. II. *Human Relations, 17*, 3–17.
Oliveri, M. E., & Reiss, D. (1987). Social networks of family members: Distinctive roles of mothers and fathers. *Sex Roles, 17*, 719–736.
Osgood, C. E., Suci, G. J., & Tannenbaum, P. H. (1957). *The measurement of meaning.* Urbana, IL: University of Illinois Press.
Pattison, P. E. (1980). *An algebraic analysis for multiple social networks.* Ph.D. Thesis, University of Melbourne.
Pattison, P. E. (1981). Equating the "joint reduction" with blockmodel common role structure: A reply to McConaghy. *Sociological Methods and Research, 9*, 286–302.
Pattison, P. E. (1982). The analysis of semigroups of multirelational systems. *Journal of Mathematical Psychology, 25*, 87–118.
Pattison, P. E. (1988). Network models: Some comments on papers in this special issue. *Social Networks, 10*, 383–411.
Pattison, P. E. (1989). Mathematical models for local social networks. In J. A. Keats, R. Taft, R. A. Heath, & S. H. Lovibond (Eds.), *Mathematical and theoretical systems* (pp. 139–149). Amsterdam: North Holland.
Pattison, P. E., & Bartlett, W. K. (1975). *An algebraic approach to group structure: The organization as a semilattice* (Melbourne Psychology Reports, No. 16). University of Melbourne, Department of Psychology.
Pattison, P. E., & Bartlett, W. K. (1982). A factorization procedure for finite algebras. *Journal of Mathematical Psychology, 25*, 51–81.
Pattison, P. E., Caputi, P., & Breiger, R. L. (1988, August). *Fitting relational models to binary relational data.* Paper presented at the Mathematical Psychology Satellite Conference, Armidale, N.S.W., Australia.
Peay, E. (1977a). Indices for consistency in qualitative and quantitative structures. *Human Relations, 30*, 343–361.
Peay, E. (1977b). Matrix operations and the properties of networks and directed graphs. *Journal of Mathematical Psychology, 15*, 89–101.
Peay, E. R. (1980). Connectedness in a general model for valued networks. *Social Networks, 2*, 385–410.
Piliksuk, M., & Froland, C. (1978). Kinship, social networks, social support and health. *Social Science and Medicine, 12*, 272–280.

Pool, I., & Kochen, M. (1978). Contacts and influence. *Social Networks, 1,* 5–51.

Press, W. H., Flannery, B. P., Teukolsky, S. A., & Vetterling, W. J. (1986). *Numerical recipes: The art of scientific computing.* New York: Cambridge University Press.

Rapoport, A. (1957). A contribution to the theory of random and biased nets. *Bulletin of Mathematical Biophysics, 19,* 257–277.

Rapoport, A. (1979). Some problems relating to randomly constructed biased networks. In P. W. Holland & S. Leinhardt (Eds.), *Perspectives on social research* (pp. 119–136). New York: Academic.

Rapoport, A. (1983). *Mathematical models in the social and behavioral sciences.* New York: Wiley.

Rapoport, A., & Horvath, W. J. (1961). A study of a large sociogram. *Behavioral Science, 6,* 279–291.

Roethlisberger, F. J., & Dickson, W. J. (1939). *Management and the worker.* Cambridge, MA: Harvard University Press.

Rogers, E. M. (1987). Progress, problems and prospects for network research: Investigating relationships in the age of electronic communication technologies. *Social Networks, 9,* 285–310.

Romney, A. K., & Faust, K. (1983). Predicting the structure of a communications network from recalled data. *Social Networks, 4,* 285–304.

Rosenthal, N., Fingrutd, M., Ethier, M., Karant, R., & McDonald, D. (1985). Social movements and network analysis: A case study of nineteenth-century women's reform in New York State. *American Journal of Sociology, 90,* 1022–1054.

Rossignol, C., & Flament, C. (1975). Décomposition de l'équilibre structural (aspects de la répresentation du groupe). *Année Psychologique, 75,* 417–425.

Runger, G., & Wasserman, S. (1979). Longitudinal analysis of friendship networks. *Social Networks, 2,* 143–154.

Rytina, S., & Morgan, D. L. (1982). The arithmetic of social relations: The interplay of category and network. *American Journal of Sociology, 88,* 88–113.

Sampson, S. F. (1969). Crisis in a cloister. Doctoral dissertation. Cornell University.

Schein, B. M. (1970). A construction for idempotent binary relations. *Proceedings of the Japanese Academy, 46,* 246–247.

Schott, T. (1987). Interpersonal influence in science: Mathematicians in Denmark and Israel. *Social Networks, 9,* 351–374.

Schwarz, S. (1970a). On idempotent binary relations on a finite set. *Czechoslovak Mathematical Journal, 20* (95), 696–702.

Schwarz, S. (1970b). On the semigroup of binary relations on a finite set. *Czechoslovak Mathematical Journal, 20* (95), 632–679.

Seidman, S. B. (1983). Internal cohesion of LS sets in graphs. *Social Networks, 5,* 97–107.

Seidman, S. B., & Foster, B. L. (1978). A graph-theoretic generalisation of the clique concept. *Journal of Mathematical Sociology, 6,* 139–154.

Seiyama, K. (1977). Personal communication, January, 1977.

Skog, O.-J. (1986). The long waves of alcohol consumption: A social network perspective on cultural change. *Social Networks, 8,* 1–32.

Snijders, T. A. B. (1981). The degree variance: An index of graph heterogeneity. *Social Networks, 3*, 163–174.

Snyder, D., & Kick, E. L. (1979). Structural position in the world system and economic growth, 1955–1970: A multiple network analysis of transnational interactions. *American Journal of Sociology, 84*, 1096–1126.

Stephenson, K., & Zelen, M. (1989). Rethinking centrality: Methods and examples. *Social Networks, 11*, 1–37.

Strauss, D., & Freeman, L. (1989). Stochastic modeling and the analysis of structural data. In L. C. Freeman, D. R. White & A. K. Romney (Eds.), *Research Methods in social network analysis* (pp. 135–183). Fairfax, VA: George Mason University Press.

Sudman, S. (1988). Experiments in measuring neighbour and relative social networks. *Social Networks, 10*, 93–108.

Thurman, B. (1979). In the office: Networks and coalitions. *Social Networks*, 2, 47–63.

Vickers, M. (1981). *Relational analysis: An applied evaluation*. Unpublished M.S. Thesis, University of Melbourne, Department of Psychology.

Vickers, M., & Chan, S. (1980). *Representing classroom social structure: A methodological study*. Unpublished manuscript, Victorian Department of Education, Melbourne.

Wang, Y. J., & Wong, G. Y. (1987). Stochastic blockmodels for directed graphs. *Journal of the American Statistical Association, 82*, 8–19.

Wasserman, S. (1980). Analysing social networks as stochastic processes. *Journal of the American Statistical Association, 75*, 280–294.

Wasserman, S., & Anderson, C. (1987). Stochastic *a posteriori* blockmodels: Construction and assessment. *Social Networks, 9*, 1–36.

Wasserman, S., & Faust, K. (1993). *Social network analysis: Methods and applications*. New York: Cambridge University Press.

Wasserman, S., & Galaskiewicz, J. (1984). Some generalisations of p_1: External constraints, interaction and non-binary relations. *Social Networks, 6*, 177–192.

Wasserman, S., & Iacobucci, D. (1986). Statistical analysis of discrete relational data. *British Journal of Mathematical and Statistical Psychology, 39*, 41–64.

Wasserman, S., & Iacobucci, D. (1988). Sequential social network data. *Psychometrika, 53*, 261–282.

Wasserman, S., & Iacobucci, D. (1991). Statistical modelling of one-mode and two-mode networks: Simultaneous analysis of graphs and bipartite graphs. *British Journal of Statistical and Mathematical Psychology, 44*, 13–43.

Wellman, B. (1979). The community question: The intimate networks of East Yorkers. *American Journal of Sociology, 84*, 1201–1231.

Wellman, B. (1982). Studying personal communities. In P. V. Marsden & N. Lin (Eds.), *Social structure and network analysis* (pp. 61–80). Beverly Hills, CA: Sage.

Wellman, B. (1983). Network analysis: Some basic principles. In R. Collins (Ed.), *Sociological theory 1983* (pp. 155–200). San Francisco: Jossey-Bass.

Wellman, B. (1988). Structural analysis: From method and metaphor to theory and substance. In B. Wellman & S. D. Berkowitz (Eds.), *Social structures: A network approach* (pp. 19–61). New York: Cambridge University Press.

White, D. R., & McCann, H. G. (1988). Cites and fights: Material entailment analysis of the eighteenth-century chemical revolution. In B. Wellman & S. D. Berkowitz (Eds.), *Social structures: A network approach* (pp. 380–400). New York: Cambridge University Press.

White, D. R., & Reitz, K. P. (1983). Graph and semigroup homomorphisms on networks of relations. *Social Networks, 5*, 193–234.

White, D. R., & Reitz, K. P. (1989). Rethinking the role concept: Homomorphisms on social networks. In L. C. Freeman, D. R. White & A. K. Romney (Eds.), *Research methods in social network analysis* (pp. 429–488). Fairfax, VA: George Mason University Press.

White, H. C. (1961). Management conflict and sociometric structure. *American Journal of Sociology, 67*, 185–199.

White, H. C. (1963). *An anatomy of kinship: Mathematical models for structures of cumulated roles*. Englewood Cliffs, NJ: Prentice-Hall.

White, H. C. (1970). *Chains of opportunity: System models of mobility in organizations*. Cambridge, MA: Harvard University Press.

White, H. C. (1977). Probabilities of homomorphic mappings from multiple graphs. *Journal of Mathematical Psychology, 16*, 121–134.

White, H. C., Boorman, S. A., & Breiger, R. L. (1976). Social structure from multiple networks: I. Blockmodels of roles and positions. *American Journal of Sociology, 81*, 730–780.

Whyte, W. F. (1943). *Street corner society* (2nd ed.). Chicago: Chicago University Press.

Wilson, T. P. (1982). Relational networks: An extension of sociometric concepts. *Social Networks, 4*, 105–116.

Winship, C. (1988). Thoughts about roles and relations: An old document revisited. *Social Networks, 10*, 209–231.

Winship, C., & Mandel, M. J. (1983). Roles and positions: A critique and extension of the blockmodelling approach. In Leinhardt, S. (Ed.), *Sociological Methodology 1983–1984* (pp. 314–344). San Francisco: Jossey-Bass.

Wish, M. (1976). Comparisons among multidimensional structures of interpersonal relations. *Multivariate Behavioral Research, 11*, 297–324.

Wish, M., Deutsch, M., & Kaplan, S. J. (1976). Perceived dimensions of interpersonal relations. *Journal of Personality and Social Psychology 33*: 409–420.

Wu, L. L. (1983). Local blockmodel algebras for analyzing social networks. In S. Leinhardt (Ed.), *Sociological Methodology 1983–1984* (pp. 272–313). San Francisco: Jossey-Bass.

Appendix A
Some basic mathematical terms

A *set* X is a collection of objects or *elements*; we use $x \in X$ to indicate that the element x is a member of the set X and we can write X as $\{x: x \in X\}$. The *intersection* $X \cap Y$ of two sets X and Y is the set of elements belonging to both X and Y; that is,

$$X \cap Y = \{z: z \in X \text{ and } z \in Y\}.$$

The *union* $X \cup Y$ of X and Y is the set of elements that are members of at least one of the sets X and Y; that is,

$$X \cup Y = \{z: z \in X \text{ or } z \in Y \text{ (or } z \in X \cap Y)\}.$$

The *null set* or *empty set* is the set $\{\ \}$ containing no elements. Two sets are *disjoint* if their intersection is null.

The *Cartesian product* $X \times Y$ of two sets X and Y is the set of all ordered pairs of elements from X and Y, that is,

$$X \times Y = \{(x, y): x \in X, y \in Y\}.$$

A *binary relation* R on a set X is a subset of the Cartesian product $X \times X$. A binary relation R on X is *reflexive* if $(x, x) \in R$, for all $x \in X$; it is *symmetric* if $(x, y) \in R$ implies $(y, x) \in R$, for any $x, y \in X$; and it is *transitive* if $(x, y) \in R$ and $(y, z) \in R$ implies $(x, z) \in R$, for any $x, y, z \in X$. A binary relation that is reflexive, symmetric and transitive is termed an *equivalence relation*. If R is an equivalence relation, we can define its *equivalence classes* as subsets of elements of X that are related by R to x, that is,

$$\pi_x = \{y \in X: (x, y) \in R\}.$$

Classes of an equivalence relation are either disjoint or identical and define a *partition* on X; that is, we can write X as the union of disjoint subsets, each of which is a class of the equivalence relation R.

A binary relation R on X is a *quasi-order* if R is reflexive and transitive. R is *antisymmetric* if $(x, y) \in R$ and $(y, x) \in R$ implies $x = y$, for any $x, y \in X$. A quasi-order that is also antisymmetric is termed a *partial order*.

Quasi-orders and partial orders are often represented as relations $\leq$ or $\geq$ such that $x \leq y$ or $x \geq y$ if and only if $(x, y) \in R$. Given a quasi-order $\geq$ defined on a set X, we can also define the following relations on X:

$x = y$ iff $x \geq y$ and $y \geq x$; $x, y \in X$;
$x > y$ iff $x \geq y$ but not $x = y$; $x, y \in X$;
$x \leq y$ iff $y \geq x$; $x, y \in X$; and
$x < y$ iff $x \leq y$ but not $x = y$; $x, y \in X$.

The *compound relation* RS formed from two binary relations R and S on a set X is the set $\{(x, z): (x, y) \in R \text{ and } (y, z) \in S \text{ for some } y \in X\}$. The *converse* of the relation R is the relation R' defined by

$$(x, y) \in R' \text{ iff } (y, x) \in R.$$

A *mapping* or *function* ϕ from a set X to a set Y is an assignment of a unique element of Y to each element of X. We write

$$\phi: X \rightarrow Y.$$

The unique element of Y associated with $x \in X$ is denoted by $\phi(x)$. The *image* of the function ϕ is

$$\text{Im } \phi = \{y \in Y: y = \phi(x), \text{ for some } x \in X\}.$$

If Im $\phi = Y$, then ϕ is termed a *surjection*. If $x \neq y$ implies $\phi(x) \neq \phi(y)$, for any $x, y \in X$, then ϕ is an *injection*. If ϕ is both a surjection and an injection, then it is termed a *bijection* or *one-to-one* mapping.

A *binary operation* f on a set X is a mapping from the set $X \times X$ to X; that is,

$$f: X \times X \rightarrow X.$$

The binary operation f is *associative* if $f(f(x, y),z) = f(x, f(y, z))$ for any $x, y, z \in X$; it is *commutative* if $f(x, y) = f(y, x)$ for all $x, y \in X$. A *semigroup* is a set S together with an associative binary operation on S. A *partially ordered semigroup* is a semigroup S with a partial ordering $\leq$ on S for which $s \leq t$ implies $us \leq ut$ and $su \leq tu$, for all $u \in S$; $s, t, \in S$.

A *lattice* is a set L and a partial order $\leq$ in which, for any $x, y \in L$, there exist unique elements glb(x, y) or meet (x, y) (the *meet* or *greatest lower bound* of x and y) and lub (x, y) or join(x, y) (the *join* or *least upper bound* of x and y) such that

1 glb $(x, y) \leq x$, glb$(x, y) \leq y$;
2 $x \leq$ lub(x, y), $y \leq$ lub(x, y);
3 if $x \leq z$ and $y \leq z$, then lub$(x, y) \leq z$; and
4 if $z \leq x$ and $z \leq y$, then $z \leq$ glb(x, y).

An equivalent definition of a lattice is as a set L together with two commutative and associative binary operations (termed meet and join) also satisfying

1 *idempotence*: $\text{glb}(x, x) = x$, $\text{lub}(x, x) = x$; and
2 *absorption*: $\text{glb}(x, \text{lub}(x, y)) = \text{lub}(x, \text{glb}(x, y)) = x$.

Appendix B   Proofs of theorems

THEOREM 3.4. *Let Ω be reflexive transitive relation defined on the partially ordered semigroup S. Then Ω is a π-relation corresponding to some homomorphism φ on* S *if and only if, for any* s, t ∈ S,

$$(s, t) \in \Omega \textit{ implies } (su, tu) \in \Omega \textit{ and } (us, ut) \in \Omega \textit{ for each } u \in S.$$

Proof: First, define the relation Ω corresponding to an isotone homomorphism ϕ by $(s, t) \in \Omega$ if and only if $\phi(s) \geq \phi(t)$. Then, Ω is clearly reflexive and transitive and $(s, t) \in \Omega$ implies $\phi(s) \geq \phi(t)$; and hence $\phi(s)\phi(u) \geq \phi(t)\phi(u)$, for any $u \in S$. Thus, $\phi(su) \geq \phi(tu)$, so that $(su, tu) \in \Omega$. Similarly $(us, ut) \in \Omega$, and so Ω has the properties stated in the theorem.

Conversely, suppose Ω is a relation on S with the stated properties. Define an equivalence relation e_Ω on S by $(s, t) \in e_\Omega$ if and only if $(s, t) \in \Omega$ and $(t, s) \in \Omega$; $s, t \in S$. Define a mapping ϕ on S by $\phi(s) = [s]$, where $[s]$ is the class of e_Ω containing s. Let $[s][t] = [st]$ and let $[s] \leq [t]$ if and only if $s \leq t$, for some $s \in [s]$, $t \in [t]$. Then the image $\phi(S)$ is well-defined, because (a) for any $s' \in [s]$, $t' \in [t]$, it follows that (s, s'), (s', s), (t, t') and $(t', t) \in \Omega$, so that $(st, s't')$ and $(s't', st) \in \Omega$, and hence $s't' \in [st]$; and (b) $s' \in [s]$, $t' \in [t]$ and $s \leq t$ implies (t', t), (t, s) and $(s, s') \in \Omega$, and hence $(t', s') \in \Omega$ and so $s' \leq t'$. Clearly also ϕ has the properties of an isotone homomorphism, so that the result is established.

THEOREM 3.6. *The collection* L(**R**) *of partially ordered semigroups on a set* **R** *of generator labels forms a lattice under the partial ordering:*

$$S_1 \leq S_2 \textit{ iff there is an isotone homomorphism from } S_2 \textit{ onto } S_1.$$

Similarly, the collection A(**R**) *of abstract semigroups with generator labels* **R** *is a lattice under the partial ordering:*

$$S_1 \leq S_2 \textit{ iff there is an (abstract) homomorphism from } S_2 \textit{ onto } S_1.$$

The set L_s *of isotone homomorphic images of a finite partially ordered semigroup S from* L(**R**) *is a finite sublattice of* L(**R**), *and the set* A_s *of abstract homomorphic images of a semigroup* S *from* A(**R**) *is a finite sublattice of* A(**R**).

Proof: Each isotone homomorphic image S_i of S is associated with an isotone homomorphism ϕ_i of S and a relation π_i on S defined by $(s, t) \in \pi_i$ if and only if $\phi_i(s) \geq \phi_i(t)$. We show that the greatest lower bound of a collection of images $\{S_1, S_2, \ldots, S_q\}$ of S is an image S' with corresponding relation $\pi' = \pi_1 \cap p_2 \cap \cdots \cap \pi_q$ and hence that $L_\pi(S)$ is a lattice (Birkhoff, 1967, p. 112). Then, because $S_1 \leq S_2$ if and only if π_2 is contained in π_1, the set $L_\pi(S)$ of relations π corresponding to isotone homomorphisms of S, partially ordered by set inclusion, is isomorphic to the dual of the partially ordered set L_s. Consequently, $L_\pi(S)$ is a lattice if and only if L_s is a lattice.

Let $\{\pi_1, \pi_2, \ldots, \pi_q\}$ be a set of relations corresponding to $\{S_1, S_2, \ldots, S_q\}$ in L_s, and define

$$\pi' = \pi_1 \cap \pi_2 \cap \cdots \cap \pi_q.$$

Then $(s, t) \in \pi'$ implies $\phi_i(s) \geq \phi_i(t)$, for each $i = 1, 2, \ldots, q$. Define a mapping $\phi'\colon S \to S'$ such that $\phi'(s) \geq \phi'(t)$ if and only if $(s, t) \in \pi'$. The image of ϕ' is the set of classes of S under the equivalence relation $E' = \{(s, t)\colon (s, t) \in \pi' \text{ and } (t, s) \in \pi'\}$; moreover,

1 $s \geq t$ implies $(s, t) \in \pi_i$ for each i and hence $\phi'(s) \geq \phi'(t)$; and
2 ϕ' is a homomorphism: if $\phi'(s_1) = \phi'(s_2)$ and $\phi'(t_1) = \phi'(t_2)$, then

$$\phi_i(s_1) = \phi_i(s_2) \text{ and } \phi_i(t_1) = \phi_i(t_2), \text{ for each } i,$$

and because each ϕ_i is a homomorphism, $\phi_i(s_1t_1) = \phi_i(s_2t_2)$, for each i, and so $\phi'(s_1t_1) = \phi'(s_2t_2)$. Thus, the equivalence relation on S associated with ϕ' has the Substitution Property, and ϕ' is a homomorphism (Birkhoff, 1967).

Thus, π' corresponds to an isotone homomorphism ϕ' on S, so that by Birkhoff's result, both $L_\pi(S)$ and L_s form a lattice. To show that $L(\mathbf{R})$ is also a lattice, it suffices to demonstrate that any two semigroups S_1 and S_2 in $L(\mathbf{R})$ possess a unique least upper bound. Define a relation π on $FS(\mathbf{R})$ by $(s, t) \in \pi$ if and only if $t \leq s$ in S_1 and $t \leq s$ in S_2. Then the semigroup S' defined on the equivalence classes e_π of the relation π, with $[t] \leq [s]$ in S' if and only if $(s, t) \in \pi$, for some $s \in [s]$, $t \in [t]$ (where $[s]$ denotes the class of e_π containing s), is indeed a semigroup. Further $S_1 \leq S'$, $S_2 \leq S'$ and S' is the least semigroup for which this is so. Thus, S' is the least upper bound of S_1 and S_2, and the result follows.

The results for $A(\mathbf{R})$ follow similarly.

THEOREM 3.11. *The collection* M(**R**) *of role algebras having the set* **R** *of generator labels forms a lattice whose partial order is given by* T ≤ Q *if and only if* T *is nested in* Q. *For any local role algebra* Q ∈ M(**R**), *the collection* $\mathrm{L_Q}$ *of role algebras nested in* Q *defines a finite sublattice of* M(**R**).

Proof: We show that any pair of role algebras P, T in $M(\mathbf{R})$ possess a unique greatest lower bound and least upper bound, given by

1 $glb(P, T) = P \cap T$; and
2 $lub(P, T) = (PT)^{\infty}$.

In fact, it suffices to prove the result for $glb(P, T)$; the result for $lub(P, T)$ then follows because it defines the least role algebra containing P and T (Birkhoff, 1967). $P \cap T$ is clearly a quasi-order; also $(s, t) \in P \cap T$ implies $(s, t) \in P$ and $(s, t) \in T$, hence $(su, tu) \in P$ and $(su, tu) \in T$ and so $(su, tu) \in P \cap T$. Thus, $P \cap T$ is a role algebra and is the greatest lower bound of P and T. By induction, $\pi_1 \cap \pi_2 \cap \cdots \cap \pi_k$ is a role algebra and is the greatest lower bound of $\{\pi_1, \pi_2, \cdots, \pi_k\}$ and so $M(\mathrm{R})$ is a lattice. Consequently, L_Q is also a finite sublattice of $M(\mathrm{R})$ for any role algebra $Q \in M(\mathbf{R})$.

THEOREM 3.12. *If* T *is a role algebra nested in the role algebra* Q, *then* π_T *is a reflexive and transitive relation on* e_Q *with the property that* $(s^*, t^*) \in \pi_T$ *implies* $(su^*, tu^*) \in \pi_T$, *for any* $u \in FS(\mathbf{R})$. *Conversely, if* π *is a transitive and reflexive relation on the classes of* e_Q *with the property that* $(s^*, t^*) \in \pi$ *implies* $(su^*, tu^*) \in \pi$, *for any* $u \in FS(\mathbf{R})$, *then* π *is the* π-*relation corresponding to some role algebra nested in* Q.

Proof: $(s^*, t^*) \in \pi_T$ implies $(s', t') \in T$ for any $s' \in s^*$, $t' \in t^*$, because T is a role algebra. Hence $(s'u, t'u) \in T$ for any $u \in FS(\mathbf{R})$, and so $(s'u^*, t'u^*) \in \pi_T$ for any u, as required. Conversely, define a relation T on $FS(\mathbf{R})$ by $(s, t) \in T$ if and only if $(s^*, t^*) \in \pi$. Clearly, T is a quasi-order nested in Q; further, $(s, t) \in T$ implies $(s^*, t^*) \in \pi$ and hence $(su^*, tu^*) \in \pi$ for any $u \in FS(\mathbf{R})$. As a result, $(su, tu) \in T$, and T is a role algebra.

THEOREM 3.13. *For any* $T \in U_n$, $S(T)$ *has distinct relations* $T, T^2, \ldots, T^n$ *and products in* $S(T)$ *defined by*

$$T^iT^j = T^{i+j} \quad \text{if} \quad i+j<n \quad \text{and} \quad T^iT^j = T^n \quad \text{if} \quad i+j \geq n.$$

The partial order in $S(T)$ *is given by*

$$T^i < T^j \text{ iff } i \geq n \text{ and } j < n.$$

(In fact, T^n *is the null relation.)*

Proof: Let x be the source of the tree T, and let $xT^i = \{y \in X: (x, y) \in T^i\}$. Then clearly $xT^i \cap xT^j = \phi$, for any i, j; further, xT^i is nonempty whenever $i < n$. Thus, $T, T^2, \ldots, T^{n-1}$ are distinct, non-null and unordered with respect to one another; also T^n is null, $T^n < T^i$ for any $i < n$, and $T^{n+j} = T^n$, for any $j > 0$. Hence the relations just stated all hold.

THEOREM 3.14. *Let* $\mathrm{T}_1, \mathrm{T}_2 \in \mathbf{T}_n$. *Then the multiplication tables of* $S(\{\mathrm{T}_1, \mathrm{T}_1^*\})$ *and* $S(\{\mathrm{T}_2, \mathrm{T}_2^*\})$ *are isomorphic, where* T^* *denotes the converse of the relation* T.

Proof: Let $L \in \mathbf{T}_n$ be a chain of length $n-1$, and label the elements of the chain by members of the set $\{1, 2, \ldots, n\}$ so that there is a path of length $j-i$ from i to j whenever $i \leq j$. Let $[a, b] = \{m: a \leq m \leq b \text{ and } m \in Z\}$, where Z is the set of integers.

Firstly, we describe the elements of the semigroup $S_L = S(\{L, L^*\})$. Any compound relation W generated by the relations L and L^* can be represented as a sequence of positive or negative integers $(i_1, i_2, \ldots, i_k)$, in which i_r denotes a compound of i_r copies of L if i_r is positive, and a compound of $-i_r$ copies of L^* if i_r is negative. We show that $(p, q) \in W$ if and only if p and q simultaneously satisfy the following three conditions:

(a) $1 \leq p \leq n, \ 1 \leq q \leq n;$

(b) $q = p + \sum_{r=1}^{k} i_r$; and

(c) $p \in \left[1 - \min_{m \in [0, k]} \sum_{r=0}^{m} i_r, \ n - \max_{m \in [0, k]} \sum_{r=0}^{m} i_r\right],$

Now, if p and q satisfy (a) through (c), then $(p, q) \in W$. Suppose conversely that $(p, q) \in W$. Then conditions (a) and (b) must clearly hold; (c) follows because p must satisfy the set of inequalities

$$
\begin{aligned}
&1 \leq p \leq n \\
&1 \leq p + i_1 \leq n \\
&1 \leq p + i_1 + i_2 \leq n \\
&\cdot \\
&\cdot \\
&\cdot \\
&1 \leq p + i_1 + i_2 + \cdots + i_k \leq n,
\end{aligned}
$$

and hence

$$1 - \sum_{r=0}^{m} i_r, \leq p \leq n - \sum_{r=0}^{m} i_r, \qquad \text{for all } m \in [0, k].$$

Secondly, we consider a compound relation V generated by the relations T and T^* for an arbitrary tree $T \in \mathbf{T}_n$; V may be represented by a sequence of integers $(i_1, i_2, \ldots, i_k)$, just as for compounds constructed from the chain L. Let x, y be any elements of the set of vertices of the tree T, and let u be the unique maximal vertex of the tree. Then, using an argument similar to the preceding, we may show that $(x, y) \in V$ if and only if

$$\text{(a)} \quad d(u, y) - d(u, x) = \sum_{r=1}^{k} i_r$$

where $d(v, w)$ is the length of the unique path from v to w, if it exists;

$$\text{(b)} \quad 1 + d(u, x) \in [1 - i_{min}, n - i_{max}],$$

where $i_0 = 0$, and where

$$i_{\min} = \min_{m \in [0, k]} \sum_{r=0}^{m} i_r, \qquad i_{\max} = \max_{m \in [0, k]} \sum_{r=0}^{m} i_r; \text{ and}$$

(c) there exists a vertex z such that

$$d(z, x) \geq 0, \qquad d(z, y) \geq 0 \qquad \text{and} \qquad i_{\min} \leq d(x, z) \leq i_{\max}.$$

Thus, connections in T are wholly determined by the collection of partial sums

$$\left\{ \sum_{r=0}^{m} i_r \right\}_{m=0}^{k}.$$

Thirdly, let W and U be compound relations represented by the sequences $(i_1, i_2, \ldots, i_k)$ and $(j_1, j_2, \ldots, j_h)$, respectively. We need to establish that $W = U$ in $S_T = S(\{T, T^*\})$ if and only if $W = U$ in S_L. Let

$$i_{\min} = \min_{m \in [0, k]} \sum_{r=0}^{m} i_r, \qquad j_{\min} = \min_{m \in [0, h]} \sum_{r=0}^{m} j_r$$

and

$$i_{\max} = \max_{m \in [0, k]} \sum_{r=0}^{m} i_r, \qquad j_{\max} = \max_{m \in [0, h]} \sum_{r=0}^{m} j_r.$$

Now we have established that $W = U$ in S_L or in S_T if and only if

$$\text{(a)} \quad \sum_{r=1}^{k} i_r = \sum_{r=1}^{h} j_r$$

(b) $i_{\min} = j_{\min}$; and

(c) $i_{\max} = j_{\max}$.

Hence, the equations in the semigroups S_L and S_T are identical, and so they are isomorphic.

THEOREM 3.18. *Let* T *be a transition graph on* n *vertices, with period sequence* $\pi(T) = (r_1, r_2, \ldots, r_n)$ *and depth sequence* $\delta(T) = (d_0, d_1, \ldots,$

d_{n-1}). *Then the monogenic semigroup generated by* T *is of type* (h + *1*, d), *where*

$$h = \begin{cases} \max\limits_{j=0,1,\ldots n-1} \{j : d_j > 0\} \\ 0 \text{ if } d_j = 0, \text{ for all } j \end{cases}$$

and

$$d = \begin{cases} \text{l.c.m.}_{i=1,2,\ldots n} \{i : r_i > 0\} \\ 1 \text{ if } r_i = 0, \text{ for all } i \end{cases}.$$

Proof: Let $S = S(\{T\})$ be the monogenic semigroup generated by T. Consider three cases:

Case I Suppose that T is weakly connected; then either (a) T is a tree: T^h is nonempty, $T^{h+1} = \phi = T^{h+2}$ and S is of type $(h+1, 1)$; or (b) T is a flower with period k, where k is minimal in satisfying $T^{(h+1)+k} = T^{h+1}$, and $d = k$. Also, if $l < h$, then T^{l+1+k} is not equal to T^{l+1} (which follows from consideration of the point(s) of maximal height of T). Thus, in both (a) and (b), S is of type $(h+1, d)$.

Case II Suppose that T has two weak components U and V. Let S of type $(h+1, d)$ and S' of type $(g+1, e)$ be the semigroups generated by U and V, respectively. Then, in S, $U^{(h+1)+d} = U^{h+1}$ whereas in S', $V^{(g+1)+e} = V^{g+1}$, and h, g, d and e are minimal integers for which these equations are true. Thus, if T is the disjoint union of U and V, then k and f are the minimum integers for which

$$T^{k+1+f} = T^{k+1},$$

where $k = \max(h, f)$ and $f = \text{l.c.m.}(d, e)$, as required.

Case III If T has m components, then it follows by induction that S is of type $(h+1, d)$, where h and d are as given.

THEOREM 4.1. *The partially ordered semigroup* S *is the direct product of partially ordered semigroups* $S_1, S_2, \ldots, S_r$ *if and only if there exist* π-*relations* $\pi_1, \pi_2, \ldots, \pi_r \in L_\pi(S)$ *such that*

1 S/π_i *is isomorphic to* S_i, *for each* i;
2 $\text{glb}(\pi_1, \pi_2, \ldots, \pi_r) = \pi_{min}$;
3 $\text{lub}(\pi_i, \text{glb}(\pi_1, \pi_2, \ldots, \pi_{i-1})) = \pi_{max}$, *for each* i; *and*
4 *the* π-*relations are* permutable, *that is,*

$$\pi_i\pi_j = \pi_j\pi_i,$$

for all $i = 1, 2, \ldots, r$; $j = 1, 2, \ldots, r$.

Proof: We establish the result for the case $r = 2$; induction may then be used to establish it for $r > 2$. If S is isomorphic to $S_1 \times S_2$, then the π-relations π_1 and π_2, associated with the mappings from S onto S_1 and S_2, satisfy (a) $glb(\pi_1, \pi_2) = \pi_{min}$, and (b) $\pi_1\pi_2 = \pi_2\pi_1 = lub(\pi_1, \pi_2) = \pi_{max}$, so that the stated conditions of the theorem hold.

Conversely, suppose that conditions (i) through (iv) hold, and let $\pi_i(s)$ denote the class of the equivalence relation associated with π_i which contains the element s. Consider the mapping ϕ from S onto $S_1 \times S_2$ given by $\phi(s) = (\pi_1(s), \pi_2(s))$. Clearly ϕ is an isotone homomorphism, because π_1 and π_2 correspond to isotone homomorphisms. Also, if $\pi_i(s) = \pi_i(t)$, for each i, then $s = t$ because $glb(\pi_1, \pi_2) = \pi_{min}$; thus, ϕ is an injection. Moreover, for any $s, t \in S$, there exists an element $u \in S$ such that $(s, u) \in \pi_1$ and $(u, t) \in \pi_2$, since $\pi_1\pi_2 = \pi_{max}$. Thus, $\phi(u) = (\pi_1(s), \pi_2(t))$, and so ϕ is a surjection. Consequently, ϕ is a bijection and S is isomorphic to $S_1 \times S_2$.

LEMMA 4.1. Each F^{**} defines a subdirect representation of S in terms of irreducible components.

Proof: Suppose that F^{**} does not define a subdirect decomposition, so that $d = glb(y_1, y_2, \ldots, y_a) > \pi_{min}$, which is the minimal element of the π-relation lattice $L_\pi(S)$. Then there exists an atom $z \leq d$ such that $glb(y_i, z) = z$, for all $i = 1, 2, \ldots, a$. But $y_i \in C_z$ for some i, which leads to a contradiction. Thus, $glb(y_1, y_2, \ldots, y_a) = \pi_{min}$, and F^{**} defines a subdirect decomposition.

We now suppose there exists some $z^* \in C_z$ that is reducible, so that

$$z^* = glb(b_1, b_2)$$

with

$$b_1 > z^*, \qquad b_2 > z^*.$$

If $b_1 \geq z$ and $b_2 \geq z$, then $glb(b_1, b_2) = z^* \geq z$, which is a contradiction. Hence, at least one of b_1 and b_2 is a meet-complement of z, contradicting membership of z^* in C_z. Thus, z^* is irreducible, and F^{**} is a subdirect representation of the partially ordered semigroup S in terms of irreducible components.

THEOREM 4.6. *F is the set of factorisations of* S.

Proof: By construction, each $F \in \mathrm{F}$ is a minimal irredundant subdirect representation of S and hence a factorisation. It suffices to show, therefore, that any factorisation F of S comprises maximal meet-complements corresponding to a subset of atoms of $L_\pi(S)$.

Let $F = \{y_1, y_2, \ldots, y_r\}$ and let $M = \{z_1, z_2, \ldots, z_a\}$ be the set of atoms of $L_\pi(S)$. Suppose that y_i does not belong to C_z for any $z \in M$. Clearly,

$glb(y_i, z) = \pi_{\min}$ for some $z_j \in M$. Now y_i is irreducible and y_i does not belong to C_z; thus, the unique element t covering y_i satisfies

$$glb(t, z) = \pi_{\min}.$$

Let $glb(y_1, y_2, \ldots, y_{i-1}, y_{i+1}, \ldots, y_r) = d$; then $glb(d, z) = \pi_{\min}$ (since F is a factorisation). Consider two cases.

Case I Suppose that $glb(t, d) = d' > \pi_{\min}$. Then there exists an atom $z' \leq d'$ for which

$$glb(t, z') = z' \qquad \text{and} \qquad glb(y_i, z') = \pi_{\min}.$$

But t covers y_i and is unique in doing so; thus, $y_i \in C_{z'}$, contradicting the supposition that y_i does not belong to C_z, for any $z \in M$.

Case II Suppose that $glb(t, d) = \pi_{\min}$. If t is irreducible, then $F' = \{y_1, y_2, \ldots, y_{i-1}, y_{i+1}, \ldots, y_r, t\}$ is an irredundant subdirect representation such that $F' < F$, in contradiction to the minimality of F. If t is reducible, then it may be expressed as the meet of irreducible elements:

$$t = glb(t_1, t_2, \ldots, t_s) \qquad \text{and} \qquad F' = \{y_1, y_2, \ldots, y_{i-1}, y_{i+1}, \ldots, y_r, t_1, t_2, \ldots, t_s\}$$

is a subdirect decomposition of S in terms of irreducible elements. If any of the t_m are redundant, they can be deleted from F', leaving an irredundant subdirect representation of S that is strictly less than F. Again the minimality of F is contradicted, and we have established that $y_i \in C_z$, for some $z \in M$. Thus, each factorisation arises as an irredundant collection of elements that are maximal meet-complements of atoms of $L_\pi(S)$, and **F** is the set of factorisations of S.

THEOREM 4.7. *If* z *has a unique maximal complement, then* $\pi(z)$ *is a* π-*relation and* $C(z) = \{\pi(z)\}$. *Conversely, if* $\pi(z)$ *is a* π-*relation, then it is the unique maximal complement of* z.

Proof: Clearly $\pi(z) \geq c$, for any $c \in C(z)$, because $(p, q) \in c$ implies

$$glb(\pi_{pq}, z) = \pi_{\min},$$

and hence $(p, q) \in \pi(z)$.

1 Suppose that $C(z) = \{c\}$. Then $c \geq \pi_{st}$ for each $(s, t) \in \pi(z)$; thus,

$$c \geq lub\{(\pi_{st}); \qquad s, t \in \pi(z)\},$$

and so $c \geq \pi(z)$. Thus, if z has a unique maximal complement, it is equal to $\pi(z)$, as required.

2 We have noted that $\pi(z) \geq c$, for any $c \in C(z)$; hence if $\pi(z)$ is a π-relation, it must be a unique maximal element for $C(z)$.

THEOREM 4.8. *Let* $M = \{z_1, z_2, \ldots, z_a\}$ *be the set of atoms of the* π-*relation lattice of a partially ordered semigroup* S. *If, for each* $z_i \in M$, z_i *has a unique maximal complement* $\pi(z_i)$, *then the factorisation of* S *is unique and* $\{\pi(z_1), \pi(z_2), \ldots, \pi(z_a)\}$ *is the factorising set.*

Proof: We show that $\{\pi(z_1), \pi(z_2), \ldots, \pi(z_a)\}$ is irredundant and hence a unique factorising set for *S*. Because $\pi(z_i) \geq z_j$, for all $i \geq j$, it follows that the omission of $\pi(z_i)$ from $\{\pi(z_1), \pi(z_2), \ldots, \pi(z_a)\}$ leads to a collection of π-relations whose meet is strictly greater than π_{min}. Further, each $\pi(z_i)$ is clearly irreducible, so that $\{\pi(z_1), \pi(z_2), \ldots, \pi(z_a)\}$ is irredundant.

THEOREM 4.9. *The association index for quotient semigroups of a partially ordered semigroup satisfies*

1 $0 \leq r(\pi_1, \pi_2) \leq 1$ *and* $r(\pi_1, \pi_2) = r(\pi_2, \pi_1)$, *for all* $\pi_1, \pi_2 \in L_\pi(S)$;
2 $r(\pi_1, \pi_2) = 0$ *iff* $S/glb(\pi_1, \pi_2)$ *is isomorphic to* $S/\pi_1 \times S/\pi_2$; *and*
3 *If* $\pi_1 = \pi_2$, *then* $r(\pi_1, \pi_2) = 1$.

Proof: Let $|S/\pi_1| = a$, $|S/\pi_2| = b$, $|S/glb(\pi_1, \pi_2)| = p$, $|S/lub(\pi_1, \pi_2)| = q$.

1 Now $glb(\pi_1, \pi_2) = \pi_1 \cap \pi_2$, so that $p \leq ab$. Also $p \geq \max(a, b)$, $q \leq \min(a, b)$ and a, b, p and q are all positive integers. Thus

$$0 = \frac{ab - ab}{ab - q} \leq \frac{ab - p}{ab - q} = r \leq \frac{ab - \max(a, b)}{ab - \min(a, b)} \leq 1.$$

2 If $\pi_1 = \pi_2$, then $lub(\pi_1, \pi_2) = glb(\pi_1, \pi_2)$ so that $p = q$ and consequently $r(\pi_1, \pi_2) = 1$.
3 Clearly, $r(\pi_1, \pi_2) = 0$ if $S/glb(\pi_1, \pi_2)$ is isomorphic to $S/\pi_1 \times S/\pi_2$; to establish the converse result, it is sufficient to demonstrate that $r(\pi_1, \pi_2) = 0$ implies that $\pi_1\pi_2 = \pi_2\pi_1 = lub(\pi_1, \pi_2)$ (by Theorem 4.1). Let the classes of $glb(\pi_1, \pi_2)$ be denoted by $C_1, C_2, \ldots, C_p$. Because $r(\pi_1, \pi_2) = 0$, $p = ab$. Thus, no π_1-class may contain more than b classes of $glb(\pi_1, \pi_2)$, nor may any π_2-class contain more than a classes of $glb(\pi_1, \pi_2)$. Therefore each π_1-class (π_2-class) contains exactly b (a) classes of $glb(\pi_1, \pi_2)$. Now any class C_i of $glb(\pi_1, \pi_2)$ is one of b distinct classes of $glb(\pi_1, \pi_2)$ contained in a single π_1-class, and each of those b classes belongs to a different π_2-class. Hence $\pi_1\pi_2 = \pi_{max}$, and because π_1 and π_2 are permutable, $\pi_1\pi_2 = \pi_2\pi_1 = lub(\pi_1, \pi_2)$, as required.

THEOREM 4.10. *Let* ϕ *be a homomorphism on a semigroup S. Then the relation* π_ϕ *on* S *given by*

$$(s, t) \in \pi_\phi \textit{ iff } \phi(s) = \phi(t); \qquad s, t \in S$$

is reflexive, symmetric and transitive and has the substitution property:

$$(s, t) \in \pi_\phi \textit{ iff } (us, ut) \in \pi_\phi \textit{ and } (su, tu) \in \pi_\phi, \textit{ for any } u \in S.$$

Conversely, any reflexive, symmetric and transitive relation having the substitution property corresponds to a homomorphism of S.

Proof: Clearly π_ϕ is reflexive, symmetric and transitive. Also, because ϕ is a homomorphism, $(s, t) \in \pi_\phi$ implies $\phi(s) = \phi(t)$, and so $\phi(su) = \phi(s)\phi(u) = \phi(t)\phi(u) = \phi(tu)$, and so $(su, tu) \in \pi_\phi$. Similarly, $(us, ut) \in \pi_\phi$, and π_ϕ has the properties asserted in the theorem. Conversely, if π is an equivalence relation possessing the substitution property, the mapping ϕ on S given by $\phi(s) = [s]$, where $[s]$ is the equivalence class of π containing s, establishes a homomorphism on S because (a) $[s][t] = [st]$ is a well-defined operation on the classes of π by virtue of the substitution property, and so (b) $\phi(st) = \phi(s)\phi(t)$.

THEOREM 5.7. *Let* C *be a partition on* X *satisfying the central representatives condition for the network* **R** *on* X, *and let* **T** *be the derived network of* **R** *induced by* C. *Then* S(**T**) *is an isotone homomorphic image of* S(**R**).

Proof: Define ϕ: $S(\mathbf{R}) \rightarrow S(\mathbf{T})$ by

$$\phi(T) = T/C,$$

where $T \in S(\mathbf{R})$ and T/C is the induced relation on the equivalence classes of C (i.e., if C_x denotes the class containing x, $(C_x, C_y) \in T/C$ if and only if, there is some $x \in C_x$, $y \in C_y$ such that $(x, y) \in T$). It is readily shown that $T_1 \leq T_2$ implies $T_1/C \leq T_2/C$, hence proving that

$$(T_1T_2)/C = (T_1/C)(T_2/C)$$

establishes that ϕ is an isotone homomorphism.

Now $(C_x, C_y) \in (T_1T_2)/C$ if and only if there exist central representatives $x_\alpha \in C_x$, $y_\alpha \in C_y$ such that $(x_\alpha, y_\alpha) \in T_1T_2$ and so if and only if there exists a central representative $z_\alpha \in C_z$ such that $(x_\alpha, z_\alpha) \in T_1$ and $(z_\alpha, y_\alpha) \in T_2$. Hence $(C_x, C_y) \in (T_1T_2)/C$ if and only if $(C_x, C_z) \in T_1/C$ and $(C_z, C_y) \in T_2/C$, that is, if and only if $(C_x, C_y) \in (T_1T_2)/C$, as required.

THEOREM 5.9. *The semigroup of the network induced by the subset* U *is an isotone homomorphic image of the semigroup of* **R**.

Proof: Let $x, y \in U$ and suppose $(x, y) \in RT$ for relations $R, T \in \mathbf{R}$. Then there exists an element $z \in X$ such that $(x, z) \in R$ and $(z, y) \in T$. By the definition of U, therefore, there exist elements $z', z'' \in U$ such that $(x, z') \in R$ and $(z'', y) \in T$. Further, because elements of U satisfy the condition

G_i, $(x, w) \in R$ and $(w, y) \in T$ for some $w \in U$. Consequently, $(x, y) \in RT$ in $S(\mathbf{U})$, where **U** is the derived network of **R** induced by U. It follows therefore that $S(\mathbf{U})$ is an isotone homomorphic image of $S(\mathbf{R})$.

THEOREM 5.10. *Let* **T** *be the derived network induced by a receiver subset* Y *of* X. *Then* S(**T**) *is an isotone homomorphic image of* S(**R**).

Proof: Define $\phi : S(\mathbf{R}) \rightarrow S(\mathbf{T})$ by

$$\phi(T) = T',$$

where $(x, y) \in T'$ if and only if $(x, y) \in T$ and $x, y \in Y$. It is sufficient to establish that $\phi(TU) = \phi(T)\phi(U)$ and hence that ϕ is a homomorphism and a surjection.

Let $(x, y) \in \phi(TU)$. Then there exists $z \in X$ such that $(x, z) \in T$ and $(z, y) \in U$. But because Y is a receiver subset, it follows that $z \in Y$ and so that $(x, y) \in \phi(T)\phi(U)$. Conversely, it is readily shown that $(x, y) \in \phi(T)\phi(U)$ implies $(x, y) \in \phi(TU)$; hence $\phi(TU) = \phi(T)\phi(U)$, as required.

THEOREM 5.11. *Let* P *be a partition on a set* X *on which a local network* **R** *is defined, and let* μ *be the partial function on* X *associated with* P. *If* P *is a regular equivalence, then the local role algebra of* μ(1) *in the derived local network* **T** *is nested in the local role algebra of node 1 in* **R**.

Proof: Let Q be the local role algebra of node 1 in **R**, and let T be the local role algebra of node $\mu(1)$ in **T**. Now if $(s, t) \in Q$, for some $s, t \in FS(\mathbf{R})$, then $(1, z) \in t$ implies $(1, z) \in s$, for all $z \in X$. Suppose now that $(\mu(1), z') \in t$ for some class z' of P. Then because P is a regular equivalence, $(1, z) \in t$ for some $z \in z'$, and hence $(1, z) \in s$. Thus, $(\mu(1), \mu(z)) \in s$, that is, $(\mu(1), z') \in s$. Thus, $(s, t) \in Q$ implies $(s, t) \in T$, and T is nested in Q.

Author index

Subject index

Printed in the USA
CPSIA information can be obtained
at www.ICGtesting.com
LVHW091952211223
767033LV00002B/93